Concepts of Combinatorial Optimization

Combinatorial Optimization

volume 1

Concepts of Combinatorial Optimization

Edited by
Vangelis Th. Paschos

ISTE

WILEY

First published 2010 in Great Britain and the United States by ISTE Ltd and John Wiley & Sons, Inc. Adapted and updated from *Optimisation combinatoire* volumes 1 to 5 published 2005-2007 in France by Hermes Science/Lavoisier © LAVOISIER 2005, 2006, 2007

ISTE Ltd
27-37 St George's Road
London SW19 4EU
UK

www.iste.co.uk

John Wiley & Sons, Inc.
111 River Street
Hoboken, NJ 07030
USA

www.wiley.com

Library of Congress Cataloging-in-Publication Data

Combinatorial optimization / edited by Vangelis Th. Paschos.
v. cm.
Includes bibliographical references and index.
Contents: v. 1. Concepts of combinatorial optimization
ISBN 978-1-84821-146-9 (set of 3 vols.) -- ISBN 978-1-84821-147-6 (v. 1)
1. Combinatorial optimization. 2. Programming (Mathematics)
I. Paschos, Vangelis Th.
QA402.5.C545123 2010
519.6'4--dc22

2010018423

British Library Cataloguing-in-Publication Data
A CIP record for this book is available from the British Library
ISBN 978-1-84821-146-9 (Set of 3 volumes)
ISBN 978-1-84821-147-6 (Volume 1)

Printed and bound in Great Britain by CPI Antony Rowe, Chippenham and Eastbourne.

MIX
Paper from responsible sources
FSC® C013604
www.fsc.org

Table of Contents

Preface

What is combinatorial optimization? There are, in my opinion, as many definitions as there are researchers in this domain, each one as valid as the other. For me, it is above all the art of understanding a real, natural problem, and being able to transform it into a mathematical model. It is the art of studying this model in order to extract its structural properties and the characteristics of the solutions of the modeled problem. It is the art of exploiting these characteristics in order to determine algorithms that calculate the solutions but also to show the limits in economy and efficiency of these algorithms. Lastly, it is the art of enriching or abstracting existing models in order to increase their strength, portability, and ability to describe mathematically (and computationally) other problems, which may or may not be similar to the problems that inspired the initial models.

Seen in this light, we can easily understand why combinatorial optimization is at the heart of the junction of scientific disciplines as rich and different as theoretical computer science, pure and applied, discrete and continuous mathematics, mathematical economics, and quantitative management. It is inspired by these, and enriches them all.

This book, *Concepts of Combinatorial Optimization*, is the first volume in a set entitled *Combinatorial Optimization*. It tries, along with the other volumes in the set, to embody the idea of combinatorial optimization. The subjects of this volume cover themes that are considered to constitute the hard core of combinatorial optimization. The book is divided into three parts:

– Part I: Complexity of Combinatorial Optimization Problems;

– Part II: Classical Solution Methods;

– Part III: Elements from Mathematical Programming.

In the first part, Chapter 1 introduces the fundamentals of the theory of (deterministic) complexity and of algorithm analysis. In Chapter 2, the context changes and we consider algorithms that make decisions by “tossing a coin”. At each stage of the resolution of a problem, several alternatives have to be considered, each one occurring with a certain probability. This is the context of probabilistic (or randomized) algorithms, which is described in this chapter.

In the second part some methods are introduced that make up the great classics of combinatorial optimization: branch-and-bound and dynamic programming. The former is perhaps the most well known and the most popular when we try to find an optimal solution to a difficult combinatorial optimization problem. Chapter 3 gives a thorough overview of this method as well as of some of the most well-known tree search methods based upon branch-and-bound. What can we say about dynamic programming, presented in Chapter 4? It has considerable reach and scope, and very many optimization problems have optimal solution algorithms that use it as their central method.

The third part is centered around mathematical programming, considered to be the heart of combinatorial optimization and operational research. In Chapter 5, a large number of linear models and an equally large number of combinatorial optimization problems are set out and discussed. In Chapter 6, the main simplex algorithms for linear programming, such as the primal simplex algorithm, the dual simplex algorithm, and the primal–dual simplex algorithm are introduced. Chapter 7 introduces some classical linear programming methods, while Chapter 8 introduces quadratic integer optimization methods. Chapter 9 describes a series of resolution methods currently widely in use, namely column generation. Chapter 10 focuses on polyhedral methods, almost 60 years old but still relevant to combinatorial optimization research. Lastly, Chapter 11 introduces a more contemporary, but extremely interesting, subject, namely constraint programming.

This book is intended for novice researchers, or even Master’s students, as much as for senior researchers. Master’s students will probably need a little basic knowledge of graph theory and mathematical (especially linear) programming to be able to read the book comfortably, even though the authors have been careful to give definitions of all the concepts they use in their chapters. In any case, to improve their knowledge of graph theory, readers are invited to consult a great, flagship book from one of our gurus, Claude Berge: *Graphs and Hypergraphs*, North Holland, 1973. For linear programming, there is a multitude of good books that the reader could consult, for example V. Chvátal, *Linear Programming*, W.H. Freeman, 1983, or M. Minoux, *Programmation mathématique: théorie et algorithmes*, Dunod, 1983.

Editing this book has been an exciting adventure, and all my thanks go, firstly, to the authors who, despite their many responsibilities and commitments (which is the lot of any university academic), have agreed to participate in the book by writing chapters in their areas of expertise and, at the same time, to take part in a very tricky exercise: writing chapters that are both educational and high-level science at the same time.

This work could never have come into being without the original proposal of Jean-Charles Pomerol, Vice President of the scientific committee at Hermes, and Sami Ménascé and Raphaël Ménascé, the heads of publications at ISTE. I give my warmest thanks to them for their insistence and encouragement. It is a pleasure to work with them as well as with Rupert Heywood, who has ingeniously translated the material in this book from the original French.

Vangelis Th. PASCHOS
June 2010

PART I

Complexity of Combinatorial Optimization Problems

Chapter 1

Basic Concepts in Algorithms and Complexity Theory

1.1. Algorithmic complexity

In algorithmic theory, a problem is a general question to which we wish to find an answer. This question usually has parameters or variables the values of which have yet to be determined. A problem is posed by giving a list of these parameters as well as the properties to which the answer must conform. An instance of a problem is obtained by giving explicit values to each of the parameters of the instanced problem.

An algorithm is a sequence of elementary operations (variable affectation, tests, forks, etc.) that, when given an instance of a problem as input, gives the solution of this problem as output after execution of the final operation.

The two most important parameters for measuring the quality of an algorithm are: its *execution time* and the *memory space* that it uses. The first parameter is expressed in terms of the number of instructions necessary to run the algorithm. The use of the number of instructions as a unit of time is justified by the fact that the same program will use the same number of instructions on two different machines but the time taken will vary, depending on the respective speeds of the machines. We generally consider that an instruction equates to an elementary operation, for example an assignment, a test, an addition, a multiplication, a trace, etc. What we call the *complexity in time* or simply the *complexity* of an algorithm gives us an indication of the time it will take to solve a problem of a given size. In reality this is a function that associates an order of

---

Chapter written by Vangelis Th. PASCHOS.

magnitude[1] of the number of instructions necessary for the solution of a given problem with the size of an instance of that problem. The second parameter corresponds to the number of memory units used by the algorithm to solve a problem. The *complexity in space* is a function that associates an order of magnitude of the number of memory units used for the operations necessary for the solution of a given problem with the size of an instance of that problem.

There are several sets of hypotheses concerning the "standard configuration" that we use as a basis for measuring the complexity of an algorithm. The most commonly used framework is the one known as "worst-case". Here, the complexity of an algorithm is the number of operations carried out on the instance that represents the worst configuration, amongst those of a fixed size, for its execution; this is the framework used in most of this book. However, it is not the only framework for analyzing the complexity of an algorithm. Another framework often used is "average analysis". This kind of analysis consists of finding, for a fixed size (of the instance) n, the average execution time of an algorithm on all the instances of size n; we assume that for this analysis the probability of each instance occurring follows a specific distribution pattern. More often than not, this distribution pattern is considered to be uniform. There are three main reasons for the worst-case analysis being used more often than the average analysis. The first is psychological: the worst-case result tells us for certain that the algorithm being analyzed can never have a level of complexity higher than that shown by this analysis; in other words, the result we have obtained gives us an upper bound on the complexity of our algorithm. The second reason is mathematical: results from a worst-case analysis are often easier to obtain than those from an average analysis, which very often requires mathematical tools and more complex analysis. The third reason is "analysis portability": the validity of an average analysis is limited by the assumptions made about the distribution pattern of the instances; if the assumptions change, then the original analysis is no longer valid.

1.2. Problem complexity

The definition of the complexity of an algorithm can be easily transposed to problems. Informally, *the complexity of a problem is equal to the complexity of the best algorithm that solves it* (this definition is valid independently of which framework we use).

Let us take a size n and a function $f(n)$. Thus:

– **TIMEf(n)** is the class of problems for which the complexity (in time) of an instance of size n is in $O(f(n))$.

1. Orders of magnitude are defined as follows: given two functions f and g: $f = O(g)$ if and only if $\exists(k,k') \in (\mathbb{R}^+, \mathbb{R})$ such that $f \leqslant kg + k'$; $f = o(g)$ if and only if $\lim_{n\to\infty}(f/g) = 0$; $f = \Omega(g)$ if and only if $g = o(f)$; $f = \Theta(g)$ if and only if $f = O(g)$ and $g = O(f)$.

– $\mathbf{SPACEf(n)}$ is the class of problems that can be solved, for an instance of size n, by using a memory space of $O(f(n))$.

We can now specify the following general classes of complexity:

– **P** is the class of all the problems that can be solved in a time that is a polynomial function of their instance size, that is $\mathbf{P} = \cup_{k=0}^{\infty}\mathbf{TIMEn^k}$.

– **EXPTIME** is the class of problems that can be solved in a time that is an exponential function of their instance size, that is $\mathbf{EXPTIME} = \cup_{k=0}^{\infty}\mathbf{TIME2^{n^k}}$.

– **PSPACE** is the class of all the problems that can be solved using a memory space that is a polynomial function of their instance size, that is $\mathbf{PSPACE} = \cup_{k=0}^{\infty}\mathbf{SPACEn^k}$.

With respect to the classes that we have just defined, we have the following relations: $\mathbf{P} \subseteq \mathbf{PSPACE} \subseteq \mathbf{EXPTIME}$ and $\mathbf{P} \subset \mathbf{EXPTIME}$. Knowing whether the inclusions of the first relation are strict or not is still an open problem.

Almost all combinatorial optimization problems can be classified, from an algorithmic complexity point of view, into two large categories. *Polynomial* problems can be solved optimally by algorithms of polynomial complexity, that is in $O(n^k)$, where k is a constant independent of n (this is the class **P** that we have already defined). *Non-polynomial* problems are those for which the best algorithms (those giving an optimum solution) are of a "super-polynomial" complexity, that is in $O(f(n)^{g(n)})$, where f and g are increasing functions in n and $\lim_{n\to\infty} g(n) = \infty$. All these problems contain the class **EXPTIME**.

The definition of any algorithmic problem (and even more so in the case of any combinatorial optimization problem) comprises two parts. The first gives the instance of the problem, that is the type of its variables (a graph, a set, a logical expression, etc.). The second part gives the type and properties to which the expected solution must conform. In the complexity theory case, algorithmic problems can be classified into three categories:

– *decision problems*;

– *optimum value calculation problems*;

– *optimization problems.*

Decision problems are questions concerning the existence, for a given instance, of a configuration such that this configuration itself, or its value, conforms to certain properties. The solution to a decision problem is an answer to the question associated with the problem. In other words, this solution can be:

– either "*yes*, such a configuration does exist";

– or "*no*, such a configuration does not exist".

Let us consider as an example the conjunctive normal form satisfiability problem, known in the literature as SAT: "Given a set U of n Boolean variables $x_1, \ldots, x_n$ and a set C of m *clauses*[2] $C_1, \ldots, C_m$, is there a model for the expression $\phi = C_1 \wedge \ldots \wedge C_m$; i.e. is there an assignment of the values 0 or 1 to the variables such that $\phi = 1$?". For an instance ϕ of this problem, if ϕ allows a model then the solution (the correct answer) is *yes*, otherwise the solution is *no*.

Let us now consider the MIN TSP problem, defined as follows: given a complete graph K_n over n vertices for which each edge $e \in E(K_n)$ has a value $d(e) > 0$, we are looking for a Hamiltonian cycle $H \subset E$ (a partial closely related graph such that each vertex is of 2 degrees) that minimizes the quantity $\sum_{e \in H} d(e)$. Let us assume that for this problem we have, as well as the complete graph K_n and the vector $\overline{d}$, costs on the edges K_n of a constant K and that we are looking not to determine the smallest (in terms of total cost) Hamiltonian cycle, but rather to answer the following question: "Does there exist a Hamiltonian cycle of total distance less than or equal to K?". Here, once more, the solution is either *yes* if such a cycle exists, or *no* if it does not.

For *optimum value calculation problems*, we are looking to calculate *the value of the optimum solution* (and not the solution itself).

In the case of the MIN TSP for example, the optimum associated value calculation problem comes down to calculating the cost of the smallest Hamiltonian cycle, and not the cycle itself.

Optimization problems, which are naturally of interest to us in this book, are those for which we are looking to establish the best solution amongst those satisfying certain properties given by the very definition of the problem. An optimization problem may be seen as a mathematical program of the form:

$$\begin{cases} \text{opt} & v(\overline{x}) \\ & \overline{x} \in C_I \end{cases}$$

where $\overline{x}$ is the vector describing the solution[3], $v(\overline{x})$ is the *objective function*, C_I is the problem's constraint set, set out for the instance I (in other words, C_I sets out both the instance and the properties of the solution that we are looking to find for this instance), and opt $\in \{\max, \min\}$. An *optimum solution* of I is a vector $\overline{x}^* \in \text{argopt}\{v(\overline{x}) : \overline{x} \in C_I\}$. The quantity $v(\overline{x}^*)$ is known as the *objective value* or *value* of the problem. A solution $\overline{x} \in C_I$ is known as a *feasible solution*.

2. We associate two *literals* x and $\overline{x}$ with a Boolean variable x, that is the variable itself and its negation; a clause is a disjunction of the literals.

3. For the combinatorial optimization problems that concern us, we can assume that the components of this vector are 0 or 1 or, if need be, integers.

Let us consider the problem MIN WEIGHTED VERTEX COVER[4]. An instance of this problem (given by the information from the incident matrix A, of dimension $m \times n$, of a graph $G(V, E)$ of order n with $|E| = m$ and a vector $\overline{w}$, of dimension n of the costs of the edges of V), can be expressed in terms of a linear program in integers as follows:

$$\begin{cases} \min & \overline{w} \cdot \overline{x} \\ & A.\overline{x} \geqslant \overline{1} \\ & \overline{x} \in \{0,1\}^n \end{cases}$$

where $\overline{x}$ is a vector from 0, 1 of dimension n such that $x_i = 1$ if the vertex $v_i \in V$ is included in the solution, $x_i = 0$ if it is not included. The block of m constraints $A.\overline{x} \geqslant \overline{1}$ expresses the fact that for each edge at least one of these extremes must be included in the solution. The feasible solutions are all the transversals of G and the optimum solution is a transversal of G of minimum total weight, that is a transversal corresponding to a feasible vector consisting of a maximum number of 1.

The solution to an optimization problem includes an evaluation of the optimum value. Therefore, an optimum value calculation problem can be associated with an optimization problem. Moreover, optimization problems always have a decisional variant as shown in the MIN TSP example above.

1.3. The classes P, NP and NPO

Let us consider a decision problem Π. If for any instance I of Π a solution (that is a correct answer to the question that states Π) of I can be found algorithmically in polynomial time, that is in $O(|I|^k)$ stages, where $|I|$ is the size of I, then Π is called a *polynomial problem* and the algorithm that solves it a *polynomial algorithm* (let us remember that polynomial problems make up the **P** class).

For reasons of simplicity, we will assume in what follows that the solution to a decision problem is:

– either "*yes*, such a solution exists, and this is it";

– or "*no*, such a solution does not exist".

In other words, if, to solve a problem, we could consult an "oracle", it would provide us with an answer of not just a *yes* or *no* but also, in the first case, a certificate

4. Given a graph $G(V, E)$ of order n, in the MIN VERTEX COVER problem we are looking to find a smallest transversal of G, that is a set $V' \subseteq V$ such that for every edge $(u, v) \in E$, either $u \in V'$, or $v \in V'$ of minimum size; we denote by MIN WEIGHTED VERTEX COVER the version of MIN VERTEX COVER where a positive weight is associated with each vertex and the objective is to find a transversal of G that minimizes the sum of the vertex weights.

proving the veracity of the *yes*. This testimony is simply a solution proposal that the oracle "asserts" as being the real solution to our problem.

Let us consider the decision problems for which the validity of the certificate can be verified in polynomial time. These problems form the class **NP**.

DEFINITION 1.1.– *A decision problem* Π *is in **NP** if the validity of all solutions of* Π *is verifiable in polynomial time.*

For example, the SAT problem belongs to **NP**. Indeed, given the assignment of the values 0, 1 to the variables of an instance ϕ of this problem, we can, with at most nm applications of the connector $\vee$, decide whether the proposed assignment is a model for ϕ, that is whether it satisfies all the clauses.

Therefore, we can easily see that the decisional variant of MIN TSP seen previously also belongs to **NP**.

Definition 1.1 can be extended to optimization problems. Let us consider an optimization problem Π and let us assume that each instance I of Π conforms to the three following properties:

1) The feasibility of a solution can be verified in polynomial time.

2) The value of a feasible solution can be calculated in polynomial time.

3) There is at least one feasible solution that can be calculated in polynomial time.

Thus, Π belongs to the class **NPO**. In other words, *the class **NPO** is the class of optimization problems for which the decisional variant is in **NP***. We can therefore define the class **PO** of optimization problems for which the decisional variant belongs to the class **P**. In other words, **PO** is the class of problems that can be optimally solved in polynomial time.

We note that the class **NP** has been defined (see definition 1.1) without explicit reference to the optimum solution of its problems, but by reference to the verification of a given solution. Evidently, the condition of belonging to **P** being stronger than that of belonging to **NP** (what can be solved can be verified), we have the obvious inclusion of : **P** $\subseteq$ **NP** (Figure 1.1).

"What is the complexity of the problems in **NP** $\setminus$ **P**?". The best general result on the complexity of the solution of problems from **NP** is as follows [GAR 79].

THEOREM 1.1.– *For any problem* $\Pi \in$ ***NP****, there is a polynomial* p_Π *such that each instance* I *of* Π *can be solved by an algorithm of complexity* $O(2^{p_\Pi(|I|)})$.

In fact, theorem 1.1 merely gives an upper limit on the complexity of problems in **NP**, but no lower limit. The diagram in Figure 1.1 is nothing more than a conjecture,

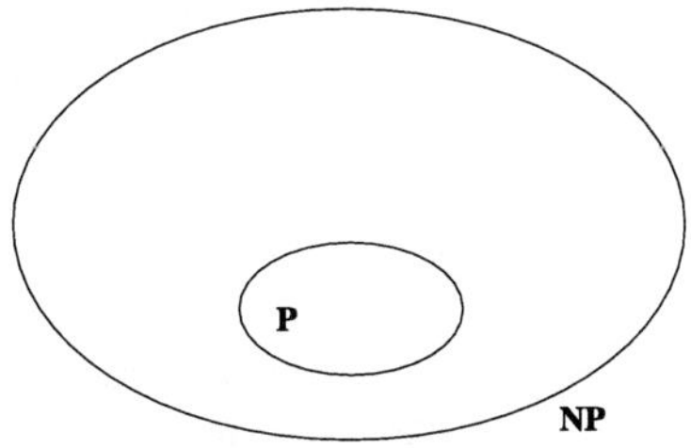

Figure 1.1. *P and NP (under the assumption P $\neq$ NP)*

and although almost all researchers in complexity are completely convinced of its veracity, it has still not been proved. The question "*Is **P** equal to or different from **NP**?*" is the biggest open question in computer science and one of the best known in mathematics.

1.4. Karp and Turing reductions

As we have seen, problems in **NP** $\setminus$ **P** are considered to be algorithmically more difficult than problems in **P**. A large number of problems in **NP** $\setminus$ **P** are very strongly bound to each other through the concept of *polynomial reduction.*

The principle of reducing a problem Π to a problem Π' consists of considering the problem Π as a specific case of Π', *modulo* a slight transformation. If this transformation is polynomial, and we know that we can solve Π' in polynomial time, we will also be able to solve Π in polynomial time. Reduction is thus a means of transferring the result of solving one problem to another; in the same way it is a tool for classifying problems according to the level of difficulty of their solution.

We will start with the classic Karp reduction (for the class **NP**) [GAR 79, KAR 72]. This links two decision problems by the possibility of their optimum (and simultaneous) solution in polynomial time. In the following, given a problem Π, let $\mathcal{I}_\Pi$ be all of its instances (we assume that each instance $I \in \mathcal{I}_\Pi$ is identifiable in polynomial time in $|I|$). Let $\mathcal{O}_\Pi$ be the subset of $\mathcal{I}_\Pi$ for which the solution is *yes*; $\mathcal{O}_\Pi$ is also known as the set of *yes*-instances (or positive instances) of Π.

DEFINITION 1.2.– *Given two decision problems Π_1 and Π_2, a Karp reduction (or polynomial transformation) is a function $f : \mathcal{I}_{\Pi_1} \rightarrow \mathcal{I}_{\Pi_2}$, which can be calculated in polynomial time, such that, given a solution for $f(I)$, we are able to find a solution for I in polynomial time in $|I|$ (the size of the instance I).*

A Karp reduction of a decision problem Π_1 to a decision problem Π_2 implies the existence of an algorithm $\mathtt{A}_1$ for Π_1 that uses an algorithm $\mathtt{A}_2$ for Π_2. Given any instance $I_1 \in \mathcal{I}_{\Pi_1}$, the algorithm $\mathtt{A}_1$ constructs an instance $I_2 \in \mathcal{I}_{\Pi_2}$; it executes the algorithm $\mathtt{A}_2$, which calculates a solution on I_2, then $\mathtt{A}_1$ transforms this solution into a solution for Π_1 on I_1. If $\mathtt{A}_2$ is polynomial, then $\mathtt{A}_1$ is also polynomial.

Following on from this, we can state another reduction, known in the literature as the *Turing reduction*, which is better adapted to optimization problems. In what follows, we define a problem Π as a couple $(\mathcal{I}_\Pi, \mathrm{Sol}_\Pi)$, where Sol_Π is the set of solutions for Π (we denote by $\mathrm{Sol}_\Pi(I)$ the set of solutions for the instance $I \in \mathcal{I}_\Pi$).

DEFINITION 1.3.– *A Turing reduction of a problem* Π_1 *to a problem* Π_2 *is an algorithm* $\mathtt{A}_1$ *that solves* Π_1 *by using (possibly several times) an algorithm* $\mathtt{A}_2$ *for* Π_2 *in such a way that if* $\mathtt{A}_2$ *is polynomial, then* $\mathtt{A}_1$ *is also polynomial.*

The Karp and Turing reductions are transitive: if Π_1 is reduced (by one of these two reductions) to Π_2 and Π_2 is reduced to Π_3, then Π_1 reduces to Π_3. We can therefore see that both reductions preserve membership of the class **P** in the sense that if Π reduces to Π' and $\Pi' \in$ **P**, then $\Pi \in$ **P**.

For more details on both the Karp and Turing reductions refer to [AUS 99, GAR 79, PAS 04]. In Garey and Johnson [GAR 79] (Chapter 5) there is also a very interesting historical summary of the development of the ideas and terms that have led to the structure of complexity theory as we know it today.

1.5. NP-completeness

From the definition of the two reductions in the preceding section, if Π' reduces to Π, then Π can reasonably be considered as at least as difficult as Π' (regarding their solution by polynomial algorithms), in the sense that a polynomial algorithm for Π would have sufficed to solve not only Π itself, but equally Π'. Let us confine ourselves to the Karp reduction. By using it, we can highlight a problem $\Pi^* \in$ **NP** such that any other problem $\Pi \in$ **NP** reduces to Π^* [COO 71, GAR 79, FAG 74]. Such a problem is, as we have mentioned, the most difficult problem of the class **NP**. Therefore we can show [GAR 79, KAR 72] that there are other problems $\Pi^{*\prime} \in$ **NP** such that Π^* reduces to $\Pi^{*\prime}$. In this way we expose a family of problems such that any problem in **NP** reduces (remembering that the Karp reduction is transitive) to one of its problems. This family has, of course, the following properties:

– It is made up of the most difficult problems of **NP**.

– A polynomial algorithm for at least one of its problems would have been sufficient to solve, in polynomial time, all the other problems of this family (and indeed any problem in **NP**).

The problems from this family are ***NP-complete*** problems and the class of these problems is called the ***NP-complete*** class.

DEFINITION 1.4.– *A decision problem* Π *is* ***NP****-complete if, and only if, it fulfills the following two conditions:*

1) $\Pi \in$ ***NP*** *;*

2) $\forall \Pi' \in$ ***NP****,* Π' *reduces to* Π *by a Karp reduction.*

Of course, a notion of **NP**-completeness very similar to that of definition 1.4 can also be based on the Turing reduction.

The following application of definition 1.3 is very often used to show the **NP**-completeness of a problem. Let $\Pi_1 = (\mathcal{I}_{\Pi_1}, \mathrm{Sol}_{\Pi_1})$ and $\Pi_2 = (\mathcal{I}_{\Pi_2}, \mathrm{Sol}_{\Pi_2})$ be two problems, and let (f, g) be a pair of functions, which can be calculated in polynomial time, where:

– $f : \mathcal{I}_{\Pi_1} \to \mathcal{I}_{\Pi_2}$ is such that for any instance $I \in \mathcal{I}_{\Pi_1}$, $f(I) \in \mathcal{I}_{\Pi_2}$;

– $g : \mathcal{I}_{\Pi_1} \times \mathrm{Sol}_{\Pi_2} \to \mathrm{Sol}_{\Pi_1}$ is such that for every pair $(I, S) \in (\mathcal{I}_{\Pi_1} \times \mathrm{Sol}_{\Pi_2}(f(I)))$, $g(I, S) \in \mathrm{Sol}_{\Pi_1}(I)$.

Let us assume that there is a polynomial algorithm $\mathtt{A}$ for the problem Π_2. In this case, the algorithm $f \circ \mathtt{A} \circ g$ is a (polynomial) Turing reduction.

A problem that fulfills condition 2 of definition 1.4 (without necessarily checking condition 1) is called **NP**-hard[5]. It follows that *a decision problem* Π *is* ***NP****-complete if and only if it belongs to* ***NP*** *and it is* ***NP****-hard.*

With the class **NP**-complete, we can further refine (Figure 1.2) the world of **NP**. Of course, if **P** = **NP**, the three classes from Figure 1.2 coincide; moreover, under the assumption **P** $\neq$ **NP**, the classes **P** and **NP**-complete do not intersect.

Let us denote by **NP**-intermediate the class $\mathbf{NP} \setminus (\mathbf{P} \cup \mathbf{NP} - complete)$. Informally, this concerns the class of problems of intermediate difficulty, that is problems that are more difficult than those from **P** but easier than those from **NP**-complete. More formally, for two complexity classes **C** and **C**$'$ such that $\mathbf{C}' \subseteq \mathbf{C}$, and a reduction R preserving the membership of **C**$'$, a problem is **C**-intermediate if it is neither **C**-complete under R, nor belongs to **C**$'$. Under the Karp reduction, the class **NP**-intermediate is not empty [LAD 75].

Let us note that the idea of **NP**-completeness goes hand in hand with *decision problems*. When dealing with optimization problems, the appropriate term, used in

5. These are the starred problems in the appendices of [GAR 79].

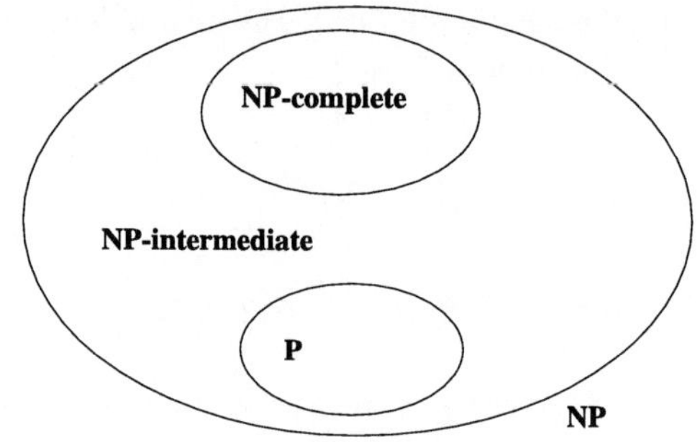

Figure 1.2. ***P**, **NP** and **NP-complete** (under the assumption **P** ≠ **NP**)*

the literature, is ***NP**-hard*[6]. *A problem of **NPO** is **NP**-hard if and only if its decisional variant is an **NP**-complete* problem.

The problem SAT was the first problem shown to be **NP**-complete (the proof of this important result can be found in [COO 71]). The reduction used (often called *generic reduction*) is based on the theory of recursive languages and Turing machines (see [HOP 79, LEW 81] for more details and depth on the Turing machine concept; also, language-problem correspondence is very well described in [GAR 79, LEW 81]). The general idea of generic reduction, also often called the "Cook–Levin technique (or theory)", is as follows: *For a generic decision (language) problem belonging to **NP**, we describe, using a normal conjunctive form, the working of a non-deterministic algorithm (Turing machine) that solves (decides) this problem (language).*

The second problem shown to be **NP**-complete [GAR 79, KAR 72] was the variant of SAT, written 3SAT, where no clause contains more than three literals. The reduction here is from SAT [GAR 79, PAP 81]. More generally, for all $k \geqslant 3$, the kSAT problems (that is the problems defined on normal conjunctive forms where each clause contains no more than k literals) are all **NP**-complete.

It must be noted that the problem 2SAT, where all normal conjunctive form clauses contain at most two literals, is polynomial [EVE 76]. It should also be noted that in [KAR 72], where there is a list of the first 21 **NP**-complete problems, the problem of linear programming in real numbers was mentioned as a probable problem from the

6. There is a clash with this term when it is used for optimization problems and when it is used in the sense of property 2 of definition 1.4, where it means that a decision problem Π is harder than any other decision problem $\Pi' \in$ **NP**.

class **NP-intermediate**. It was shown, seven years later ([KHA 79] and also [ASP 80], an English translation of [KHA 79]), that this problem is in **P**.

The reference on **NP**-completeness is the volume by Garey and Johnson [GAR 79]. In the appendix, *A list of NP-complete problems*, there is a long list of **NP**-complete problems with several commentaries for each one and for their limited versions. For many years, this list has been regularly updated by Johnson in the *Journal of Algorithms* review. This update, supplemented by numerous commentaries, appears under the title: *The NP-completeness column: an ongoing guide.*

"What is the relationship between optimization and decision for **NP**-complete problems?". The following theory [AUS 95, CRE 93, PAZ 81] attempts to give an answer.

THEOREM 1.2.– *Let* Π *be a problem of* ***NPO*** *and let us assume that the decisional version of* Π*, written* Π_d*, is* ***NP****-complete. It follows that a polynomial Turing reduction exists between* Π_d *and* Π.

In other words, the decision versions (such as those we have considered in this chapter) and optimization versions of an **NP**-complete problem are of equivalent algorithmic difficulty. However, the question of the existence of a problem **NPO** for which the optimization version is more difficult to solve than its decisional counterpart remains open.

1.6. Two examples of NP-complete problems

Given a problem Π, the most conventional way to show its **NP**-completeness consists of making a Turing or Karp reduction of an **NP**-complete Π' problem to Π. In practical terms, the proof of **NP**-completeness for Π is divided into three stages:

1) proof of membership of Π to **NP**;

2) choice of Π';

3) building the functions f and g (see definition 1.3) and showing that they can both be calculated in polynomial time.

In the following, we show that MIN VERTEX COVER$(G(V,E),K)$ and MAX STABLE$(G(V,E),K)$, the decisional variants of MIN VERTEX COVER and of MAX STABLE[7], respectively, are **NP**-complete. These two problems are defined as follows. MIN VERTEX COVER$(G(V,E),K)$: given a graph G and a constant $K \leqslant |V|$, does there exist in G a transversal $V' \subseteq V$ less than or equal in size to K? MAX

7. Given a graph $G(V,E)$ of magnitude n, we are trying to find a stable set of maximum size, that is a set $V' \subseteq V$ such that $\forall (u,v) \in V' \times V'$, $(u,v) \notin E$ of maximum size.

STABLE SET$(G(V,E),K)$: given a graph G and a constant $K \leqslant |V|$, does there exist in G a stable set $V' \subseteq V$ of greater than or equal in size to K?

1.6.1. MIN VERTEX COVER

The proof of membership of MIN VERTEX COVER$(G(V,E),K)$ to **NP** is very simple and so has been omitted here. We will therefore show the completeness of this problem for **NP**. We will transform an instance $\phi(U,\mathcal{C})$ from 3SAT, with $U = \{x_1,\ldots,x_n\}$ and $\mathcal{C} = \{C_1,\ldots,C_m\}$, into an instance $(G(V,E),K)$ of MIN VERTEX COVER.

This graph is made up of two component sets, joined by edges. The first component is made up of $2n$ vertices $x_1,\bar{x}_1,\ldots,x_n,\bar{x}_n$ and n edges $(x_i,\bar{x}_i)$, $i=1,\ldots,n$, which join the vertices in pairs. The second is made up of m vertex-disjoint triangles (that is of m cliques with three vertices). For a clause C_i, we denote the three vertices of the corresponding triangle by c_{i1}, c_{i2} and c_{i3}. In fact, the first set of components, for which each vertex corresponds to a literal, serves to define the truth values of the solution for 3SAT; the second set of components corresponds to the clauses, and each vertex is associated with a literal of its clause. These triangles are used to verify the satisfaction of the clauses. To finish, we add $3m$ "unifying" edges that link each vertex of each "triangle-clause" to its corresponding "literal-vertex". Let us note that exactly three unifying edges go from (the vertices of) each triangle, one per vertex of the triangle. Finally, we state $K = n + 2m$. It is easy to see that the transformation of $\phi(U,\mathcal{C})$ to $G(V,E)$ happens in polynomial time in $\max\{m,n\}$ since $|V| = 2n + 3m$ and $|E| = n + 6m$.

As an example of the transformation described above, let us consider the instance $\phi = (x_1 \vee \bar{x}_2 \vee x_3) \wedge (\bar{x}_1 \vee x_2 \vee \bar{x}_3) \wedge (x_2 \vee x_3 \vee \bar{x}_4) \wedge (\bar{x}_1 \vee \bar{x}_2 \vee x_4)$ from 3SAT. The graph $G(V,E)$ for MIN VERTEX COVER is given in Figure 1.3. In this case, $K = 12$.

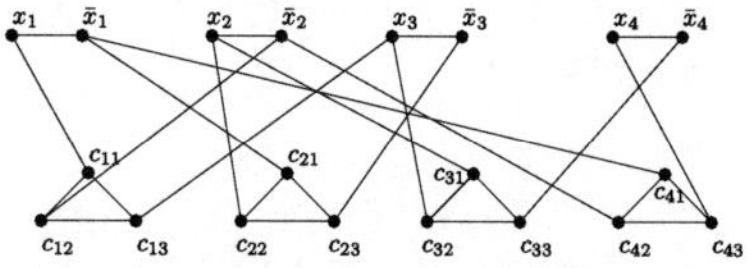

Figure 1.3. *The graph associated with the expression* $(x_1 \vee \bar{x}_2 \vee x_3) \wedge (\bar{x}_1 \vee x_2 \vee \bar{x}_3) \wedge (x_2 \vee x_3 \vee \bar{x}_4) \wedge (\bar{x}_1 \vee \bar{x}_2 \vee x_4)$

We will now show that G *allows a transversal less than or equal in size to* $n + 2m$ *if and only if the expression* ϕ *can be satisfied.*

Let us first show that the condition is necessary. Let us assume that there is a polynomial algorithm A that answers the question "Is there in G a transversal $V' \subseteq V$ of size $|V'| \leqslant K$?", and, if so, returns V'. Let us execute it with $K = n + 2m$. If the answer from A is *yes*, then the transversal must be of a size *equal* to $n+2m$. In fact, any transversal needs at least n vertices in order to cover the n edges corresponding to the variables of ϕ (one vertex per edge) and $2m$ vertices to cover the edges of m triangles (two vertices per triangle). As a result, if A answers *yes*, it will have calculated a transversal of exactly $n + 2m$ vertices.

In the light of the previous observation, given such a transversal V', we state that $x_i = 1$ if the extremity x_i of the edge $(x_i, \bar{x}_i)$ is taken in V'; if the extremity $\bar{x}_i$ is included, then we state that $\bar{x}_i = 1$, that is $x_i = 0$. We claim that this assignment of the truth values to the variables satisfies ϕ. Indeed, since only one extremity of each edge $(x_i, \bar{x}_i)$ is taken in V', *only one literal is set to 1 for each variable* and, in consequence, the assignment in question is coherent (one, and only one, truth value is assigned to each literal). Moreover, let us consider a triangle T_i of G corresponding to the clause C_i; let us denote its vertices by c_{i1}, c_{i2} and c_{i3}, and let us assume that the last two belong to V'. Let us also assume that the unifying edge having as an extremity the vertex c_{i1} is the edge (c_{i1}, ℓ_k), ℓ_k being one of the literals associated with the variable x_k. Since $c_{i1} \notin V'$, ℓ_k belongs to it, that is $\ell_k = 1$, and the existence of the edge (c_{i1}, ℓ_k) means that the literal ℓ_k belongs to the clause C_i. This is proved by setting ℓ_k to 1. By iterating this argument for each clause, the need for the condition is proved. Furthermore, let us note that obtaining the assignment of the truth values to the variables of ϕ is done in polynomial time.

Let us now show that the condition is also good enough. Given an assignment of truth values satisfying the expression ϕ, let us construct in G a transversal V' of size $n + 2m$. To start with, for each variable x_i, if $x_i = 1$, then the extremity x_i of the edge $(x_i, \bar{x}_i)$ is put in V'; otherwise, the extremity $\bar{x}_i$ of the edge $(x_i, \bar{x}_i)$ is put there. We thereby cover the edges of type $(x_i, \bar{x}_i)$, $i = 1, \ldots, n$, and one unifying edge per triangle. Let T_i (corresponding to the clause C_i), $i = 1, \ldots, m$, be a triangle and let (ℓ_k, c_{i1}) be the unifying edge covered by the setting to 1 of ℓ_k. We therefore put the vertices c_{i2} and c_{i3} in V'; these vertices cover both the edges of T_i and the two unifying edges having as extremities c_{i2} and c_{i3}, respectively. By iterating this operation for each triangle, a transversal V' of size $n + 2m$ is eventually constructed in polynomial time.

1.6.2. MAX STABLE

The proof of membership of MAX STABLE$(G(V, E), K)$ is so simple that it is omitted here.

Let us consider a graph $G(V, E)$ of magnitude n having m edges and let us denote by A its incidence matrix[8]. Let us also consider the expression of MIN VERTEX COVER as a linear program in whole numbers and the transformations that follow:

$$\left\{\begin{array}{ll}\min & \bar{1}\cdot\bar{y}\\ & A\cdot\bar{y}\geqslant\bar{1}\\ & \bar{y}\in\{0,1\}^n\end{array}\right. \Leftrightarrow \left\{\begin{array}{ll}\min & \bar{1}\cdot\bar{y}\\ & 2.\bar{1}-A\cdot\bar{y}\leqslant\bar{1}\\ & \bar{y}\in\{0,1\}^n\end{array}\right.$$

$$\Leftrightarrow \left\{\begin{array}{ll}\min & \bar{1}\cdot\bar{y}\\ & A\cdot(\bar{1}-\bar{y})\leqslant\bar{1}\\ & \bar{y}\in\{0,1\}^n\end{array}\right. \overset{\bar{y}=\bar{1}-\bar{x}}{\Longleftrightarrow} \left\{\begin{array}{ll}\min & \bar{1}\cdot(\bar{1}-\bar{x})\\ & A\cdot\bar{x}\leqslant\bar{1}\\ & \bar{x}\in\{0,1\}^n\end{array}\right.$$

$$\Leftrightarrow \left\{\begin{array}{ll}\max & \bar{1}\cdot\bar{x}\\ & A\cdot\bar{x}\leqslant\bar{1}\\ & \bar{x}\in\{0,1\}^n\end{array}\right.$$

We note that the last program in the series is nothing more than the linear program (in whole numbers) of MAX STABLE. Furthermore, this series of equivalents indicates that if a solution vector $\bar{x}$ for MAX STABLE is given, then the vector $\bar{y} = \bar{1} - \bar{x}$ (that is the vector $\bar{y}$ where we interchange the "1" and the "0" regarding $\bar{x}$) is a feasible solution for MIN VERTEX COVER. Furthermore, if $\bar{x}$ contains at least K "1" (that is the size of the stable set is at least equal to K), then the solution vector deduced for MIN VERTEX COVER contains at most $n - K$ "1" (that is the size of the transversal is at most equal to $n - K$). Since the function $\bar{x} \mapsto \bar{1} - \bar{x}$ is polynomial, then so is the described transformation.

1.7. A few words on strong and weak NP-completeness

Let Π be a problem and I an instance of Π of size $|I|$. We denote by $\max(I)$ the highest number that appears in I. Let us note that $\max(I)$ can be exponential in n. An algorithm for Π is known as *pseudo-polynomial* if it is polynomial in $|I|$ and $\max(I)$ (if $\max(I)$ is exponential in $|I|$, then this algorithm is exponential for I).

DEFINITION 1.5.– *An optimization problem is **NP**-complete in the strong sense (strongly **NP**-complete) if the problem is **NP**-complete because of its structure and not because of the size of the numbers that appear in its instances. A problem is **NP**-complete in the weak sense (weakly **NP**-complete) if it is **NP**-complete because of its valuations (that is* $\max(I)$ *affects the complexity of the algorithms that solve it).*

Let us consider on the one hand the SAT problems, or MIN VERTEX COVER problems, or even the MAX STABLE problems seen previously, and on the other hand the KNAPSACK problem for which the decisional variant is defined as: "given a maximum cost b, n objects $\{1, \ldots, n\}$ of respective values a_i and respective costs $c_i \leqslant b$,

8. This matrix is of dimensions $m \times n$.

$i = 1, \ldots, n$, and a constant K, is there a subset of objects for which the sum of the values is at least equal to K without the sum of the costs of these objects exceeding b?". This problem is the most well-known weakly **NP**-complete problem. Its intrinsic difficulty is not due to its structure, as is the case for the previous three problems where no large number affects the description of their instances, but rather due to the values of a_i and c_i that do affect the specification of the instance of KNAPSACK.

In Chapter 4 (see also Volume 2, Chapter 8), we find a dynamic programming algorithm that solves this problem in linear time for the highest value of a_i and in logarithmic time for the highest value of c_i. This means that if this value is a polynomial of n, then the algorithm is also polynomial, and if the value is exponential in n, the algorithm itself is of exponential complexity.

The result below [GAR 79, PAS 04] follows the borders between strongly and weakly **NP**-complete problems. *If a problem* Π *is strongly* ***NP****-complete, then it cannot be solved by a pseudo-polynomial algorithm, unless* ***P*** $=$ ***NP****.*

1.8. A few other well-known complexity classes

In this section, we briefly present a few supplementary complexity classes that we will encounter in the following chapters (for more details, see [BAL 88, GAR 79, PAP 94]). Introductions to some complexity classes can also be found in [AUS 99, VAZ 01].

Let us consider a decision problem Π and a generic instance I of Π defined on a data structure S (a graph, for example) and a decision constant K. From the definition of the class **NP**, we can deduce that if there is a solution giving the answer *yes* for Π on I, and if this solution is submitted for verification, then the answer for any correct verification algorithm will always be *yes*. On the other hand, if such a solution does not exist, then any solution proposal submitted for verification will always bring about a *no* answer from any correct verification algorithm.

Let us consider the following decisional variant of MIN TSP, denoted by co-MIN TSP: "given K_n, $\bar{d}$ and K, is it true that there is no Hamiltonian cycle of a total distance less than or equal to K?". How can we guarantee that the answer for an instance of this problem is *yes*? This questions leads on to that of this problem's membership of the class **NP**. We come across the same situation for a great many problems in **NP** $\setminus$ **P** (assuming that **P** $\neq$ **NP**).

We denote by $\mathcal{I}_\Pi$ the set of all the instances of a decision problem $\Pi \in$ **NP** and by $\mathcal{O}_\Pi$ the subset of $\mathcal{I}_\Pi$ for which the solution is *yes*, that is the set of *yes*-instances (or positive instances) of Π. We denote by $\bar{\Pi}$ the problem having as *yes*-instances the set $\mathcal{O}_{\bar{\Pi}} = \mathcal{I}_\Pi \setminus \mathcal{O}_\Pi$, that is the set of *no*-instances of Π. All these problems make

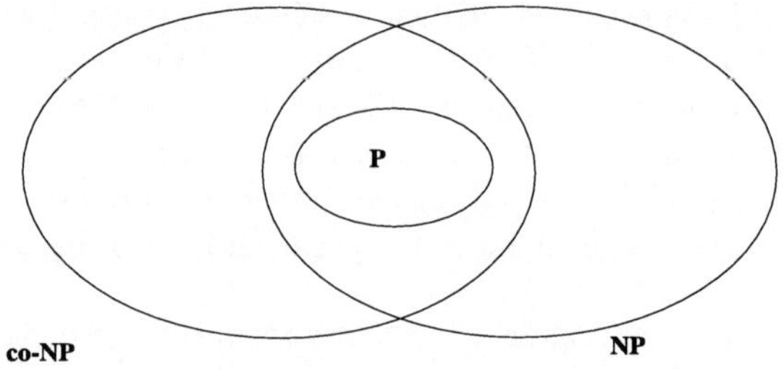

Figure 1.4. *P, NP and co-NP (under the conditions P ≠ NP and NP ≠ co-NP)*

up the class **co-NP**. In other words, **co-NP** $=\{\bar{\Pi} : \Pi \in$ **NP**$\}$. It is surmised that **NP** $\neq$ **co-NP**. This surmise is considered as being "stronger" than **P** $\neq$ **NP**, in the sense that it is possible that **P** $\neq$ **NP**, even if **NP** $=$ **co-NP** (but if **P** $=$ **NP**, then **NP** $=$ **co-NP**).

Obviously, for a decision problem $\Pi \in$ **P**, the problem $\bar{\Pi}$ also belongs to **P**; as a result, **P** $\subseteq$ **NP** $\cap$ **co-NP**.

A decision problem Π belongs to **RP** if there is a polynomial p and an algorithm $\mathtt{A}$ such that, for any instance I: if $I \in \mathcal{O}_\Pi$, then the algorithm gives a decision in polynomial time for at least half of the candidate solutions (certificates) S, such that $|S| \leqslant p(|I|)$, that are submitted to it for verification of their feasibility; if, on the other hand, $I \notin \mathcal{O}_\Pi$ (that is if $I \in \mathcal{I}_\Pi \setminus \mathcal{O}_\Pi$), then for any proposed solution S with $|S| \leqslant p(|I|)$, $\mathtt{A}$ rejects S in polynomial time. A problem $\Pi \in$ **co-RP** if and only if $\bar{\Pi} \in$ **RP**, where $\bar{\Pi}$ is defined as before. We have the following relationship between **P**, **RP** and **NP**: **P** $\subseteq$ **RP** $\subseteq$ **NP**. For a very simple and intuitive proof of this relationship, see [VAZ 01].

The class **ZPP** (an abbreviation of *zero-error probabilistic polynomial time*) is the class of decision problems that allow a randomized algorithm (for this subject see [MOT 95] and Chapter 2) that always ends up giving the correct answer to the question "$I \in \mathcal{O}_\Pi$?", with, on average, a polynomial complexity. A problem Π belongs to **ZPP** if and only if Π belongs to **RP** $\cap$ **co-RP**. In other words, **ZPP** $=$ **RP** $\cap$ **co-RP**.

1.9. Bibliography

[ASP 80] ASPVALL B., STONE R.E., "Khachiyan's linear programming algorithm", *J. Algorithms*, vol. 1, p. 1–13, 1980.

[AUS 95] AUSIELLO G., CRESCENZI P., PROTASI M., "Approximate solutions of NP optimization problems", *Theoret. Comput. Sci.*, vol. 150, p. 1–55, 1995.

[AUS 99] AUSIELLO G., CRESCENZI P., GAMBOSI G., KANN V., MARCHETTI-SPACCAMELA A., PROTASI M., *Complexity and Approximation. Combinatorial Optimization Problems and their Approximability properties*, Springer-Verlag, Berlin, 1999.

[BAL 88] BALCÀZAR J.L., DIAZ J., GABARRÒ J., *Structural Complexity*, vol. 1 and 2, Springer-Verlag, Berlin, 1988.

[COO 71] COOK S.A., "The complexity of theorem-proving procedures", *Proc. STOC'71*, p. 151–158, 1971.

[CRE 93] CRESCENZI P., SILVESTRI R., "Average measure, descriptive complexity and approximation of minimization problems", *Int. J. Foundations Comput. Sci.*, vol. 4, p. 15–30, 1993.

[EVE 76] EVEN S., ITAI A., SHAMIR A., "On the complexity of timetable and multicommodity flow problems", *SIAM J. Comput.*, vol. 5, p. 691–703, 1976.

[FAG 74] FAGIN R., "Generalized first-order spectra and polynomial-time recognizable sets", KARP R.M., Ed., *Complexity of Computations*, p. 43–73, American Mathematics Society, 1974.

[GAR 79] GAREY M.R., JOHNSON D.S., *Computers and Intractability. A Guide to the Theory of NP-completeness*, W. H. Freeman, San Francisco, 1979.

[HOP 79] HOPCROFT J.E., ULLMAN J.D., *Introduction to Automata Theory, Languages and Computation*, Addison-Wesley, 1979.

[KAR 72] KARP R.M., "Reducibility among combinatorial problems", MILLER R.E., THATCHER J.W., Eds., *Complexity of computer computations*, p. 85–103, Plenum Press, New York, 1972.

[KHA 79] KHACHIAN L.G., "A polynomial algorithm for linear programming", *Dokladi Akademiy Nauk SSSR*, vol. 244, p. 1093–1096, 1979.

[LAD 75] LADNER R.E., "On the structure of polynomial time reducibility", *J. Assoc. Comput. Mach.*, vol. 22, p. 155–171, 1975.

[LEW 81] LEWIS H.R., PAPADIMITRIOU C.H., *Elements of the Theory of Computation*, Prentice-Hall, New Jersey, 1981.

[MOT 95] MOTWANI R., RAGHAVAN P., *Randomized Algorithms*, Cambridge University Press, Cambridge, 1995.

[PAP 81] PAPADIMITRIOU C.H., STEIGLITZ K., *Combinatorial Optimization: Algorithms and Complexity*, Prentice-Hall, New Jersey, 1981.

[PAP 94] PAPADIMITRIOU C.H., *Computational Complexity*, Addison-Wesley, 1994.

[PAS 04] PASCHOS V.TH., *Complexité et approximation polynomiale*, Hermes, Paris, 2004.

[PAZ 81] PAZ A., MORAN S., "Non deterministic polynomial optimization problems and their approximations", *Theoret. Comput. Sci.*, vol. 15, p. 251–277, 1981.

[VAZ 01] VAZIRANI V., *Approximation Algorithms*, Springer-Verlag, Berlin, 2001.

Chapter 2

Randomized Complexity

A deterministic algorithm is "a method of solution, by the application of a rule, a finite number of times" (Chapter 1). This definition extends to probabilistic algorithms by allowing probabilistic rules, or by giving a distribution of probability over a set of deterministic algorithms. By definition, probabilistic algorithms are potentially more *powerful* than their deterministic counterparts: the class of probabilistic algorithms contains the class of deterministic algorithms. It is therefore natural to consider probabilistic algorithms for solving the hardest combinatorial optimization problems, all the more so because probabilistic algorithms are often simpler than their deterministic counterparts.

A deterministic algorithm is correct if it solves each instance in a valid way. In the context of probabilistic algorithms for which the execution depends both on the instance and on the randomization of the algorithm, we consider the correct complexity of algorithms for any instance and any randomization, but also the complexity of algorithms such that for each instance I, the probability that the algorithm correctly solves I is higher than a constant (typically $1/2$).

In the context of combinatorial optimization, we consider algorithms for which the result *approximates* the solution of the problem set. We then examine both the complexity of the algorithm and the *quality* of the approximations that it gives. We can, without loss of generality, limit ourselves to the problems of minimizing a cost function, in which case we look to minimize both the complexity of the algorithm and the cost of the approximation that it produces. Depending on whether we examine the complexity or the cost of the approximation generated, the terms are different but the

---

Chapter written by Jérémy BARBAY.

techniques are similar: in both cases, we are looking to minimize a measure of the performance of the algorithm. This performance, for a probabilistic algorithm and a given instance I, is defined as the average of the performance of the corresponding deterministic algorithm on I.

The aim of this chapter is to show that the analysis techniques used to study the complexity of probabilistic algorithms can be just as easily used to analyze the approximation quality of combinatorial optimization algorithms. In section 2.1, we give a more formal definition of the concepts and notations generally used in the study of the complexity of probabilistic algorithms, and we introduce a basic problem that is used to illustrate the most simple results of this chapter. In section 2.2, we give and prove the basic results that allow us to prove lower bounds for the performance of probabilistic algorithms for a given problem. These results can often be used as they stand, but it is important to understand their causes in order to adapt them to less appropriate situations. The most common technique for proving an upper bound to the performance of the best probabilistic algorithm for a given problem is to analyze the performance of the probabilistic algorithm, and for this purpose there are as many techniques as there are algorithms: for this reason we do not describe a general method, but instead give an example of analysis in section 2.3.

2.1. Deterministic and probabilistic algorithms

An *algorithm* is a method for solving a problem using a finite number of rule applications. In the computer science context, an algorithm is a precise description of the stages to be run through in order to carry out a calculation or a specific task. A *deterministic algorithm* is an algorithm such that the choice of each rule to be applied is deterministic. A *probabilistic algorithm* can be defined in a similar way by a probabilistic distribution over the deterministic algorithms, or by giving a probabilistic distribution over the rules of the algorithm. In both cases, the data received by the algorithm make up the *instance* of the problem to be solved, and the choice of algorithm or rules makes up the *randomization* of the execution.

Among deterministic algorithms, generally only the algorithms that *always* give a correct answer are considered, whereas for probabilistic algorithms, algorithms that may be incorrect are also considered, with some constraints. In the context of decision problems, where the answer to each instance is Boolean (accepted or refused), we consider "Monte Carlo" *probabilistic algorithms* [PAP 94, section 11.2] such that for a fixed positive instance, the probability that the algorithm accepts the instance is at least $1/2$, and for a fixed negative instance, the algorithm always refuses the instance (whatever its randomization). This value of $1/2$ is arbitrary: any value ε that is strictly positive is sufficient, since it is enough to repeat the algorithm k times independently to reduce the probability that the algorithm accepts a negative instance to $(1-\varepsilon)^k$.

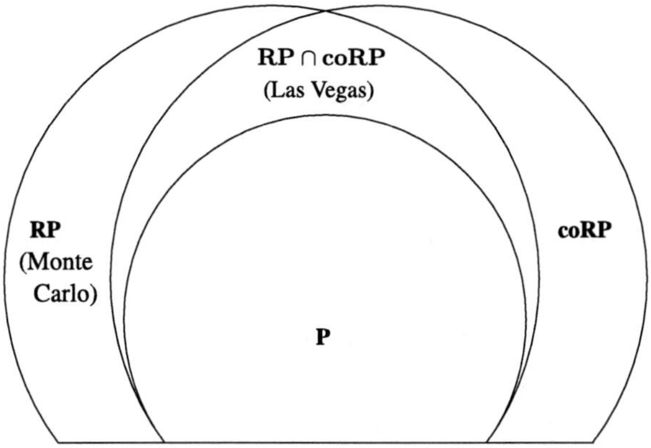

Figure 2.1. *Complexity classes relative to probabilistic algorithms that run in polynomial time*

Problems that can be solved using Monte Carlo algorithms running in polynomial time make up the complexity class **RP**. Among these, problems that can be resolved by deterministic algorithms running in polynomial time make up the complexity class **P**. In a similar way, the class **co-RP** is made up of the set of problems that can be solved in polynomial time by a probabilistic algorithm that always accepts a positive instance, but refuses a negative instance with a probability of at least $1/2$. If a problem is in **ZPP** = **RP**∩**co-RP**, it allows an algorithm of each kind, and so allows an algorithm that is a combination of both, which always accepts positive instances and refuses negative instances, but in an unlimited time. The execution time of algorithms of this type can be random, but they always find a correct result. These algorithms are called "Las Vegas": their result is certain, but their execution time is random, a bit like a martingale with a sufficiently large initial stake.

Another useful complexity class concerns probabilistic algorithms that can make mistakes both by accepting and by refusing instances of the class **BPP**, formed by problems that can be solved in polynomial time by a probabilistic algorithm refusing a positive instance or accepting a negative instance with a probability of at most $1/4$. Just as for **RP**, the value of $1/4$ is arbitrary and can be replaced by $1/2 - p(n)$ for any polynomial $p(n)$ without losing the important properties of **RP** (but the value of $1/2$ would not be suitable here, see [MOT 95, p. 22]).

Despite the fact that these complexity classes are generally defined in terms of decision problems, they can equally well be used to order complexities of a larger class of problems, such as research problems, or combinatorial problems [MOT 95, p. 23].

EXAMPLE.– A classic problem is the *bin-packing* problem: given a list of objects of heights $L = \{x_1, \ldots, x_n\}$, between 0 and 1, we must make vertical stacks of objects in such a way that the height of each stack does not exceed 1, and such that there is a minimum number of stacks. This problem is **NP**-complete and the approximation of the optimal number μ of stacks is a classic combinatorial optimization problem. Even if it is not a decision problem, it can be categorized in the same classes. It is enough to consider the following variant: given a list of the weights of objects $L = \{x_1, \ldots, x_n\}$, and a number of stacks M, can these n objects be organized into M stacks, without any of them exceeding one unit of height? Any algorithm approximating μ by a value m such that $\Pr[m = \mu] \geqslant 3/4$ (if need be by iterating the same algorithm several times) allows us to decide whether M stacks will suffice ($m \leqslant M$) without ever being wrong when M is not sufficient ($M < \mu \leqslant m$), and with a probability of being wrong of at most $1/4$ if M is sufficient ($\Pr[\mu < m = M] \leqslant 1/4$): the decision problem is in **RP**!

2.1.1. *Complexity of a Las Vegas algorithm*

Given a cost function over the operations carried out by the algorithm, the complexity of an algorithm A on an instance I is the sum of the costs of the instructions corresponding to the execution of A on I. The algorithms solving this same problem are compared by their complexity. The *time complexity* $C(A, I)$ of an algorithm A on an instance I corresponds to the number of instructions carried out by A when it is executed to solve I.

EXAMPLE.–

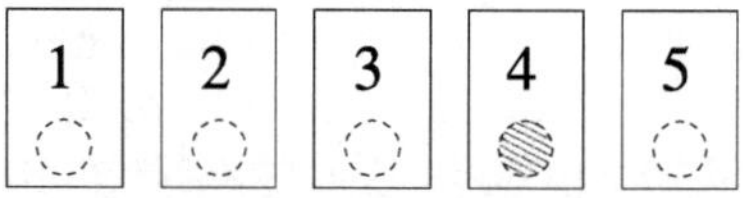

Figure 2.2. *An instance of the hidden coin problem: one silver coin amongst four copper coins, the coins being hidden by cards*

The hidden coin problem is another abstract problem that we will use to illustrate the different ideas of this chapter. We have a row of n cards. Each card hides a coin, which can be a copper coin or a silver coin. The *hidden coin problem* is to decide whether the row contains at least one silver coin.

In this particular case, an algorithm must indicate which coins to uncover, and in which order, depending on which type of coin is revealed. For such a simple problem, we can limit the study to the algorithms that stop uncovering coins as soon as they have discovered a silver coin: any coin uncovered after this would be superfluous. Each of these algorithms is defined by the order σ in which it uncovers the coins as long as there is no silver coin:

– A deterministic algorithm uncovers coins in a fixed order.

– A potential algorithm chooses half of the coins randomly and uniformly, uncovers them, accepts the instance if there is a silver coin, and otherwise refuses. This algorithm always refuses an instance that does not contain a silver coin (therefore a negative instance), and accepts any instance containing at least one silver coin (therefore positive) with a probability of at least $1/2$: therefore it is a Monte Carlo algorithm.

– Another potential algorithm uncovers coins in a random order until it has found a silver coin, or it has uncovered all of the coins: this algorithm always gives the right answer, but the number of coins uncovered is a random variable that depends on the chosen order: it is therefore a Las Vegas algorithm.

In a configuration that has only one silver coin hidden under the second card from the right, the algorithm that uncovers coins from left to right will uncover $n-1$ coins, the algorithm that uncovers coins from right to left will only uncover 2 coins.

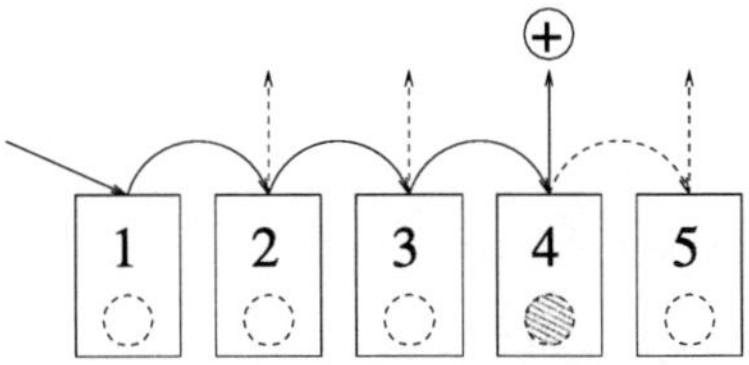

Figure 2.3. *The algorithm choosing coins from left to right: dotted lines show the possible executions, and solid lines show the executions on this instance. The answer of the algorithm is positive because the row contains a silver coin*

Let $F = \{I_1, \ldots, I_{|F|}\}$ be a finite set of instances, and A an algorithm for these instances. The complexity $C(A, F)$ of A on F can be defined in several ways. The set of values taken by the complexity of A on the instances of F is $\{C(A, I_1), \ldots, C(A, I_{|F|})\}$:

– their maximum $C(A, F) = \max_{I \in F} C(A, I)$ corresponds to the *worst-case complexity*;

– and the average $C(A, F) = \sum_{I \in F} C(A, I)p(I)$ corresponds to the *average complexity* according to a probability distribution $p(I)$ on the instances.

EXAMPLE.– The worst-case complexity of the algorithm choosing the coins from left to right is n.

The complexity of an algorithm on an infinite number of instances cannot be defined in the same way: the number of values of complexity to be considered is potentially infinite. To define this complexity, the set of instances is partitioned into subsets of a finite cardinality indexed by $\mathbb{N}$, for example the instances that can be coded in n machine words, for any integer n. For any integer n, the complexity $f(n)$ (in the worst case, or on average) of the algorithm on the subset of index n is therefore well defined. The complexity $C(A)$ of the algorithm A on the problem is defined as the function f which for each integer n connects $f(n)$.

EXAMPLE.– The set of configurations is finite. For the hidden coin problem with n coins, where n is fixed, algorithm 2.1 will uncover all n coins in the worst case: its complexity in the worst case is therefore $f(n) = n$. Its average complexity on the uniform distribution of the instances containing only one silver coin is $(n+1)/2$.

Algorithm 2.1 Listing of the algorithm that uncovers coins from left to right

```
while there are still coins to uncover, and no silver coin has been found do
  uncover the coin the furthest to the left that has not been uncovered already;
end while
if a silver coin has been uncovered then
  answer positively
else
  answer negatively.
end if
```

2.1.2. *Probabilistic complexity of a problem*

To prove a lower bound to the complexity of a problem, we must be able to consider all the possible algorithms. To this end, the algorithms are represented in the form of trees, where each node is an instruction, each branch is an execution, and each leaf is a result: this is the *decision tree* model.

DEFINITION 2.1.– *A* questionnaire *is a tree whose leaves are labeled by classes and whose internal nodes are labeled by tests. If the test has k possible results, the internal node has k threads, and the k arcs linking the node to its threads are labeled by these results. The questionnaire allows us to decide to which class a given instance belongs. The instance is subjected to a series of tests, starting with the test contained in the root node. Following the result of the test, the series of tests continues in the corresponding sub-branch. If the sub-branch is a leaf, the label of this leaf is the class associated with the instance.*

DEFINITION 2.2.– *A* decision tree *is a questionnaire where each internal node corresponds to a deterministic operation that can be executed on any instance in a finite time.*

All deterministic algorithms ending in a finite time can be expressed as a decision tree. Each leaf then corresponds to a possible result of the algorithm. The number of operations carried out by the algorithm on this instance is thus the length of the corresponding branch.

DEFINITION 2.3.– *A* comparison tree *is a decision tree for which the tests are comparisons between internal elements of the instance.*

Knuth [KNU 73] uses comparison trees to obtain a lower bound on the complexity of comparison-based sorting algorithms. His analysis excludes algorithms such as the sorting by counting algorithm, which directly access the values being sorted, and hence are not in the comparison model.

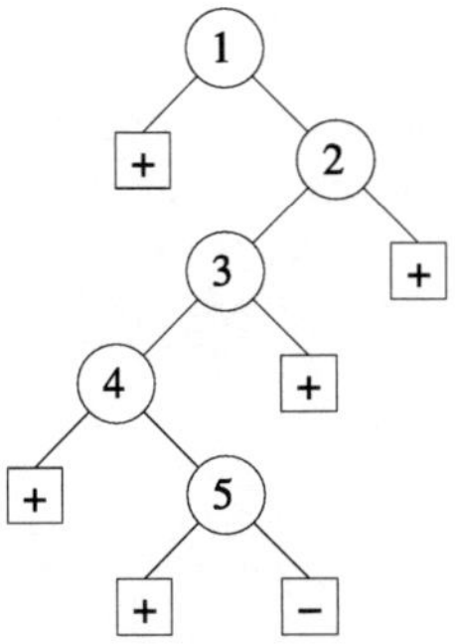

Figure 2.4. *A decision tree for the hidden coin problem when there are five coins*

EXAMPLE.– Any decision tree corresponding to an algorithm for the hidden coin problem must contain at least $n + 1$ leaves: one for each possible position for the first silver coin uncovered, and one for the configuration not containing any silver coins. Each test allows us to eliminate exactly one potential position for the silver coin. In this particular case, any decision tree corresponds to a chain. Its height is equal to n. The worst-case complexity of any algorithm for this problem is therefore $\Omega(n)$.

The decision tree model only concerns deterministic algorithms. It can be extended to probabilistic algorithms by distribution over deterministic decision trees.

DEFINITION 2.4.– *For a fixed problem, a* probabilistic algorithm *is defined by a distribution over deterministic algorithms. In the same way, a* probabilistic decision tree *is a distribution over decision trees.*

The complexity of a probabilistic algorithm R on an instance I is the average of the complexities of the deterministic algorithms A on I according to the distribution associated with R; $C(R, I) = \sum_A \Pr\{A\} C(A, I)$.

The model of probabilistic algorithms is more general than that of deterministic algorithms, and allows better performances.

EXAMPLE.– If an instance of the hidden coins problem contains $n/2$ silver coins and $n/2$ copper coins, for each deterministic algorithm there is a configuration of the coins such that it uncovers $n/2$ coins before uncovering a silver coin. The probabilistic algorithm choosing an order σ of positions at random will possibly also uncover up to $n/2$ coins, but with a very low probability of $1/2^{\,n/2}$. For the worst instance containing $n/2$ silver coins and $n/2$ copper coins, the probabilistic algorithm will uncover less than two coins on average: a lot less than a deterministic algorithm.

DEFINITION 2.5.– *The probabilistic complexity of a problem is equal to the average complexity of the best deterministic algorithm on the worst distribution.*

The complexity of any probabilistic algorithm for a given problem gives an upper bound to the probabilistic complexity of this problem: an example is given in section 2.3.1. Moreover, the minimax theorem allows us to obtain a lower bound to the probabilistic complexity of a given problem: this is presented and proved in section 2.2.

2.2. Lower bound technique

The minimax theorem is a fundamental theoretical tool in the study of probabilistic complexity. It is presented in the context of game theory, in particular for games for two players with a sum total of zero (games where the sum total of the winnings of the two players is always equal to zero). The Yao principle applies this theorem in the context of probabilistic algorithms; the first player applies the algorithm and the second creates an instance in a way that maximizes the complexity of the algorithm: this is related to the min–max algorithms (see Volume 2, Chapter 4).

2.2.1. *Definitions and notations*

Let Γ be a zero sum game for two players Alice and Bernard such that:

– the sets $A = \{a_1, a_2, \dots, a_m\}$ and $B = \{b_1, b_2, \dots, b_n\}$ of possible deterministic strategies for Alice and Bernard are finite;

– the winnings for Alice when she applies strategy a_i and when Bernard applies strategy b_j, are denoted by the element $M_{i,j}$ of the matrix M.

The set of probabilistic strategies (or mixed strategies) is denoted by $\mathcal{A}$ for Alice, $\mathcal{B}$ for Bernard. These probabilistic strategies are obtained by combining the deterministic strategies (or pure strategies) according to a probability vector:

$$\begin{aligned}\mathcal{A} = \quad & \{\alpha = (\alpha_1, \alpha_2, \ldots, \alpha_m) \in [0,1]^m \text{ such that } \textstyle\sum_{i=1}^m \alpha_i = 1 \\ & \quad \text{and the strategy } a_i \text{ is used with probability } \alpha_i \; \} \\ \mathcal{B} = \quad & \{\beta = (\beta_1, \beta_2, \ldots, \beta_n) \in [0,1]^n \text{ such that } \textstyle\sum_{j=1}^n \beta_j = 1 \\ & \quad \text{and the strategy } b_j \text{ is used with probability } \beta_i\}\end{aligned}$$

Of course, the pure strategies are included in the set of mixed strategies, as the probability vectors for which the full weight is on one component: a_1 corresponds to the mixed strategy of the probability vector $(1, 0, \ldots, 0)$. A mixed strategy is a linear combination of pure strategies: $\alpha = \alpha_1 a_1 + \alpha_2 a_2 + \ldots + \alpha_n a_n$.

COMMENT 2.1.– The performance of a mixed strategy α for Alice against a mixed strategy β for Bernard is calculated by the following formula:

$$\alpha^T M \beta = \sum_{i=1}^{m} \sum_{j=1}^{n} \alpha_i M_{i,j} \beta_j$$

A couple of strategies (α^*, β^*) is a *Nash equilibrium* of the game if, for all strategies α and β, $\alpha^T M \beta^* \leqslant {\alpha^*}^T M \beta^* \leqslant {\alpha^*}^T M \beta$. These configurations have the specificity that each player reduces his winnings if he is the only one to change strategy. For certain problems, there is a Nash equilibrium among the pure strategies. Von Neumann's minimax theorem shows that there is always a Nash equilibrium among the mixed strategies. Loomis's lemma shows that this equilibrium is attained by a couple formed from a pure strategy and a mixed strategy. The Yao principle is a reformulation of the minimax theorem which shows that the average complexity of the best deterministic algorithm over the worst distribution is the worst-case complexity of the best probabilistic algorithm. These results are proved in the following sections.

2.2.2. *Minimax theorem*

The minimax theorem is proved using the fixed point theorem, and lemmas 2.1 and 2.3. The lemmas are proved here; for the proof of the fixed point theorem, refer to [BOR 98, p. 112]. The proof of the minimax theorem introduced here is drawn from the one presented by Borodin and El-Yaniv [BOR 98], but slightly extended for clarification. Lemma 2.1 is a basic result used to show the double inequality of the minimax theorem (theorem 2.2).

LEMMA 2.1.– *The existence of $\tilde{\alpha}$ and $\tilde{\beta}$. Let ϕ and ψ be defined in $\mathbb{R}^m$ and $\mathbb{R}^n$ by:*

$$\phi(\alpha) = \sup_{\beta} \alpha^T M\beta \text{ and } \psi(\beta) = \inf_{\alpha} \alpha^T M\beta$$

Thus:

1) $\phi(\alpha) = \max_\beta \alpha^T M\beta$ and $\psi(\beta) = \min_\alpha \alpha^T M\beta$.

2) There are mixed strategies $\tilde{\alpha}$ for Alice and $\tilde{\beta}$ for Bernard such that ϕ reaches its minimum in $\tilde{\alpha}$ and ψ reaches its maximum in $\tilde{\beta}$.

Proof. The first point follows on from the bilinearity of M. If α is fixed, the function which associates $\alpha^T M\beta$ with β is linear, and thus iscontinuous, and, $\mathcal{B}$ being compact, it reaches its upper bound on $\mathcal{B}$ (theorem 29 in [SCH 81]): $\phi(\alpha) = \sup_\beta \alpha^T M\beta = \max_\beta \alpha^T M\beta$. The same logic applies to ψ.

To demonstrate this second point, it is sufficient to show that ϕ and ψ are continuous: the existence of $\tilde{\alpha}$ and $\tilde{\beta}$ is implied by the compactness of $\mathcal{A}$ and $\mathcal{B}$. To show the continuity of ϕ (the reasoning is equally valid for ψ), let $\alpha \in \mathcal{A}$ and $\varepsilon = (\varepsilon_1, \varepsilon_2, \ldots, \varepsilon_m)$:

$$\begin{aligned} \phi(\alpha + \varepsilon) &= \max_\beta (\alpha + \varepsilon)^T M\beta \\ &\leqslant \max_\beta \alpha^T M\beta + \max_\beta \varepsilon^T M\beta \\ &= \phi(\alpha) + \max_\beta \varepsilon^T M\beta. \end{aligned}$$

But for any norm $|.|$, $|\varepsilon^T M\beta| \leqslant |\varepsilon| \times |M\beta|$. Since $\mathcal{B}$ is of finite dimension n, $|M\beta|$ is finite, and $|\varepsilon^T M\beta|$ tends towards zero when $|\varepsilon|$ tends towards zero. Therefore $\lim_{|\varepsilon| \to 0} (\phi(\alpha + \varepsilon) - \phi(\alpha)) = 0$, and ϕ is continuous in α. ∎

The minimax theorem states that $\min_\alpha \max_\beta \alpha^T M\beta$ is equal to $\max_\beta \min_\alpha \alpha^T M\beta$. This is proved by showing a double inequality, the first of which is valid even for pure strategies. Lemma 2.2 below is the "easy" part of the minimax theorem.

LEMMA 2.2.– *Simple inequality.*

$$\min_{\alpha} \max_{\beta} \alpha^T M \beta \geqslant \max_{\beta} \min_{\alpha} \alpha^T M \beta$$

Proof. Let $\tilde{\alpha}$ and $\tilde{\beta}$ be expressed by lemma 2.1:

$$\min_{\alpha} \max_{\beta} \alpha^T M \beta = \max_{\beta} \tilde{\alpha}^T M \beta \geqslant \tilde{\alpha}^T M \tilde{\beta} \geqslant \min_{\alpha} \alpha^T M \tilde{\beta} = \max_{\beta} \min_{\alpha} \alpha^T M \beta$$

∎

The lemma below is used to prove the minimax theorem.

LEMMA 2.3.– *Minimax lemma. In the game Γ the following two conditions are equivalent:*

(i) $\min_{\alpha} \max_{\beta} \alpha^T M \beta = \max_{\beta} \min_{\alpha} \alpha^T M \beta$.

(ii) *There is a real number v and mixed strategies $\tilde{\alpha}$ and $\tilde{\beta}$ such that:*

$$\tilde{\alpha}^T M b_j \leqslant v \quad \forall j = 1, 2, \ldots, n$$

$$a_i^T M \tilde{\beta} \geqslant v \quad \forall i = 1, 2, \ldots, m$$

Proof. $(ii) \Rightarrow (i)$ Let us assume the existence of the real number v and mixed strategies $\tilde{\alpha}$ and $\tilde{\beta}$ of (ii): by the linearity of M, for all $\beta \in \mathcal{B}$, we obtain $\tilde{\alpha}^T M \beta \leqslant v$. Therefore, in particular,

$$\max_{\beta} \tilde{\alpha}^T M \beta \leqslant v$$

which implies by the minimum definition, $\min_{\alpha} \max_{\beta} \alpha^T M \beta \leqslant v$. The same applies for $\tilde{\beta}$, $v \geqslant \max_{\beta} \min_{\alpha} \alpha^T M \beta$. So $\min_{\alpha} \max_{\beta} \alpha^T M \beta \leqslant v \leqslant \max_{\beta} \min_{\alpha} \alpha^T M \beta$. Moreover, lemma 2.2 implies the inverse inequality:

$$\min_{\alpha} \max_{\beta} \alpha^T M \beta \geqslant \max_{\beta} \min_{\alpha} \alpha^T M \beta$$

hence point (i).

$(i) \Rightarrow (ii)$ Let us assume that we have the equality (i). Let $v = \min_{\alpha} \max_{\beta} \alpha^T M \beta$. Using lemma 2.1, $\tilde{\alpha} \in \mathcal{A}$ exists such that $\max_{\beta} \tilde{\alpha}^T M \beta = v$. By definition, $\forall \beta \in \mathcal{B}$ $\tilde{\alpha}^T M \beta \leqslant v$, and the inequality $\tilde{\alpha}^T M b_j \leqslant v$ of (ii) is confirmed for all values of j from 1 to n. In the same way, there is a $\tilde{\beta} \in \mathcal{B}$ such that $\min_{\alpha} \alpha^T M \tilde{\beta} = v$, therefore $\forall \alpha \in \mathcal{A}$ $\alpha^T M \tilde{\beta} \geqslant v$, and as a consequence the inequality $a_i^T M \tilde{\beta} \leqslant v$ of (ii) is confirmed for all values of i from 1 to m. ∎

The proof of the minimax theorem given here relies on Brouwer's fixed point theorem.

THEOREM 2.1.– *Brouwer's fixed point [BOR 98, p. 112]. Let X be a compact and convex set in $\mathbb{R}^n$. Any continuous function $\phi : X \to X$ allows a fixed point $x \in X$ such that $\phi(x) = x$.*

THEOREM 2.2.– *von Neumann's minimax theorem. The game Γ is defined by the matrix M:*

$$\min_{\alpha} \max_{\beta} \alpha^T M\beta = \max_{\beta} \min_{\alpha} \alpha^T M\beta$$

Proof. Given the mixed strategies of each player α and β, let:

$$\begin{array}{lllll} p_i & = & p_i(\alpha, \beta) & = & a_i^T M\beta - \alpha^T M\beta \quad \forall i = 1, \ldots, m \\ q_j & = & q_j(\alpha, \beta) & = & \alpha^T M\beta - \alpha^T M b_j \quad \forall j = 1, \ldots, n \end{array}$$

Let the function Φ be: $\mathcal{A} * \mathcal{B} \to \mathcal{A} * \mathcal{B}$ such that each couple (α, β) matches the couple of mixed strategies (ξ, η) expressed by:

$$\begin{array}{ll} \forall i = 1, \ldots, m & \xi_i = \dfrac{\alpha_i + \max\{p_i, 0\}}{1 + \sum_{k=1}^{n} \max\{p_k, 0\}} \\ \forall j = 1, \ldots, n & \eta_j = \dfrac{\beta_j + \max\{q_j, 0\}}{1 + \sum_{k=1}^{m} \max\{q_k, 0\}} \end{array}$$

The function Φ is the sum of continuous functions: it is continuous. $\mathcal{A}$ and $\mathcal{B}$ are compact and convex vector spaces. Hence theorem 2.1 implies that Φ allows a fixed point $(\tilde{\alpha}, \tilde{\beta})$. Let p_i be considered at the point $(\tilde{\alpha}, \tilde{\beta})$: $p_i = p_i(\tilde{\alpha}, \tilde{\beta})$ for all i. We have:

$$\begin{aligned} \sum_i \tilde{\alpha}_i p_i &= \sum_i \tilde{\alpha}_i (a_i^T M\tilde{\beta} - \tilde{\alpha}^T M\tilde{\beta}) \\ &= (\sum_i \tilde{\alpha}_i a_i)^T M\tilde{\beta} - (\sum_i \tilde{\alpha}_i)\tilde{\alpha}^T M\tilde{\beta} \\ &= \tilde{\alpha}^T M\tilde{\beta} - \tilde{\alpha}^T M\tilde{\beta} \\ &= 0 \end{aligned}$$

The terms of a zero sum are either all zero, or include at least one negative and one positive term:

– If the terms are all zero, for all i $\tilde{\alpha}_i = 0$ or $p_i = 0$. Let us assume $p_i \neq 0$. From the fixed point definition $\tilde{\alpha}_i \sum_{k=1}^{n} \max\{p_k, 0\} = \max\{p_i, 0\}$: we have $\max(p_i, 0) = 0$ and $p_i \leqslant 0$. So $p_i \leqslant 0$ for all i.

– If not all the terms are zero, there is a strictly negative term $\tilde{\alpha}_i p_i$. $\tilde{\alpha}$ being a distribution of probability, $\tilde{\alpha}_i$ is positive, which implies that p_i is strictly negative. From the fixed point definition, $\tilde{\alpha}_i \sum_{k=1}^{n} \max\{p_k, 0\} = \max\{p_i, 0\}$, which is zero since $p_i < 0$. The terms of the zero sum of positive terms $\max\{p_k, 0\}$ are all zero, so $p_k \leqslant 0$ for all k.

The q_j are all shown to be negative or zero in a similar way. The point (ii) of lemma 2.3 is confirmed by noting that $v = \tilde{\alpha}^T M \tilde{\beta}$, which proves the theorem. ∎

2.2.3. *The Loomis lemma and the Yao principle*

The Loomis lemma is the essential argument for extending the minimax theorem to the Yao principle. The proof given here is the one given by Borodin and El-Yaniv [BOR 98].

LEMMA 2.4.– *The Loomis lemma [BOR 98, Lemma 8.2]. Given a fixed mixed strategy α for Alice, there is an optimal deterministic strategy b_j for Bernard when Alice applies the strategy α:*

$$\max_{\beta} \alpha^T M \beta = \alpha^T M b_j$$

In the same way, given a fixed mixed strategy β for Bernard, there is an optimal deterministic strategy a_i for Alice when Bernard applies the strategy β:

$$\min_{\alpha} \alpha^T M \beta = a_i^T M \beta$$

Proof. For a fixed α, $\alpha^T M$ is a linear function on $\mathcal{B}$, and for this reason reaches its maximum in at least one pure strategy b_j, corresponding to a maximum coefficient of $\alpha^T M$. Therefore, $\max_{\beta} \alpha^T M \beta = \alpha^T M b_j$. The same reasoning holds for Bernard. ∎

The equality of the minimax theorem can be rewritten by applying the Loomis lemma to optimal strategies $\tilde{\alpha}$ and $\tilde{\beta}$:

$$\min_{\alpha} \max_{j} \alpha^T M b_j = \max_{\beta} \min_{i} a_i^T M \beta \qquad [2.1]$$

These game theory results are applied to the analysis of algorithmic complexity. Let B be the *finite* set of instances of size s, $\mathcal{B}$ the set of random distributions over these instances, A a *finite* set of deterministic algorithms that solve the problem for a fixed data size s, and $\mathcal{A}$ the set of probabilistic algorithms defined by a distribution of probability over A.

Thus $\alpha^T M b_j$ is the average complexity of the probabilistic algorithm $\alpha \in \mathcal{A}$ over the instance $b_j \in B$, and $\max_j \alpha^T M b_j$ is the average complexity of this algorithm over the worst instance. Symmetrically, $a_i^T M \beta$ is the average complexity of the deterministic algorithm $a_i \in A$ over the distribution of instances $\beta \in \mathcal{B}$ and $\max_\beta a_i^T M \beta$ is the complexity of this algorithm over the worst imaginable distribution of instances.

The following inequality, known as "Yao's inequality", is obtained by applying this interpretation to equation [2.1]. This inequality allows us to obtain lower bounds to the complexity of probabilistic algorithms.

THEOREM 2.3.– *Yao principle [YAO 77]. The complexity of the best deterministic algorithm over the worst distribution of instances $\beta \in \mathcal{B}$ is equal to the complexity of the best probabilistic algorithm over the worst instance:*

$$\max_\beta \min_i a_i^T M \beta = \min_\alpha \max_j \alpha^T M b_j$$

This approach allows us to obtain lower bounds for *finite* sets of deterministic algorithms over sets of finite instances. This is not a difficult bound to reach, for a finite number of instances, once the size of the instance has been fixed. For example, the computational model based on comparisons is reduced to the algorithms that do not carry out the same comparison twice. The number of such algorithms is polynomial in the number of elements to be compared. There is therefore a finite number of algorithms on instances of fixed size.

COMMENT 2.2.– The fact that $\mathcal{A}$ and $\mathcal{B}$ are of a finite dimension is crucial in all the extensions of the minimax principle [SIO 58].

EXAMPLE.– In the hidden coins problem, a single silver coin being uniformly hidden, any deterministic algorithm uncovers an average of $n/2$ cards. The Yao principle implies that the best Las Vegas algorithm uncovers on average $n/2$ cards for at least one configuration.

This factor of $\frac{1}{2}$ is not very important, but more important differences in performance appear between deterministic and probabilistic algorithms when complexity is analyzed more closely.

EXAMPLE.– In the hidden coin problem, if half the coins are silver, and if they are evenly distributed, the best deterministic algorithm uncovers $\lceil n/2 \rceil$ coins in the worst case, while the best probabilistic algorithm uncovers less than 2 coins, on average, in the worst instance.

2.3. Elementary intersection problem

The analysis of the hidden coin problem can be adapted to other, less trivial problems, which can be decomposed as a conjunction of independent subproblems. The elementary intersection problem, which is relatively close to the hidden coins problem used throughout this chapter, is given as an applied example.

Formally, an instance of the elementary intersection problem is made up of an element x and k sorted tables, of respective sizes $n_1 \leqslant \ldots \leqslant n_k$. The problem therefore consists of deciding whether or not x belongs to the intersection of the tables, that is whether there is a table that does not contain x. The analogy with the hidden coins problem can be seen if we make a copper coin correspond to the tables containing x, and a silver coin to the tables not containing x. As for the hidden coin problem, the instance is harder if almost all the tables contain x, both for the deterministic and the probabilistic algorithms, and a probabilistic algorithm will end more quickly than a deterministic algorithm if only half the tables contain x.

We will show here how to obtain a lower bound on the complexity of the elementary intersection problem using the Yao principle, and how to obtain an upper bound by analyzing a simple randomized algorithm.

2.3.1. *Upper bound*

The complexity of any probabilistic algorithm for a given problem forms an upper bound on the probabilistic complexity of this problem. The bound is all the better since the complexity of the algorithm is reduced. By definition, this complexity is equal to the average of the complexities of the deterministic algorithms making up the probabilistic algorithm on a fixed worst instance: $\max_I \sum_i p_i C(A_i, I)$. This corresponds to the last example in section 2.1.2, in the case where half the coins are made of silver.

To calculate this complexity, we draw upon the Yao principle, adapting the randomization of the algorithm on the instance, in such a way as to define a worst distribution of instances, from the worst instance: then the complexity in the worst case of the probabilistic algorithm corresponds to the average complexity of a deterministic algorithm over a worst distribution.

EXAMPLE.– In the example of the hidden coins problem, any probabilistic algorithm can be defined as a distribution $(p_i)_{i \in \{1,\ldots,n!\}}$ over the permutations of $\{1, \ldots, n\}$. For any instance I, and in particular for the worst instance, $(p_i)_{i \in \{1,\ldots,n!\}}$ defines a distribution over the permutations of I. The complexity of a probabilistic algorithm then corresponds to the average complexity of any deterministic algorithm on this distribution.

The complexity of any probabilistic algorithm that solves the elementary intersection problem forms an upper bound to the probabilistic complexity of the problem.

Let us consider a deterministic algorithm that is looking for element x in each table using the dichotomous search algorithm. The cost of the search, in numbers of comparisons in a table of n_i elements, is $O(\log n_i)$. Thus this type of deterministic algorithm carries out at most $O(\sum_{i=1}^{k} \log n_i)$ comparisons.

Therefore, the probabilistic complexity of the elementary intersection problem of an instance of signature $(k, n_1, \ldots, n_k)$ must be less than or equal to, give or take a multiplicative constant, $\sum_{i=1}^{k} \log n_i$. We will show that this upper bound corresponds to the upper bound, and that therefore this deterministic algorithm is optimal amongst all the deterministic or probabilistic algorithms that solve the elementary intersection problem.

2.3.2. *Lower bound*

To prove a lower bound on the worst-case complexity of any *deterministic* algorithm, it is sufficient to define an opponent's strategy, which constructs, for each deterministic algorithm, its worst instance. Such a method provides a lower bound of $\sum_{i=1}^{k} \log n_i$, which proves that the type of algorithm described in the previous section is optimal among the deterministic algorithms, but does not prove anything about probabilistic algorithms.

To prove a lower bound on the worst-case complexity of any *probabilistic* algorithm, we use the Yao principle, which allows us to obtain this bound from a worst distribution for all the deterministic algorithms. The following lemma defines such a worst distribution.

LEMMA 2.5.– *For any set of integers $k > 1$, $n_1, \ldots, n_k > 0$, there is a distribution of instances $(x, A_1, \ldots, A_k)$ of signature $(k, n_1, \ldots, n_k)$ such that any deterministic algorithm that decides whether x is in the intersection $A_1 \cap \ldots \cap A_k$ carries out on average at least $\Omega(\sum_{i=1}^{k} \log n_i)$ comparisons.*

The distribution is defined by choosing a table A_w where x may be absent with probability $\log n_i / \sum \log n_i$ for each table A_i; for each other table, we choose a position where x may be placed with probability $1/n_i$ for each table A_i. Any algorithm looking for x will find it in $k/2$ tables on average before finding the table A_w that does not contain x, and will carry out, in each of these tables, on average $\Omega(\sum_{i=1}^{k} \log n_i)$ comparisons. The rigorous proof of the lower bound is not relevant to us, and is given elsewhere [BAR 08].

By directly applying the Yao principle we obtain the desired lower bound.

THEOREM 2.4.– *For any set of integers $k > 1$, $n_1, \ldots, n_k > 0$, and for any probabilistic algorithm A that solves the elementary intersection problem, there is an instance $(x, A_1, \ldots, A_k)$ of signature $(k, n_1, \ldots, n_k)$ such that A carries out on average at least $\Omega(\sum_{i=1}^{k} \log n_i)$ comparisons on the instance $(x, A_1, \ldots, A_k)$.*

The probabilistic complexity of the elementary intersection problem is therefore greater than or equal to $\sum_{i=1}^{k} \log n_i$, give or take a multiplicative constant.

2.3.3. *Probabilistic complexity*

The previous results give lower and upper bounds on the probabilistic complexity accurate to one multiplicative constant. This is sufficient to know the order of magnitude of the probabilistic complexity of the elementary intersection problem, and to prove that the type of deterministic algorithm proposed in section 2.3.1 is optimal.

2.4. Conclusion

The analysis of probabilistic algorithms is not generally considered as combinatorial optimization, while heuristics, particular probabilistic algorithms, are often essential to solve **NP**-difficult problems, and in particular for optimization problems. This is certainly because in practice it is difficult to analyze the algorithms used, and rare to be able to prove close lower and upper bounds. Let us hope that this chapter helps to bridge the gap between the theoretical analysis of probabilistic complexity and the study of heuristics.

2.5. Bibliography

[BAR 08] BARBAY J., KENYON C., "Alternation and redundancy analysis of the intersection problem", *ACM Trans. Algorithms*, vol. 4, num. 1, p. 1–18, 2008.

[BOR 98] BORODIN A., EL-YANIV R., *Online Computation and Competitive Analysis*, Cambridge University Press, Cambridge, 1998.

[KNU 73] KNUTH D.E., *The Art of Computer Programming, vol. 3: Sorting and Searching*, Addison-Wesley, Reading, MA, 1973.

[MOT 95] MOTWANI R., RAGHAVAN P., *Randomized Algorithms*, Cambridge University Press, New York, 1995.

[PAP 94] PAPADIMITRIOU C.H., *Computational Complexity*, Addison-Wesley, Reading, MA, 1994.

[SCH 81] SCHWARTZ L., *Cours d'analyse*, Hermann, Paris, 1981.

[SIO 58] SION M., "On general Minimax theorems", *Pacific Journal of Mathematics*, p. 171–176, 1958.

[YAO 77] YAO A.C., "Probabilistic computations: Toward a unified measure of complexity", *Proc. FOCS'77*, p. 222–227, 1977.

PART II

Classical Solution Methods

Chapter 3

Branch-and-Bound Methods

3.1. Introduction

This chapter describes the principles of *branch and bound* in combinatorial optimization and gives details of one of its applications. These methods are also known by other names: *implicit enumeration methods*; SSE or $PSSE$ (for *procedures using separation and sequential evaluation*); *divide and conquer*; or $A*$ *algorithms* in artificial intelligence (see for example [LAU 86]). A name such as *tree enumeration by separation and pruning* would perhaps be a more accurate description of the characteristics of these methods, without limiting pruning to the fruits of the evaluation, but is not one of the common names. For bibliographical references to these particular methods and combinatorial optimization in general, see [CHA 96, DEL 97, GON 84, KOR 02, MIN 86, SAK 84, SCH 03, WER 03, WOL 98]. We assume that the maximization (respectively minimization) problems that we solve are of the following form:

$$\max_{\omega \in \Omega} f(\omega) \text{ (respectively } \min_{\omega \in \Omega} f(\omega))$$

where Ω is a finite set, the elements of which will be called *feasible solutions*, or simply *solutions*, and f is a function, called the *objective function*, for Ω in the set of real numbers $\mathbb{R}$; an element that gives its optimal value to f will be called the *optimal solution* or *exact solution* of the problem. We will assume in what follows that it is easily possible to generate the elements of Ω and that it is easy, for a given element

---

Chapter written by Irène CHARON and Olivier HUDRY.

ω of Ω, to calculate $f(\omega)$[1]. We denote by f_{max} (respectively f_{min}) the maximum (respectively the minimum) of f.

The relation given by the theorem below, true whatever the form of f and whatever the set Ω, allows us to solve maximization problems if we know how to solve the minimization problems with the opposite sign for the objective function and *vice versa*.

THEOREM 3.1.– *For any set Ω and any real function f on Ω, we have:*

$$\max_{\omega \in \Omega} f(\omega) = -\min_{\omega \in \Omega} [-f(\omega)]$$

We also note that the finiteness of Ω guarantees the existence of an optimal solution that we can compute, at least in theory, by an exhaustive enumeration: we inspect all the solutions of the problem one by one and we memorize, as we go along, the best solution found since the beginning; the solution retained in this way gives an optimal solution at the end of this enumeration.

In practice, such a method is only conceivable if the cardinality $|\Omega|$ of Ω is not too big. For many problems, $|\Omega|$ increases very quickly with the size of the instance being considered, which leads very quickly to prohibitive calculation times (greater than the life expectancy of the solar system, for example). For this reason we try to avoid an exhaustive enumeration in favor of more efficient algorithms; when possible, we prefer algorithms whose complexity is upper bounded by a polynomial function of the size of the variables. There are, however, problems, most notably **NP**-hard problems (see Chapter 1 for a discussion of algorithmic complexity theory and a classification of problems according to their difficulty), for which such polynomial algorithms are not known. Several approaches are then possible (see the other chapters in this book): abandoning an exact solution to use an approximate solution that we will be able to compute in a "reasonable" time with the help of a heuristic (i.e. an approximate method, general or specific, with or without a performance guarantee); applying probabilistic algorithms that will give an optimal solution with a certain probability; designing an exact method, at the risk of being confronted with high or even unacceptable calculation times, etc.

Branch-and-bound methods belong to the category of exact methods: they provide one or all of the optimal solutions of the considered instance for various optimization

1. For certain problems, it can be difficult to generate even one element of Ω; this is the case, for example, if Ω is the set of Hamiltonian cycles of a weighted graph, not necessarily complete, for which we seek a Hamiltonian cycle of minimum weight. In the same way, for a given element ω of Ω, it can be difficult to evaluate $f(\omega)$; this is the case for instance if ω is an integer and the calculation of $f(\omega)$ requires prime factors of ω.

problems[2]. The global principle of branch-and-bound methods often involves, at least implicitly, an exhaustive enumeration (with all the prohibitive calculation time problems that this is likely to involve), with the fundamental difference that we will try to make this enumeration the least exhaustive possible. The idea of exploiting this principle to solve combinatorial optimization problems with the help of computers goes back to the 1950s, with the pioneering work of G.B. Dantzig, D.R. Fulkerson and S.M. Johnson [DAN 54, DAN 59], then of C. Tompkins [TOM 56], M.J. Rossman and R.J. Twery [ROS 58], and W.L. Eastman [EAS 58, EAS 59]. Branch-and-bound methods appeared explicitly for the first time in 1960, in an article by A.H. Land and A.G. Doig [LAN 60]. J.D.C. Little, K.G. Murty, D.W. Sweeney and C. Karel [LIT 63] named this type of method *branch and bound* in an article in 1963 devoted to the exact solution of the traveling salesman problem, while P. Bertier and B. Roy developed, from 1964 onwards (see [BER 64] and [BER 65]for several combinatorial problems). Other articles published in the 1960s include [AGI 66, BAL 65, BAL 68, DAK 65, GEO 67, GEO 69, GOL 65, HER 67, LAW 66, ROY 65, ROY 69, WAL 60]. Since then, this method has been applied to countless combinatorial optimization problems and its applications have benefited from various improvements (see the references cited above).

Practical use of a branch-and-bound method requires the specification of several ingredients; a general description is given in section 3.2, while section 3.3 describes them for the binary knapsack problem. Section 3.4 concludes by giving some variants of branch-and-bound methods.

3.2. Branch-and-bound method principles

Branch-and-bound methods have a common feature: constructing a *rooted tree*[3] A (sometimes called the *search tree* of the branch-and-bound method); each vertex S of A represents a subset $\Omega(S)$ of the set Ω of the solutions of the problem. The root R

2. For certain problems, an approximate solution will suffice if the quality of the solution, measured by the difference (relative or absolute) between the obtained value and the optimal value, is judged sufficient. A variant of branch-and-methods methods, which we mention briefly later on, which simply consists of interrupting the method during execution, allows us, in some cases, to obtain such an approximate solution, which will not necessarily be exact. This is not, however, the main purpose of branch-and-bound methods.

3. Let us remember that a rooted tree in graph theory (see [BER 85] on graph theory) is an oriented graph with a particular vertex R, called the *root*, such that there is only one path from R to any other vertex of A; in what follows, we will call such a path going from R to a vertex of A a *branch*. The *level* (or *depth*) of a vertex S of A is the number of arcs (i.e. directed edges) of the branch going from R to S. There are examples of trees in section 3.3. For an arc (s, t) of A, we say that s is the *predecessor* or the *parent* of t and that t is a *successor* or a *child* of s.

of A is associated with Ω itself: $\Omega(R) = \Omega$. The development of A from the root is done with the help of three main ingredients:

– a principle of separation;

– principles of pruning, generally the use of evaluation functions and bounds;

– a development strategy for the tree.

3.2.1. *Principle of separation*

The principle of separation conditions practical applications of the method. This is what specifies how the vertices S of A and the associated sets $\Omega(S)$ are constructed. In general, separation is obtained by an exhaustive enumeration principle, ensuring that no solution is missed. As we will see, however, there can be considerations other than pure enumeration.

The role of separation consists of replacing the set on which we seek to optimize f with several smaller sets. Let S be the current vertex of A to which we wish to apply the principle of separation, and let $\Omega(S)$ be the set of solutions associated with it. Separating S creates a certain number $q(S)$ which depends on S of subsets Ω_1, Ω_2, ..., $\Omega_{q(S)}$ of $\Omega(S)$, the union of which is equal to $\Omega(S)$:

$$\Omega(S) : \bigcup_{i=1}^{q(S)} \Omega_i = \Omega(S)$$

With respect to the tree under construction, this comes down to creating (if they are not eliminated by pruning criteria) $q(S)$ successors $S_1, \ldots, S_{q(S)}$ of S, respectively associated with $\Omega_1, \ldots, \Omega_{q(S)}$: for $1 \leq i \leq q(S)$, $\Omega(S_i) = \Omega_i$. Any solution appearing in the set associated with S must end up in the subset of at least one successor S_i of S (possibly in several) and, *vice versa*, only the solutions of $\Omega(S)$ can end up in the successors of S. In other words, we must neither lose any solutions on the way (which could be catastrophic for the exactness of the method), nor add any solution that has been eliminated previously (which would be clumsy and could reduce the efficiency of the method). Actually, it is generally preferable for the efficiency of the method to ensure that each element of $\Omega(S)$ ends up in one and only one of the sets $\Omega(S_i)$, which is expressed by: $i \neq j \Rightarrow \Omega_i \cap \Omega_j = \emptyset$. Therefore we seek to conceive a principle of separation allowing us to partition the set $\Omega(S)$ into $q(S)$ subsets.

In practice, the principle of separation can be obtained in various ways, for example by testing one after the other all the values that we can assign to the variables that describe the problem, which means that the number of possible values has to be finite and the values must be easy to determine (one variant consists of forbidding certain values, another in partitioning the set of all the possible values of a variable into subsets that are not necessarily singletons). Thus, in the case of the binary knapsack problem, which is discussed later, the variables can take only two values, which gives two possible branches going from each vertex in the tree[4]. We can be led to change the formulation of the problem and progressively create new problems (in this case we must check that we can obtain an optimal solution for the original problem with the help of these new problems). This way of proceeding, which can be quite different from the principles related to an exhaustive enumeration, will be illustrated later using the binary knapsack problem (see section 3.3.8, where the principle of separation has the effect of progressively adding new constraints). Furthermore, when the set Ω comes from structural constraints, the principle of separation can also be guided by the structure to be obtained, and can seek to build progressively an optimal solution according to this structure. Depending on the problem and the principle of separation used, the number $q(S)$ of successors of S may be independent of S. This can sometimes simplify the way to program the method. However, the algorithm can be more natural with a variable number of successors.

In order to build the search tree A, we separate the vertices of A as long as necessary, starting at the root. Section 3.2.2 will allow us to define what we mean by "as long as necessary".

3.2.2. *Pruning principles*

If we only applied the principle of separation as described above, we might have to develop A until its leaves[5] are associated with singletons of Ω, which can require considerable calculation time (exponential growth). It is therefore essential to prune A by eliminating unnecessary vertices.

To define the conditions under which a vertex of A can be considered to be unnecessary, we consider two cases: in the first one, we are seeking only one optimal solution; in the second, we want them all. In the first case, a vertex S of A is unnecessary if it cannot contain an optimal solution or if we have already found an element of

4. An integer variable α, not necessarily binary but positive and upper bounded by α_0, can be decomposed into powers of two (we are thus considering the binary decomposition of α): $\alpha = \sum_{j=0}^{\lceil \log_2(\alpha_0+1) \rceil - 1} \beta_j \, 2^j$, where $\lceil a \rceil$ denotes the ceiling of a and where the terms β_j are each equal to 0 or 1. Thus, we come back to a linear programming problem with binary variables (the variables β_j), and we can apply the principle of separation into the two branches that we just referred to (one for $\beta_j = 1$, one for $\beta_j = 0$) for each vertex of A. This process is more theoretical, by showing that binary trees can be more general than we might think, than practical, due to the great number of variables of type β_j that we are led to introduce, and the even greater number (since this can increase exponentially with the number of variables) of vertices of A. It is often preferable to use one of the other possibilities previously mentioned.

5. Let us remember that a *leaf* of a tree is a vertex without a successor.

Ω at least as good as the best element of $\Omega(S)$. In the second case, S is unnecessary if $\Omega(S)$ cannot contain an optimal solution.

We will consider two causes of pruning: the first uses a bound and an evaluation function to estimate the optimum of f in a set $\Omega(S)$ when considering S; the second takes into account the combinatorial or structural properties of the solutions that are sought.

3.2.2.1. *Bound*

The definition of a bound depends on the nature of the problem to be solved:

– If the problem consists of maximizing f, we define the *bound B* as any lower bound of $f_{max} : B \leqslant f_{max}$.

– If the problem consists of minimizing f, we define the *bound B* as any upper bound of $f_{min} : B \geqslant f_{min}$.

Thanks to the hypotheses stated at the start of this chapter, it is easy to calculate a bound: any element ω of Ω provides one in the shape of $f(\omega)$. However, the efficiency of the bound will generally be higher if the bound is better, that is closer to the optimum (this efficiency also depends on the tree-development strategy, see section 3.2.3). It might therefore be in our interest to calculate a good bound, found by an efficient heuristic. Anyway, we can always consider an element of Ω, chosen at random or by more relevant criteria, or even, if necessary, take $+\infty$ for a minimization problem and $-\infty$ for a maximization problem (sometimes 0, depending on the sign of the values taken by f).

We often compute certain elements of Ω during execution of the branch-and-bound method. In this case, we update the bound by keeping the best value among those that might serve as the bound.

3.2.2.2. *Evaluation function*

The definition of an evaluation function also depends on the nature of the problem to be solved. Evaluating a vertex S of the tree consists of ascertaining a quantity that we will express as $E(S)$, the *evaluation of S*, that satisfies the following property:

– If the problem consists of maximizing f, $E(S)$ must be an upper bound of the maximum for f when we are restricted to the set $\Omega(S)$ associated with S: $E(S) \geq \max\limits_{\omega \in \Omega(S)} f(\omega)$.

– If the problem consists of minimizing f, $E(S)$ must be a lower bound of the minimum for f when restricted to the set $\Omega(S)$ associated with S:$E(S) \leq \min\limits_{\omega \in \Omega(S)} f(\omega)$.

In particular, when we evaluate the root R of A, $E(R)$ provides an upper bound (respectively a lower bound) of the maximum (respectively the minimum) sought for the initial maximization (respectively minimization) problem. Because the set of solutions associated with the vertices of a particular branch of A becomes smaller and smaller as we go along the branch away from the root, the evaluations (for a given evaluation function) of the vertices of any branch of A generally make up a monotonous series (in the broad sense) as we get further away from R (more precisely, this series is generally decreasing for a maximization problem or increasing for a minimization problem).

We can often split the evaluation $E(S)$ into two additive parts: the first part, $E_1(S)$, is the outcome of the choices that have been made to end up at S from the root R of A; the second, $E_2(S)$, predicts what is yet to be done to complete the decisions that have been made to end up at S in a way that gives an optimal element of $\Omega(S)$; the sum of the two partial evaluations gives $E(S)$: $E(S) = E_1(S) + E_2(S)$. Obviously, since no decision has yet been made when we are at R, unavoidably we have $E_1(R) = 0$; other vertices of A can give a zero value for E_1, depending on the separation principle. But let us consider the fairly frequent case where the separation principle consists of setting the value of certain variables that enable the problem to be described. The tree branch that goes from R to S thus corresponds to a series of decisions, each one setting the value of such variables. Let $V(S)$ be the set of variables with values set to end up at S and let $W(S)$ be the set of other variables. Since the evaluation of S consists of estimating the optimum of f on $\Omega(S)$ (which is, in this case, the set of elements of Ω such that the variables belonging to $V(S)$ have exactly the values that have been set), we can, for this, replace the variables of $V(S)$ in f by the values that have been assigned to them. The optimization of f on $\Omega(S)$ amounts to the optimization of a new function g_S defined on $\Omega(S)$, the expression of which depends only on the variables of $W(S)$. It is possible (this is in particular the case if f is linear) that this substitution causes a constant (in the sense that it does not depend on the variables of $W(S)$) term to emerge: this term comes exclusively from the decisions taken step by step to construct S and affects the expression of f for all the elements of $\Omega(S)$. It is this term that gives the first part, $E_1(S)$, of the evaluation of S. The second part of the evaluation, $E_2(S)$, corresponds to the optimum of g_S on $\Omega(S)$. As we go further from R in A, we can expect $E_1(S)$ to play an increasingly important part in the evaluation of S, since the number of variables for which the value is fixed increases as we go further from R. This decomposition can lead to a non-negligible gain in calculation time when evaluating a successor S' of a vertex S of A: instead of calculating $E_1(S')$ from nothing, we update $E_1(S)$ with the help of the contribution of the variable for which we have fixed the value to go from S to S', provided that, of course, we have easy access to $E_1(S)$. (For example, we can store this information with the other characteristics of S, at the expense of an increase in the memory space used; see also section 3.2.3.2). In general, such a "heredity" is unfortunately not applicable to the E_2 part of the evaluation because of the greater independence of E_2 with respect to the separation principle. The application of branch-and-bound methods to the binary knapsack problem will illustrate the decomposition of E into two components E_1 and E_2.

A classic way of obtaining $E(S)$ (or, when appropriate, $E_2(S)$) consists of immersing the set $\Omega(S)$ on which the optimization is being performed in a set $\Phi(S)$ containing $\Omega(S)$. Using the inequalities obtained from theorem 3.2, we see that such an embedding gives us the required result.

THEOREM 3.2.– *For any set Y, any set Z containing Y as a subset, and any function g defined on Z, we have the following relations:*

$$\max_{y \in Y} g(y) \leq \max_{z \in Z} g(z) \text{ and } \min_{z \in Z} g(z) \leq \min_{y \in Y} g(y)$$

This embedding is often achieved by eliminating or modifying the constraints of the problem that make the problem difficult to solve. This can most notably be the case for *integer linear programming problems* or *mixed linear programming problems*. These problems, met in for many practical situations, can be expressed in the following way: $\max c.x$ with $Mx \leq b$ and $x \in \mathbb{N}^p \times \mathbb{R}^q$, where $\mathbb{N}$ and $\mathbb{R}$ represent, respectively, the set of natural integers and the set of real numbers, where c is a given (line) vector, M a given matrix, b a given (column) vector, x an unknown (column) vector, and where p and q are integers defining the dimension of the a problem, with $q = 0$ for an integer linear programming problem. Classically, for these problems, we consider *integrity constraints relaxation* (or *continuous relaxation*), for which $\mathbb{N}$ is replaced by $\mathbb{R}$ in the previous expression. In other words, instead of looking for integers for the components of x that must belong to $\mathbb{N}$, we accept real numbers. This allows us to apply continuous linear optimization techniques (see Chapter 7), for example the simplex algorithm. In the same way, a binary constraint of the form $\alpha \in \{0, 1\}$ can be replaced by a constraint of the form $\alpha \in [0, 1]$.

Another classic method consists of applying the *Lagrangian relaxation*. Without going into the theory of Lagrangian relaxation here, let us recall the general principles. Let us assume that we want to solve a problem, known as the *primal problem*, of the following form:

$$\max_{y \in Y} g(y) \text{ with } m \text{ constraints of the form } h_j(y) \geq 0, \text{ for } 1 \leq j \leq m$$

where the set Y and the functions g and h_j $(1 \leq j \leq m)$ are given. The relaxation of the m constraints $h_j(y) \geq 0$ (let us note that we could relax only a part of them) consists of no longer taking into account the functions h_j as constraints, but rather aggregating them into g, each one with a variable weight known as the *Lagrange multiplier*. More exactly, we define the *Lagrange function* L as follows: for $y \in Y$ and for $\Lambda = (\lambda_j)_{1 \leq j \leq m} \in (\mathbb{R}^+)^m$:

$$L(y, \Lambda) = g(y) + \sum_{j=1}^{m} \lambda_j h_j(y)$$

Thanks to the sign that must be satisfied for the constraints h_j and to the positive sign imposed on the Lagrange multipliers λ_j, we may interpret $\sum_{j=1}^{m} \lambda_j h_j(y)$ as a way of penalizing a solution y not conforming to the constraints (if the constraint given by h_j is not satisfied by y, the term $\lambda_j h_j(y)$ is negative, which goes in the opposite direction from the maximization of g) or, on the other hand, as a means of "giving a bonus" to a solution that conforms to them (the term $\lambda_j h_j(y)$ is positive for each constraint h_j satisfied). We now define the *dual function* w: for $\Lambda = (\lambda_j)_{1 \leq j \leq m} \in (\mathbb{R}^+)^m$, $w(\Lambda) = \max_{y \in Y} L(y, \Lambda)$, and the *dual problem* by: $\min_{\Lambda \in (\mathbb{R}^+)^m} w(\Lambda)$. We can now establish a link between the maximum g_{max} of the primal problem and the minimum w_{min} of the dual problem by proving the following theorem.

THEOREM 3.3.– *With the previous notation, we have:* $w_{min} \geq g_{max}$.

As a consequence, if we know how to solve a dual problem (or even if we only know how to produce an upper bound of w_{min}), we can use w_{min} (or its upper bound) to design the E_2 part of the evaluation function.

Let us note, finally, that a good evaluation function is a function that is both quick to calculate (polynomial time) but also provides values that are close to the optimum that we seek to calculate.

3.2.2.3. *Use of the bound and of the evaluation function for pruning*

Simultaneous use of the bound and of the evaluation function E will enable, in certain cases, the elimination of vertices from the search tree A[6].

More precisely, let S be a vertex of A and let B be the value of the bound as we consider S (let us remember that the value of B may evolve during the run of the algorithm). Let us assume that the problem to be solved is a maximization problem. If we have the relationship $E(S) < B$, then $\Omega(S)$ cannot contain an optimal solution. Indeed, for a maximization problem, $E(S)$ provides an upper bound of f when we restrict to $\Omega(S)$ the set on which we maximize f, and B is a lower bound of f_{max}. We therefore have the series of inequalities $\max_{\omega \in \Omega(S)} f(\omega) \leq E(S) < B \leq f_{max}$, from which we deduce the strict inequality $\max_{\omega \in \Omega(S)} f(\omega) < f_{max}$, which shows that no element of $\Omega(S)$ can reach the maximum sought. We can therefore, under these conditions, eliminate S from the tree.

In the same way we can eliminate S if there is an equality between $E(S)$ and B and if we are only looking for one optimal solution, and if, furthermore, we know an

6. We will see in section 3.2.3 that the evaluation function can also play a role in the way in which A develops, by showing vertices which seem more promising than others.

element of Ω which gives f the value of the bound B (or if we are only looking for the optimal value, and not the structure of an optimal solution). Indeed, in this case, it is possible that $\Omega(S)$ contains an optimal solution, but $\Omega(S)$ cannot contain a solution that is strictly better than the one associated with B. Since the objective is to improve on what we already know, if that is possible, the vertex S cannot contain a solution of interest and therefore we can eliminate S from A.

We can recapitulate this using the following rules. For a maximization problem (respectively minimization problem), when we evaluate a vertex S of A:

– If we have $E(S) < B$ (respectively $E(S) > B$), we eliminate S.

– If we have $E(S) = B$, are seeking only one optimal solution, and know an element ω of Ω satisfying $f(\omega) = B$, then we can eliminate S.

– In the other cases, we keep S (unless another pruning principle allows us to eliminate it) in order to apply the separation principle to it at a later stage.

3.2.2.4. *Other pruning principles*

Occasionally, in certain cases, we can eliminate the vertex S from the tree for reasons other than those resulting from the comparison between the bound B and from evaluation $E(S)$. Generally, as stated above, this is the case when we can show that S does not contain any interesting solutions.

In particular, we avoid (if possible) creating a vertex S associated with an empty set $\Omega(S)$. Therefore, if a decision made at the time of a separation leads to not being able to simultaneously satisfy all the constraints of the problem (in other words, there is an incompatibility between this decision and anterior decisions), we prune the branch associated with this choice, since otherwise it ends in a vertex S associated with an empty set. For example, if the problem contains constraints in the form $\alpha + \beta \leq 1$, $\alpha \in \{0, 1\}$ and $\beta \in \{0, 1\}$, and if the first separation after the root of the tree is performed according to the two values of α (1 on one side, 0 on the other), then it is not necessary to think about the branch consisting of assigning the value 1 to β behind the branch obtained by setting α as 1, since the two assignments ($\alpha = 1$ and $\beta = 1$) are incompatible with the constraint $\alpha + \beta \leq 1$. Another way of considering the same thing consists of observing that it is sometimes possible to draw conclusions from a decision made at a certain level of the tree A for deeper levels of A; thus, in the previous example, we could deduce that β can only take a zero value if we assign the value 1 to α.

In other cases, we can establish that such a decision leads to a vertex S for which the set $\Omega(S)$ is not empty but does not contain an optimal solution because, for example, certain properties required by the optimal solutions are not satisfied by the elements of $\Omega(S)$, or because it is possible to produce another vertex S' of the tree which dominates S, in the sense that with any solution s of $\Omega(S)$ we can associate a solution s' of $\Omega(S')$ better than s: $\forall\ s \in \Omega(S)$, $\exists\ s' \in \Omega(S')$ with $f(s) < f(s')$ if we are dealing with a maximization problem, or with $f(s) > f(s')$ if we are dealing with a minimization problem. We deduce from this pairing that the optimum of f on $\Omega(S')$ is strictly better than that of f on $\Omega(S)$ and that we can therefore eliminate S from A. (If we are only seeking one optimal solution, we can sometimes replace the strict inequalities by non-strict inequalities; but we must take care, however, in the extreme case where S and S' dominate each other mutually, in the sense of non-strict inequalities, not to eliminate S because of S' and simultaneously S' because of S!)

3.2.2.5. *Pruning order*

Let us note, to conclude these considerations linked to the different possibilities of reducing the size of the search tree, that the order in which we apply the various pruning principles affects the performance of the method. Of course, we favor first applying a pruning principle that is efficient and fast to apply, followed by a pruning principle that eliminates fewer vertices and necessitates more CPU time, rather than the other way round (but it is not so easy to order a slow but efficient principle and a fast principle that eliminates few vertices). Because evaluation functions often necessitate more calculation time than other pruning principles, we usually resort to them only after the other principles have been tried.

We will therefore try to retain pruning principles that are efficient and quick, but also complementary if possible, in the sense that some are capable of eliminating vertices without any optimal solutions that the others would not have been able to eliminate. Nevertheless, because of the possible exponential growth of the tree size, leading to large CPU times for dealing with it in its entirety, it may be worth keeping a pruning principle that does not seem very efficient. The compromise between the gain obtained by using a pruning principle and the consumption of resources needed to put it into practice (in particular the CPU time, but problems of available memory space can also arise) is not always easy to establish. In these cases, simulations can show the pruning principles that are worth retaining. Let us note, however, that a good evaluation function, which is good at predicting what will have to be done (in other words, using the expressions used above, a function such that the part $E_2(S)$ of the evaluation gets very close to the optimum of g_S on $\Omega(S)$), almost always becomes increasingly indispensable as we want to deal with instances that become bigger and bigger or, more generally, increasingly difficult. This is why the evaluation function, along with the separation principle, appears to be the most important ingredient when we develop a branch-and-bound method.

3.2.3. *Developing the tree*

3.2.3.1. *Description of development strategies*

The last ingredient of a branch-and-bound method consists of defining the order in which the vertices of the tree A are to be considered in order to develop it. The most common strategies are the *depth first search strategy* and the *best first strategy*, to which we can also add the *breadth strategy*. It is also possible to imagine "hybrid" strategies, combining the previous ones.

In the depth first search strategy, we go as far away from the root of A as possible, with separations creating branches of increasing length, considering only one branch at a time. When we can no longer increase the length of the branch of A under inspection (because, for example, the current leaf is associated with a subset of Ω of cardinality small enough for us to easily find the maximum of f on it, or because we have been able to eliminate the branch as of no interest), we go back up a level in A (returning therefore to a vertex of A already encountered) and we seek a new branch allowing us to descend A: if such a branch exists, we follow it; otherwise we go back up a level and apply the same process again. The method terminates when we come back to the root of A and there are no more unexplored branches branching off from the root. In such a strategy, we only need to know explicitly the branch linking the root to the leaf being examined.

In the best first search strategy, we consider, among all the current leaves of the tree being constructed, the leaf F that has the best evaluation, i.e. the highest evaluation if we are dealing with a maximization problem, and the lowest evaluation in the case of a minimization problem; we apply the principle of separation to F in order to extend A. We therefore create all the successors (not eliminated by a pruning principle) of F simultaneously and we calculate their evaluations: F is no longer a leaf of A and is replaced by its successors, which provides many leaves to take into account for the following separation. The development of A is therefore, in this case, more erratic than in the previous case, and we can be led to jump from one branch of A to another or from one level of A to another. Note that evaluation plays a supplementary role to that of pruning A, which was the role assigned to it originally: evaluation is also used here as a guide to indicate the most promising leaves, in the sense that, intuitively (no more), we can expect to have a greater chance of finding an optimal solution.

In the breadth strategy, we first of all create all the successors of the root (at least those that are not eliminated by pruning), then all the successors of the successors of the root (with the same condition), and so on, constructing A from level to level.

3.2.3.2. *Compared properties of the depth first and best first strategies*

The two principal strategies do not offer the same advantages. The depth-first strategy, which does not require knowledge of the whole tree being constructed, but only of the current branch in use, allows us to save on memory space. As well as this, always extending the vertex of A at which we have just arrived, or always going back to the vertex where we were in the previous iteration, means that the depth-first strategy makes programming of the method in a recursive way easier, which can ease considerably the conception and programming of the method. Finally, for the same reasons, we can exploit certain properties established for the vertex to which we are applying the principle of separation and thus reduce the (amortized) complexity of certain calculations (for example, the updating of the part E_1 of the evaluation; see section 3.2.2.2).

On the other hand, the depth-first strategy is much more sensitive to the quality of the bound than is the best first search strategy. In fact, while B has not attained the optimum sought or if we seek all the optimal solutions, the influence of the bound in the best first search strategy is limited to elimination of vertices that, however the value of B differs from the optimal value sought, would not have been developed by this strategy anyway. To establish this property, let us consider a maximization problem for which we are only seeking one optimal solution and let us use the best first search strategy. Note first of all that the evaluation $E(F)$ of a leaf F to which we apply the principle of separation satisfies the relationship $E(F) \geq f_{max}$. In fact, when we separate F, this leaf is of maximum evaluation $E(F)$ among the leaves that have not been eliminated from the current tree A. If $E(F)$ were less than f_{max}, then all the leaves that have not been eliminated would be of an evaluation strictly less than f_{max} and the optimal solutions would therefore be associated with previously explored vertices of A. Because of this prior exploration, the current bound would be f_{max} and all the leaves of A would be eliminated because of evaluations that would be too weak: in the end there would be no leaf to which the separation principle could be applied. Let us now consider a leaf F of A that we eliminate because of the bound. We therefore have the inequality $E(F) \leq B$. If we assume that B is strictly less than f_{max}, $E(F)$ is also strictly less than f_{max} and F does not satisfy the previous condition ($E(F) \geq f_{max}$): F cannot therefore be a leaf to which the separation principle is applied by the best first search strategy. Hence the result stated above. As a consequence, if, at the moment of calculating the initial value of B, we do not expect to reach f_{max} or f_{min}, depending on the case in question, it is not necessarily vital to devote much time to the initialization of B. A good value of B will certainly allow us to eliminate more leaves than would a less good initial value, which is probably not negligible, but as these vertices would not be developed anyway, this gain often does not radically change the size of the tree. On the other hand, if we obtain the optimum as the initial value of B, and if we are only looking for one optimal solution, the gain can become substantial.

For a depth-first strategy, it is preferable to have a good bound from the start, to eliminate vertices with a bad evaluation which, without pruning because of the bound, could create non-negligible subtrees. Because of this, the size of the tree created by the best first search strategy is often (although not systematically) smaller than the one produced by the depth-first strategy. On the other hand, comparing the CPU times is not easy to make: if, on the one hand, having fewer vertices to consider contributes to a smaller calculation time, dealing with a vertex can take longer using the best first search strategy, because, unfortunately, the best first search strategy has other drawbacks. In particular, we can no longer be satisfied with just knowing the branch being developed: on the contrary, we need to know a large part of the tree being constructed[7], which can quickly cause insurmountable problems connected to memory overload. Even if the available space is large enough, establishing the best leaf can slow down the method when the number of leaves becomes high[8]. Furthermore, as the leaves developed successively are not necessarily on the same branch, it is harder, when we develop a leaf F, to exploit the properties established for the predecessor of F (unless we keep these properties in memory for each vertex of the tree, but this increases the double problem of available space and the time necessary to find these interesting properties), and this makes programming based on recursivity difficult (or even almost impossible).

Let us finally note that we can also adopt the best first search strategy to develop an approximate method, the quality of which we will be able to evaluate. Indeed, assuming we have a bound B that we know how to reach with the help of a feasible solution, premature interruption of a branch-and-bound method (whichever strategy is retained to develop the tree) allows us to obtain bounds for the optimum sought, and thus to estimate the distance from B to the optimum sought. In the case of a maximization problem, the largest of the evaluations $E(S)$ of the vertices S of A to which we have not entirely applied the separation principle (in other words, new branches are likely to come from such a vertex; the best first search strategy concerns the biggest evaluation of the leaves of A) gives an upper bound of f_{max}, while B gives a lower bound. The error $f_{max} - B$ that we make by using B instead of f_{max} is then upper bounded by $\max_S E(S) - B$ when S goes through all the vertices of A to which we have not entirely applied the separation principle. We can then decide to stop the process when the error does not exceed a given threshold. The best first search strategy then seems appropriate for most quickly decreasing the previous maximum (but it is possible that another strategy will make B increase more quickly). As mentioned above, this is not, however, the main objective of branch-and-bound methods.

3.3. A detailed example: the binary knapsack problem

The binary knapsack problem (see [KEL 04]) is a classic example of an **NP**-hard problem and branch-and-bound methods belong to the usual methods applied to solve this problem (another being dynamic programming; see Chapter 4, as well as the references cited at the beginning of the chapter, for this optimization technique). It takes its name from a situation with which a hiker is often confronted when he fills his

7. At the least, we must keep the part of the tree likely to lead to leaves that have not yet been explored.

8. In this case it is advisable to store the leaves separately in an appropriate structure that allows us to calculate the best leaf quickly, for example a heap structure (see [COR 01] for the definition of a heap and other, more sophisticated, structures).

knapsack: he has n objects i $(1 \leq i \leq n)$, each one characterized by a strictly positive utility u_i and a strictly positive weight p_i; taking into account a weight P fixed in advance, which must not be exceeded (because the seams of the bag would split above this weight, or because the hiker does not believe that he can carry a greater weight, etc.) and assuming that the utilities are additive (the objects therefore cannot be substituted for one another), the hiker wishes to know which objects he must take with him in order to maximize the total utility of the objects taken without exceeding the prescribed limit P for the weight. Formally, a binary knapsack problem is a problem (Π) of the following form:

$$\max z = \sum_{i=1}^{n} u_i \, x_i$$

with the constraints $\sum_{i=1}^{n} p_i \, x_i \leq P$ and, for $1 \leq i \leq n$, $x_i \in \{0, \ 1\}$.

For this problem we will interpret a value equal to 1 for a variable x_i $(1 \leq i \leq n)$ as the decision to take the object i, and a value 0 as the decision not to take the object i. We will assume that the following relationships are satisfied (which simplifies the problem):

$$\forall \, i \in \{1, \ 2, \ ..., \ n\}, 0 < p_i \leq P \text{ and } \sum_{i=1}^{n} p_i > P$$

To illustrate the application of branch-and-bound methods to this problem, we will consider the following instance (Π_0):

$$\max z = 40x_1 + 40x_2 + 30x_3 + 18x_4 + 20x_5 + x_6$$

with $21x_1 + 22x_2 + 17x_3 + 13x_4 + 16x_5 + 2x_6 \leq 55$ and, for $1 \leq i \leq 6$, $x_i \in \{0, \ 1\}$.

3.3.1. *Calculating the initial bound*

A method sometimes used to establish an initial bound (in this case a bound of z) for integer linear programming problems (in this case, with binary variables) consists of relaxing the integrity constraints (in this case, binary constraints), and then solving the continuous problem thus obtained (for example, with the help of the simplex algorithm), and, finally, rounding off in an appropriate way (which is not always easy) the values provided by this solution in such a way as to build an integer solution that satisfies all the constraints.

For a binary knapsack problem (Π), the relaxation of binary constraints is obtained by replacing the constraints $x_i \in \{0, \ 1\}$ for $1 \leq i \leq n$ by the constraints $x_i \in [0, \ 1]$ for $1 \leq i \leq n$. The relaxed problem (Π^r) associated with (Π) is then expressed:

$$\max z^r = \sum_{i=1}^{n} u_i \, x_i$$

with the constraints $\sum_{i=1}^{n} p_i\, x_i \leq P$ and, for $1 \leq i \leq n$, $x_i \in [0,\ 1]$.

For the example below, this gives for the instance (Π_0^r):

$$\text{maximize} z^r = 40x_1 + 40x_2 + 30x_3 + 18x_4 + 20x_5 + x_6$$

with $21x_1 + 22x_2 + 17x_3 + 13x_4 + 16x_5 + 2x_6 \leq 55$ and, for $1 \leq i \leq 6$, $x_i \in [0,\ 1]$. It is not hard to establish the following theorem.

THEOREM 3.4.– *Let us assume that the variables of the problem* (Π^r) *are numbered according to the ratios* $\frac{u_i}{p_i}$ *(for* $1 \leq i \leq n$*) decreasing:* $\frac{u_1}{p_1} \geq \frac{u_2}{p_2} \geq \ldots \geq \frac{u_n}{p_n}$. *Let k be the index between 1 and* $n-1$ *defined by* $\sum_{i=1}^{k} p_i \leq P$ *and* $\sum_{i=1}^{k+1} p_i > P$. *So an optimal solution* $(x_i^*)_{1 \leq i \leq n}$ *of* (Π^r) *is given by the following relationships:*

– *for i satisfying* $1 \leq i \leq k, x_i^* = 1$;

– $x_{k+1}^* = \left(P - \sum_{i=1}^{k} p_i\right) / p_{k+1}$;

– *for i satisfying* $k+2 \leq i \leq n$, $x_i^* = 0$.

By rounding off the value of x_{k+1}^* to 0, we obtain a feasible solution of (Π) (because the coefficient p_{k+1} is positive) and therefore a lower boubd (that we could have obtained directly, but in this way we illustrate a general procedure) of the maximum of (Π). For the binary knapsack problem, we can however propose another heuristic solution, at least as good (with respect to the value taken by z) as the one provided by the previous rounding off. It is specified by proposition 3.1.

PROPOSITION 3.1.– *Let* z_{max} *be the maximum of* (Π). *Let us assume that the variables of the problem* (Π) *are numbered according to the ratios* $\frac{u_i}{p_i}$ *(for* $1 \leq i \leq n$*) decreasing:* $\frac{u_1}{p_1} \geq \frac{u_2}{p_2} \geq \ldots \geq \frac{u_n}{p_n}$. *Let* $(x_i^*)_{1 \leq i \leq n}$ *be the solution of* (Π) *defined step by step by the following relationships: for i satisfying* $1 \leq i \leq n, x_i^* = 1$ *if* $P - \sum_{j=1}^{i-1} p_j x_j^* \geq p_i$ *and* $x_i^* = 0$ *otherwise. We therefore have* $\sum_{i=1}^{n} u_i x_i^* \leq z_{max}$.

For the example (Π_0), the relaxation of the binary constraints gives the following solution (in continuous variables), because the variables are numbered according to the ratios u_i/p_i decreasing: $x_1 = 1$, $x_2 = 1$, $x_3 = 12/17$, $x_4 = 0$, $x_5 = 0$, $x_6 = 0$; the integer solution obtained by rounding x_3 to its floor value is therefore: $x_1 = 1$, $x_2 = 1$, $x_3 = 0$, $x_4 = 0$, $x_5 = 0$, $x_6 = 0$, which gives z the value 80 and a total weight of 43. Now, the heuristic of proposition 3.1 gives: $x_1 = 1$, $x_2 = 1$, $x_3 = 0$, $x_4 = 0$, $x_5 = 0$, $x_6 = 1$, giving a value of 81 for z and a total weight of 45. We will therefore take 81 as a lower bound of the maximum of (Π_0), a lower bound that, furthermore, we know how to reach with the help of a feasible solution.

3.3.2. *First principle of separation*

The first principle of separation is a classic; it consists of carrying out the separation according to the possible values of x_1, then of x_2, etc. (here we assume that the variables are numbered according to the ratios u_i/p_i decreasing), this being made possible by the finiteness of the number of possible values for each variable. Here, the variables being binary even allows us to predict that the tree will have no more than $2^6 = 64$ leaves, but we will see that most of them can be eliminated. (Such a tree with 2^6 leaves is shown in Figure 3.1; for readability reasons, we have only specified the variables according to which the separations are made for the first two levels: the others are associated with x_3, then with x_4, then with x_5, and finally with x_6. For the same reasons, we have not specified the values on the arcs leading to the leaves: as for the other arcs, they are associated with the values 1 to the left and 0 to the right, alternately.)

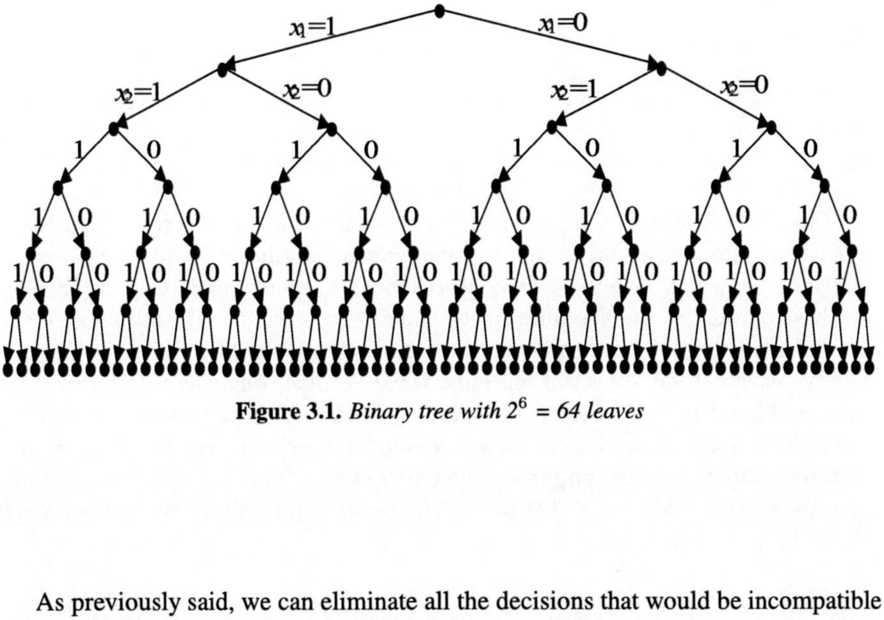

Figure 3.1. *Binary tree with* 2^6 *= 64 leaves*

As previously said, we can eliminate all the decisions that would be incompatible with the constraint related to the weight. For instance, the successive decisions $x_1 = 1$, $x_2 = 1$ and $x_3 = 1$ are incompatible with the constraint $21x_1 + 22x_2 + 17x_3 + 13x_4 + 16x_5 + 2x_6 \leq 55$. In other words, the decisions $x_1 = 1$ and $x_2 = 1$ force the value of x_3 to 0. As a result, we can eliminate the subtree to which the branch associated with the successive decisions $x_1 = 1$, $x_2 = 1$ and $x_3 = 1$ leads from the tree in Figure 3.1. Other eliminations of the same kind are possible. If we perform all these eliminations, and if we do nothing else, we obtain the tree in Figure 3.2, which enumerates all the feasible solutions, in this case 44 of them.

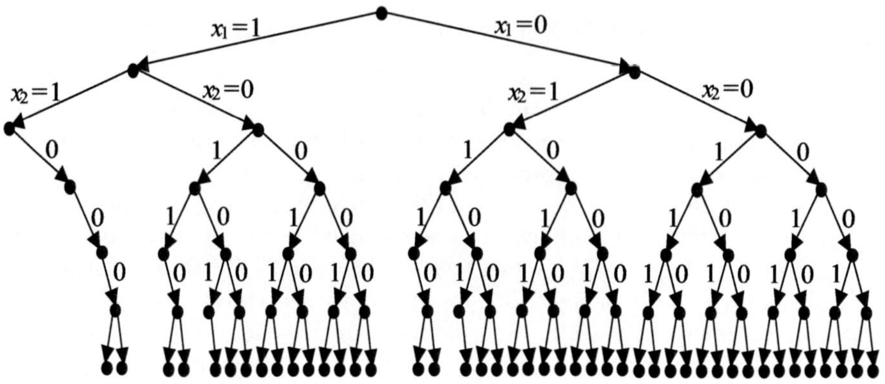

Figure 3.2. *Tree of feasible solutions of* (Π_0)

3.3.3. *Pruning without evaluation*

It is still possible to reduce the size of the rooted tree A in Figure 3.2 by eliminating branches that cannot lead to vertices associated with sets of solutions containing an optimal solution, even without using an evaluation function.

Most of the leaves of A appear in couples from the same parent, one of the leaves of each couple (to the left) therefore corresponding to the value $x_6 = 1$ and the other (to the right) to $x_6 = 0$. Now, since it concerns the last (in relation to the order in which we have considered the variables for the successive separations) variable according to which we carry out the separation (all the other variables already have a value), it is obvious that it is useless to keep some weight for the following variables as there are none! In other words, each feasible solution obtained in the previous enumeration with $x_6 = 1$ dominates the identical solution with $x_6 = 0$ instead of $x_6 = 1$. It is therefore pointless creating the branch associated with $x_6 = 0$ when the similar branch associated with $x_6 = 1$ is present. In the example, this allows us to eliminate 21 of the 44 leaves of A.

We can generalize this pruning principle to a feasible solution in which certain variables have the value 1. All solutions obtained by replacing the value 1 with 0 for one or more of the variables of value 1 are in fact dominated by the solution under consideration before these replacements (because of the strictly positive values of the coefficients of the function z). Thus, in the example, the feasible solution (0, 1, 0, 1, 1, 1) dominates the fifteen solutions: (0, 1, 0, 1, 1, 0), (0, 1, 0, 1, 0, 1), (0, 1, 0, 1, 0, 0), (0, 1, 0, 0, 1, 1), (0, 1, 0, 0, 1, 0), (0, 1, 0, 0, 0, 1), (0, 1, 0, 0, 0, 0), (0, 0, 0, 1, 1, 1), (0, 0, 0, 1, 1, 0), (0, 0, 0, 1, 0, 1), (0, 0, 0, 1, 0, 0), (0, 0, 0, 0, 1, 1), (0, 0, 0, 0, 1, 0), (0, 0, 0, 0, 0, 1), (0, 0, 0, 0, 0, 0). On the other hand, we cannot draw anything from this point of view between (0, 1, 0, 1, 1, 1) and, for example, (0, 1, 1, 0, 0, 0) or (0, 0, 1, 1, 1, 1)[9]. By applying this pruning principle to the tree in Figure 3.2, we obtain for (Π_0) the tree in Figure 3.3, which now has only eight leaves.

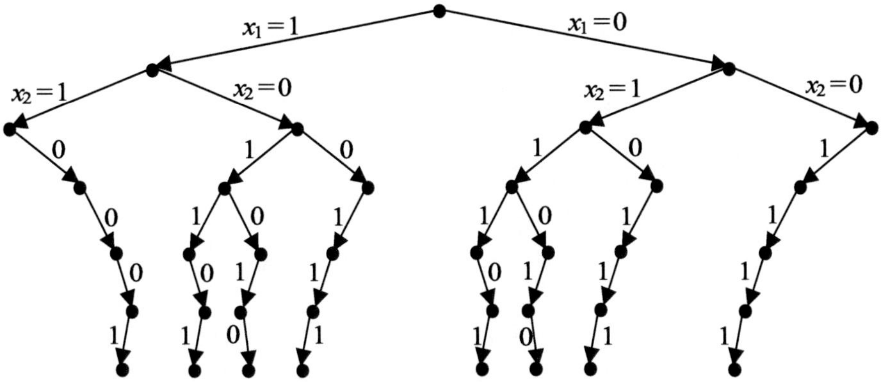

Figure 3.3. *Tree of non-dominated solutions of (Π_0) for finding all the optimal solutions*

We can complete this domination principle with the help of a comment on the variables x_1 and x_2 of (Π_0), but only in the case where we seek only one optimal solution. In fact, we notice that u_1 and u_2 are equal, while p_1 is strictly smaller than p_2: we can associate with any feasible solution in the form $(0, 1, x_3, x_4, x_5, x_6)$, a feasible solution that gives the same value to z (and that weighs less); this is a solution of the form $(1, 0, x_3, x_4, x_5, x_6)$. If we seek only one optimal solution, we can then ignore all solutions of the form $(0,\ 1,\ x_3,\ x_4,\ x_5,\ x_6)$ (this does not, however, mean that a solution of the form $(0,\ 1,\ x_3,\ x_4,\ x_5,\ x_6)$ cannot be optimal, as the full solution of the example will show). With respect to the rooted tree in Figure 3.3, we can therefore eliminate three leaves and obtain the tree in Figure 3.4, with only five leaves, if we seek only one optimal solution. In general, if we seek only one optimal solution, we can apply this pruning principle when confronted with two variables x_i and x_j, for two appropriate indices i and j, for which we have simultaneously $u_i \geq u_j$ and $p_i \leq p_j$: in this case, we can systematically ignore solutions for which x_i is equal to 0 and x_j to 1.

9. We can indeed eliminate the solution (0, 0, 1, 1, 1, 1), because of (0, 1, 0, 1, 1, 1), by remarking that u_2 is greater than u_3. We will not consider this extension of the domination relationship here.

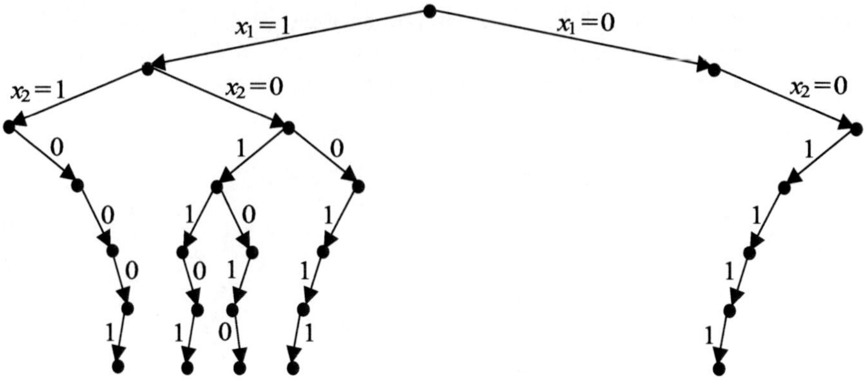

Figure 3.4. *Tree of non-dominated solutions of (Π_0) for finding only one optimal solution*

3.3.4. *Evaluation*

Thanks to theorem 3.2, theorem 3.4 provides us with an evaluation[10]. Indeed, the relaxation of the binary constraints ($x_i \in \{0,\ 1\}$ becomes $x_i \in [0,\ 1]$) has the effect of enlarging the set on which we seek to optimize z: the optimal value obtained after relaxation of the binary constraints provides an upper bound of the maximum value of z sought, and can therefore be used as the evaluation of the root of the tree.

Applying this relaxation to (Π_0) gives an optimal solution with continuous variables equal to (1, 1, 12/17, 0, 0, 0), giving the evaluation of the root of the tree as equal to $80 + 30 \times 12/17$, i.e. about 101.18. The coefficients of z in (Π_0) being integers, we deduce from the previous value that 101 is an upper bound of the maximum of (Π_0). This, and the lower bound calculated above, allow us to conclude that the maximum of (Π_0) is between 81 and 101.

We can evaluate the vertices of the tree in Figure 3.4 in the same way. Let us, for example, evaluate the vertex S associated with the set of solutions in the form $(1,\ 0,\ x_3,\ x_4,\ x_5,\ x_6)$ obtained after the separations $x_1 = 1$ and $x_2 = 0$. By carrying forward these values to the expression of (Π_0), we obtain the following new problem:

$$\max 40 + 30x_3 + 18x_4 + 20x_5 + x_6$$

with $17x_3 + 13x_4 + 16x_5 + 2x_6 \leq 34$ and, for $3 \leq i \leq 6$, $x_i \in \{0,\ 1\}$.

10. Let us point out that the application of the Lagrangian relaxation consisting of relaxing the constraint on the weight and keeping the binary constraints would give the same evaluations here.

Theorem 3.4 allows us once again to obtain an upper bound for this problem, that is an evaluation $E(S)$ of S: $40 + 30 + 18 + 20 \times 4/16$, i.e. 93 (the value reached by $x_3 = 1$, $x_4 = 1$, $x_5 = 4/16 = 1/4$ and $x_6 = 0$). As stated above, we could decompose this evaluation into two parts $E_1(S)$ and $E_2(S)$: $E_1(S)$ represents what has been established for z because of the decisions $x_1 = 1$ and $x_2 = 0$, i.e. 40, and $E_2(S)$ is an upper bound on what we can do with the help of the other variables, i.e. 53. We can do the same thing with the other vertices of the tree in Figure 3.4, but, in fact, some of these vertices will not be created.

3.3.5. *Complete execution of the branch-and-bound method for finding only one optimal solution*

We can apply everything that has already been discussed in order to see at last how a branch-and-bound method works. We must still, however, decide on the tree-development strategy. What follows corresponds to the tree obtained using a depth first strategy. We will assume for the moment that we seek only one optimal solution. Figure 3.5 shows the execution of the branch-and-bound method. The vertices S that are under evaluation are represented by a circle containing the value of $E(S)$; the other vertices, represented by a black dot, are not evaluated because the decisions that lead to them involve the attribution of forced valuesto some variables, as seen above (in the case in question, because the evaluation can be calculated with a weak complexity, a variant would consist of evaluating these vertices as well). The branching-off is done as above, with respect to the values of the variables; the variables are considered according to the decreasing utility/weight ratios. The Roman numerals beside the evaluated vertices give the order in which the vertices have been created.

Let us remember that before constructing the tree, we apply a heuristic computing a bound B, that is a lower bound of the maximum sought. What precedes shows that 81 is such a bound: $B = 81$.

As the evaluation of the root of the tree (vertex I), equal to 101, is greater than the current value of B, we apply a first branch-off, provisionally setting the value of x_1 to 1 (we will consider the value 0 later on), which leads to vertex II, which also evaluates to 101 (which was predictable, since the evaluation of the root corresponded to the solution $(1, 1, 12/17, 0, 0, 0)$, which is compatible with the value 1 for x_1). We now set x_2 to 1, which forces x_3, x_4 and x_5 to 0. Evaluation of vertex III assigns the value 1 to x_6 and thus gives a feasible solution: (1, 1, 0, 0, 0, 1), which is none other than the feasible solution computed by the second heuristic. The phenomenon is not due to chance and comes from the order in which we consider the variables to make our branch-offs, which is the same as the one in which the variables are inspected in the second heuristic. We could, therefore, in this case, have refrained from applying a preliminary heuristic and waited for the first branch setting the value of all the variables to be constructed before assigning an initial value to B. This coincidence is, however, not noticeable if we consider another order of inspection of the variables for the branch-off, or if we invert the values 0 and 1 at the branch-offs, or even, of course, if we apply another heuristic or another separation principle. As the evaluation is equal to the bound, and since we seek only one optimal solution, the set of solutions associated with vertex III does not contain any better solution than that which we already know (indeed, here it is the same) and it is therefore pointless to extend the tree from this vertex. This is represented in Figure 3.5 by the cross located under vertex III. This symbolizes the fact that potential branches from vertex III are not created.

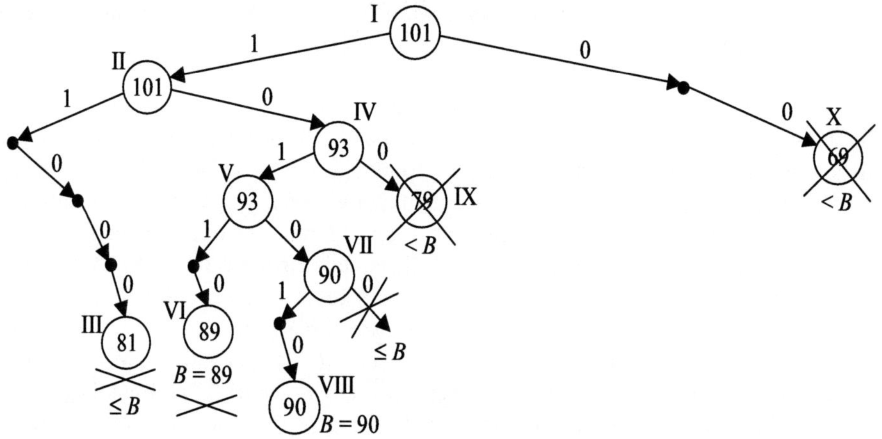

Figure 3.5. *Development of the tree according to the depth first search strategy for finding only one optimal solution of* (Π_0)

The depth-first search now goes back up to vertex II to explore the branch corresponding to the value 0 assigned to x_2. We end up at vertex IV, evaluated as 93. The separation principle applied to x_3 with the value 1 leads to the creation of vertex V, which also evaluates to 93. The following branch-off, assigning the value of 1 to x_4, is prolonged by the forced attribution of the value 0 to x_5, giving vertex VI. Evaluation of this latter (equal to 89) assigns the value 1 to x_6, which shows a feasible solution of value 89. We therefore update the bound B, which thus becomes 89, the value achieved by the solution (1, 0, 1, 1, 0, 1). We are therefore in a similar situation to that of vertex III: evaluation of vertex VI being henceforth equal to the bound, and since we seek only one optimal solution, we do not create the branches leaving from vertex VI. We therefore go back up to vertex V to create vertex VII, which evaluates to 90, then, following a mechanism similar to the aforementioned, we create vertex VIII, which has a set of associated solutions containing only one solution: (1, 0, 1, 0, 1, 0). This solution being feasible by construction and being of value 90, we update B once more: $B = 90$. This update renders pointless the exploration of the branch corresponding to $x_5 = 0$ leaving from vertex VII (hence the cross on the arc in Figure 3.5), since we seek only one optimal solution: vertex VII is evaluated as 90, so it is not possible to find, in the set associated with vertex VII, a better solution than that associated with the new value of B, particularly in the branch $x_5 = 0$ leaving from vertex VII.

In general, we notice that evaluations decrease (in a large sense) when we go from a vertex S to one of its successors. In the same way, the order of the variables according to which we carry out the separations means that, for vertices with the same parent, evaluation is decreasing (in a large sense) from left to right. We then go back to vertex IV (of which the evaluation of 93 gives us hope of finding a better solution than the best currently known solution) to explore the branch $x_3 = 0$. This one leads to vertex IX, which evaluates to 79. This value being less than B, we eliminate vertex IX (this is equally applicable if we seek all the optimal solutions), which we symbolize in Figure 3.5 by crossing out vertex IX. To end with, we go back to the root to explore the branch $x_1 = 0$. Since we seek only one optimal solution, the nullity of x_1 causes that of x_2. We therefore come to vertex X; its evaluation to 69 is too small with respect to B to keep this vertex. We have thus finished: the maximum of (Π_0) is 90 and an optimal solution is (1, 0, 1, 0, 1, 0), of total weight 54.

3.3.6. *First variant: finding all the optimal solutions*

It is easy to change this application in order to obtain all the optimal solutions. For this, we must not eliminate the branches that have been cut just because they could not lead to a better solution than the best known solution (on the other hand, we continue to eliminate branches that lead to vertices associated with sets that cannot contain an optimal solution). In our example, we should on the one hand develop the vertices that have an evaluation equal to the bound, and on the other hand we should not eliminate the branch defined by $x_1 = 0$ and $x_2 = 1$. Exploration of these branches can, of course, be done at the same time as the others, but we can also proceed in two stages (which, as we have already mentioned, would be less useful for the best first search strategy): firstly, calculate the optimal value and an optimal solution as we have just done, then apply the method once more to compute all the optimal solutions by assigning the now known optimal value to B. If, for example, we use this second possibility (so with $B = 90$ right from the start), the tree that we obtain is that given in Figure 3.6 (the numbers in Roman numerals again show the order in which the vertices have been created). We note that there are two optimal solutions of (Π_0): the one found above, that is (1, 0, 1, 0, 1, 0), of total weight 54, and the solution (0, 1, 1, 0, 1, 0), of total weight 55[11].

11. Let us note that on the face of it there is no reason why an optimal solution should saturate the constraint, or even get as close as possible to the constant term of the constraint. If necessary,

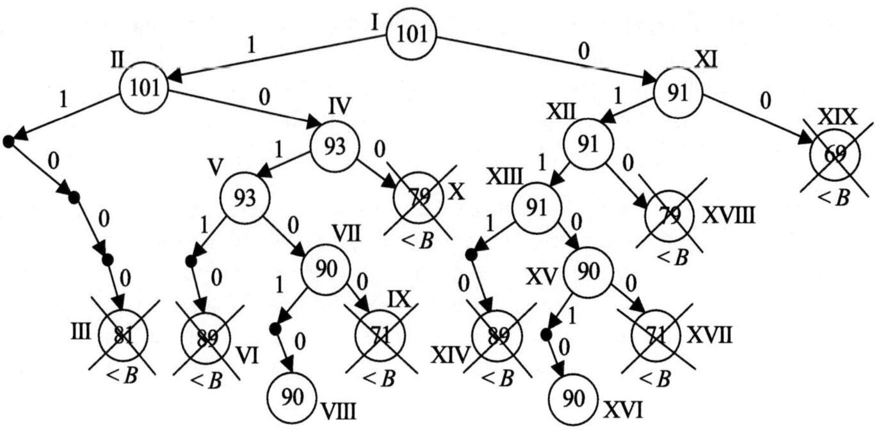

Figure 3.6. *Developing the tree to find all the optimal solutions of* (Π_0)

3.3.7. *Second variant: best first search strategy*

Let us assume that we again wish to find all the optimal solutions and that we again start with the bound initially found by the second heuristic: $B = 81$ (in fact, as mentioned above, the initial value of B is not fundamental to the best first search strategy if it is not equal to the optimum sought or if, as is the case here, we seek all the optimal solutions; here we could start from $B = 0$ or from $B = 90$ without changing the final tree). Applying the best first search strategy gives the same tree here as does the depth-first strategy (this coincidence is not general), that is the tree in Figure 3.6, but the vertices are created in a different order and are not necessarily eliminated at the same time.

More precisely, after the root (vertex I), we create and evaluate its two children (vertices II and XI). The evaluation of vertex II being bigger than that of vertex XI, we apply the separation principle to vertex II: we create and evaluate vertices III and IV. The evaluation of vertex IV being the best among the current leaves (vertices III, IV and XI), we separate vertex IV into vertices V and X (vertex X is eliminated because it evaluates to less than B). The leaf of the current tree with the biggest evaluation being

---

the reader will be easily convinced of this with the help of the following example, in which a is a given parameter greater than or equal to 1: maximize $z = 2x_1 + x_2$ with the constraints $x_1 + ax_2 \leq a$, $x_1 \in \{0, 1\}$ and $x_2 \in \{0, 1\}$. The only optimal solution being (1, 0), the difference between the constant term of the constraint, equal to a, and the value of $x_1 + ax_2$ taken by the optimal solution, equal to 1, can be made as big as we like by a suitable choice of a, while the constraint is saturated for (0, 1).

vertex V, we separate it into vertices VI and VII. Here we have to make a "jump" from the branch on which we are to vertex XI, since this is currently the vertex that makes up the leaf with the biggest evaluation. The separation of vertex XI gives vertices XII and XIX (eliminated); separation of XII gives vertices XIII and XVIII (eliminated); separation of XIII gives vertices XIV and XV. Here, two leaves, VII and XV, have the same maximum evaluation among the leaves and are therefore candidates for separation. Let us imagine that we continue from XV: we thus obtain XVI (allowing us to change B to 90, which in turn leads to the elimination of vertices III, VI and XIV) and XVII (eliminated), before jumping to VII, separated into VIII and IX (eliminated). The order in which the vertices have been created is therefore as follows: I; II, XI; III, IV; V, X; VI, VII; XII, XIX; XIII, XVIII; XIV, XV; XVI, XVII; VIII, IX. We check in the example that only the vertices that evaluate to greater than or equal to the maximum sought (here, 90) are subject to separation (which is not generally the case for the depth-first strategy).

3.3.8. *Third variant: second principle of separation*

The previous example highlights a defect in the separation principle used above, based on the values assigned to x_1, then x_2, etc.: the optimal solutions of the continuous problem may not evolve when we go from a vertex to one if its children, which leads to stagnation of the evaluations. This is the case, for example, in the tree in Figure 3.5, when we create vertex II from the root: the attribution of the value 1 to x_1 being compatible with the values obtained from the evaluation of the root, vertex II keeps the evaluation of its parent and we must wait until we go down to the lower level before the situation changes.

We can increase the dynamics of the system by adopting another separation principle, less directly linked to an enumeration principle and more subtle than the first. For this, note that the evaluation of a vertex can end up in one of the following situations:

– Calculation of the evaluation does not lead to a feasible solution because one of the variables x_i takes a fractional value; in this case, we can carry out a separation on x_i. This type of separation can be generalized to any integer linear programming problem when the evaluation function is obtained by relaxing the integrity constraints. More exactly, if evaluation of the current vertex S assigns a fractional value a to a variable α supposed to be an integer, we separate S into two branches: in the first, we add the constraint of the problem inherited from the parent of S, the constraint $\alpha \leq \lfloor a \rfloor$ (where $\lfloor a \rfloor$ denotes the floor of a) to the constraints of the problem inherited from the parent of S; in the second, we add the constraint $\alpha \geq \lceil a \rceil$ (where $\lceil a \rceil$ represents the ceiling of a) to the constraints of the same problem inherited from the parent of S. These new constraints will be passed on to all the descendants of S. This separation principle is frequently applied, most notably when we cannot (or we do not know how to) upper bound the number of values that the variable on which the separation is carried out can take.

– Calculation of the evaluation shows a solution for which all the values are integers and which gives the value of the evaluation to the objective function; in this case, and if we seek only one optimal solution, we may not, because of what has been explained above, develop the vertex under consideration (if we want all the optimal solutions we come back to the first principle of separation).

Thus, for example, the evaluation of the root gave the solution (in continuous variables) (1, 1, 12/17, 0, 0, 0). The third variable not being an integer, we carry out a separation according to x_3. The complete application of this separation principle to establish only one optimal solution is shown in Figure 3.7, with a bound initially equal to 81 (the Roman numerals show the order in which the vertices are created by a depth-first search; the best first search strategy would create the same vertices, but in the order I; II, X; III; XI, XIII; XII; XIV, XVI; IV, IX; XV; V, VI; VII, VIII).

Some observations can be made about the tree in Figure 3.7. First of all, we notice that the separations on the same level do not necessarily correspond to the same variable. For example, the separation that follows vertex II relies on x_2 because the evaluation of vertex II attributes a fractional value to x_2 (in this case $17/22$), while the one that follows vertex X relies on x_4, because this is the variable that takes a fractional value (in this case $12/13$ in the evaluation of vertex X). We notice also that, for the example, there are fewer vertices which evaluate to the same as their parents (in fact, here they are all unique if we do not take rounding into account).

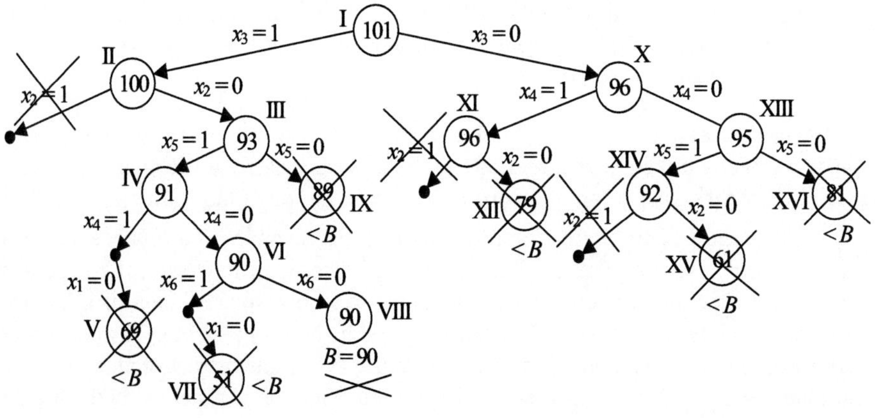

Figure 3.7. *Tree for the second principle of separation applied to* (Π_0)

The branches $x_2 = 1$ which leave from vertices II, XI and XIV, have been shown for memory's sake, but they can be pruned as soon as we consider them. Indeed, the decisions that precede them mean that, in order to satisfy the constraint, it is necessary to set x_1 to 0 when we give the value 1 to x_2. As we have seen above, we can eliminate this configuration when we seek only one optimal solution.

Concerning the updating of the bound, B changes directly from 81 (initial value) to 90 (thanks to vertex VIII), without going through the step 89. This update explains the elimination of vertex IX.

3.4. Conclusion

We could multiply the number of examples of applications as much as we like. As soon as we can show a principle of separation, branch-and-bound methods are potentially applicable. Their efficiency, however, will largely depend on the principles that will allow pruning of the search tree. Among these is almost always the use of an evaluation function applied conjointly with the calculation and updating of a bound. But it would be too simplistic to consider that this is the only way to prune the search tree: other pruning principles, independent of the evaluation function and sometimes much quicker to implement, can considerably reduce the size of the tree and thus substantially speed up the finding of an optimal solution.

Nonetheless, these methods are of high (exponential) complexity, because they implicitly rely on an enumeration principle. Let us remember, however, that we do not know any qualitatively better methods for **NP**-hard problems, which are the ones to which we usually apply these branch-and-bound methods.

As often happens to general methods, branch-and-bound methods have seen various developments and variants. As well as the classic *branch-and-bound* methods that have been the subject of this chapter, we find the following variants: *branch and cut*, resorting to polyhedral techniques (use of intersecting planes; see Chapter 10); *branch and price*, making use of the column generation technique (see Chapter 9); *branch and cut and price*, the result of a combination of the *branch and cut* and *branch and price* methods; *augment and branch and cut* (see for example [LET 03]), adding a feasible solution improvement phase to a method of the *branch and cut* type; *branch and peg* [GOL 03], aiming to set the value of certain variables describing the problem to treat it independently of the principle of separation. These diverse variants, most notably the *branch and cut* method and its derivatives (see for example [JÜN 01, JÜN 03], as well as certain references cited at the beginning of this chapter) allow us to deal with increasingly large instances. It is not impossible that new variants will increase the efficiency of these methods even further and allow us, in the future, to solve instances that are currently too hard to be solved exactly.

3.5. Bibliography

[AGI 66] AGIN N., "Optimum seeking with branch-and-bound", *Management Science*, vol. 13, p. 176–185, 1966.

[BAL 65] BALAS E., "An additive algorithm for solving linear programs with 0-1 variables", *Operations Research*, vol. 13, p. 517–546, 1965.

[BAL 68] BALAS E., "A note on the branch-and-bound principle", *Operations Research*, vol. 15, p. 442–445, 1968.

[BER 64] BERTIER P., ROY B., "Une procédure de résolution pour une classe de problèmes pouvant avoir un caractère combinatoire", *Cahiers du CERO*, vol. 6, p. 202–208, 1964.

[BER 65] BERTIER P., ROY B., Trois exemples numériques d'application de la procédure SEP, Note de travail de la direction scientifique de la SEMA num. 32, SEMA, 1965.

[BER 85] BERGE C., *Graphs*, North-Holland Publishing Co., Amsterdam, 1985.

[CHA 96] CHARON I., GERMA A., HUDRY O., *Méthodes d'optimisation combinatoire*, Masson, Paris, 1996.

[COR 01] CORMEN T.H., LEISERSON C.E., RIVEST R.L., STEIN C., *Introduction to Algorithms*, 2nd edn., MIT Press/McGraw-Hill, 2001.

[DAK 65] DAKIN R.J., "A tree search algorithm for mixed integer programming problems", *The Computer Journal*, p. 250–255, 1965.

[DAN 54] DANTZIG G.B., FULKERSON D.R., JOHNSON S.M., "Solution of a large-scale traveling-salesman problem", *Operations Research*, vol. 2, p. 393–410, 1954.

[DAN 59] DANTZIG G.B., FULKERSON D.R., JOHNSON S.M., "On a linear-programming, combinatorial approach to the traveling-salesman problem", *Operations Research*, vol. 7, p. 58–66, 1959.

[DEL 97] DELL'AMICO M., MAFFIOLI F., MARTELLO S., Eds., *Annotated Bibliographies in Combinatorial Optimization*, Wiley, Chichester, 1997.

[EAS 58] EASTMAN W.L., Linear programming with pattern constraints, PhD thesis, Harvard, 1958.

[EAS 59] EASTMAN W.L., "A solution to the traveling-salesman", *American Summer Meeting of the Econometric Society*, Cambridge, MA, 1959.

[GEO 67] GEOFFRION A.M., "Integer programming by implicit enumeration and Balas method", *SIAM Review*, vol. 9, p. 178–190, 1967.

[GEO 69] GEOFFRION A.M., "An improved implicit enumeration approach for integer programming", *Operations Research*, vol. 17, p. 437–454, 1969.

[GOL 65] GOLOMB S.W., BAUMERT L.D., "Backtrack programming", *Journal of the ACM*, vol. 12, p. 516–524, 1965.

[GOL 03] GOLDENGORIN B., GHOSH D., SIERKSMA G., "Branch and peg algorithms for the simple plant location problem", *Computers & Operations Research*, vol. 30, num. 7, p. 967–981, 2003.

[GON 84] GONDRAN M., MINOUX M., *Graphs and Algorithms*, Wiley, New York, 1984.

[HER 67] HERVÉ P., "Résolution des programmes linéaires à variables mixtes par la procédure SEP", *METRA*, vol. 6, num. 1, 1967.

[JÜN 01] JÜNGER M., NADDEF D., Eds., *Computational Combinatorial Optimization*, Springer, Berlin, 2001.

[JÜN 03] JÜNGER M., REINELT G., RINALDI G., Eds., *Combinatorial Optimization – Eureka, You Shrink!*, vol. 2570 of *LNCS*, Springer, Berlin, 2003.

[KOR 02] KORTE B., VYGEN J., *Combinatorial Optimization*, Springer, Berlin, 2002.

[KEL 04] KELLERER H., PFERSCHY U., PISINGER D., *Knapsack Problems*, Springer, Berlin, 2004.

[LAN 60] LAND A.H., DOIG A.G., "An automatic method for solving discrete programming problems", *Econometrica*, vol. 28, p. 497–520, 1960.

[LAU 86] LAURIÈRE J.-L., *Intelligence artificielle. Résolution de problèmes par l'Homme et la machine*, Eyrolles, Paris, 1986.

[LAW 66] LAWLER E.L., WOOD D.E., "Branch and bound methods: a survey", *Operations Research*, vol. 14, p. 699–719, 1966.

[LET 03] LETCHFORD A.N., LODI A., "An augment and branch and cut framework for mixed 0-1 programming", JÜNGER M., REINELT G., RINALDI G., Eds., *Combinatorial Optimization – Eureka, You Shrink!*, vol. 2570 of *LNCS*, Springer, Berlin, 2003.

[LIT 63] LITTLE J.D.C., MURTY K.G., SWEENEY D.W., KAREL C., "An algorithm for the traveling salesman problem", *Operations Research*, vol. 11, p. 972–989, 1963.

[MIN 86] MINOUX M., *Mathematical Programming: Theory and Algorithms*, Wiley, New York, 1986.

[ROS 58] ROSSMAN M.J., TWERY R.J., "A solution to the travelling salesman problem by combinatorial programming", *Operations Research*, vol. 6, 1958.

[ROY 65] ROY B., BERTIER P., NGHIEM P.T., "Programmes linéaires en nombres entiers et procédure SEP", *METRA*, vol. 4, num. 3, p. 441–460, 1965.

[ROY 69] ROY B., "Procédures d'exploration par séparation et évaluation (PSEP et PSES)", *RAIRO-OR*, vol. 1, p. 61–90, 1969.

[SAK 84] SAKAROVITCH M., *Optimisation combinatoire, Tome 1: Graphes et programmation linéaire, et Tome 2: Programmation discrète*, Hermann, Paris, 1984.

[SCH 03] SCHRIJVER A., *Combinatorial Optimization*, Springer, Berlin, 2003.

[TOM 56] TOMPKINS C., "Machine attacks on problems whose variables are permutations", *Numerical Analysis*, p. 195–211, 1956.

[WAL 60] WALKER R.S., "An enumerative technique for a class of combinatorial problems", *Proceedings of the AMS Symposium on Applied Mathematics*, vol. 10, p. 91–94, 1960.

[WER 03] DE WERRA D., LIEBLING T.M., HÈCHE J.-F., *Recherche opérationnelle pour ingénieurs*, vol. 1, Presses polytechniques et universitaires romandes, Lausanne, 2003.

[WOL 98] WOLSEY L., *Integer Programming*, Wiley, New York, 1998.

Chapter 4

Dynamic Programming

4.1. Introduction

Dynamic programming is an optimization method that proceeds by implicit enumeration of the solutions. Although already in use, it was elevated to the rank of a general solution method by the work of Bellman [BEL 54, BEL 57], who formalized and named the approach. This approach allows us to solve sequential decision problems efficiently. More generally, it consists of tackling optimization problems with a strategy consisting of two essential parts:

– breaking down the problem into a sequence of problems;

– establishing a recurrence link between the optimal solutions of the problems.

The name "dynamic programming" is justified more by fashion (the domain developed in parallel with that of linear programming) than by purely scientific reasons. The following is an extract from the autobiography of Bellman [BEL 84, DRE 02]: "My first task was to find a name for multistage decision processes. In the first place I was interested in planning, in decision making, in thinking. But planning is not a good word for various reasons. I decided therefore to use the word "programming". I wanted to get across the idea that this was dynamic, this was multistage, this was time-varying – I thought, let's kill two birds with one stone. Let's take a word that has an absolutely precise meaning, namely dynamic, in the classical physical sense."

In other words, the word *programming* must be understood here as an action plan, while the word *dynamic* refers to the splitting up of the problem into phases. Let us

Chapter written by Bruno ESCOFFIER and Olivier SPANJAARD.

make clear from the start that the paradigm of dynamic programming is imprecisely defined, and that several formalisms coexist in the abundant literature dedicated to the subject [BEL 57, KAR 67, LAU 79, MIT 64, MIT 74, MOR 82]. Here we have adopted an intermediate formalism that has its advantages and disadvantages, and can therefore certainly be disputed. In the context of this book, we have concerned ourselves above all with the method's structuring virtues. After having introduced the idea of dynamic programming intuitively with a simple and playful example (section 4.2), we introduce a fairly general formal framework (section 4.3) based on the concepts of state spaces, the structure of transitions (between states) and preference structures (between policies). We will then show how various combinatorial problems can be brought into into this framework (section 4.4), and introduce two solution procedures (forwards and backwards) for which we will establish the admissibility (section 4.5). In section 4.6, we will solve the problems introduced in section 4.4 by dynamic programming. Lastly, in section 4.7, we introduce some extensions.

4.2. A first example: crossing the bridge

We consider four individuals a, b, c, d who are coming back from a hike at nightfall. They come across a suspension bridge crossing a river. A sign indicates that a maximum of two people may cross the bridge simultaneously for safety reasons. The four companions decide to cross the bridge two by two. However, the group only having one torch, one person having crossed the bridge must go back to the other side to take back the torch after each crossing. Finally, let us specify that individual a takes one minute to cross, b takes two minutes, c takes five minutes and d takes 10 minutes. If two members of the group cross the bridge together, they walk at the speed of the slowest member (for example, a and c take five minutes to cross together). In what order must the group cross to minimize the total crossing time?

This problem is properly solved by a dynamic programming approach. Firstly, we are easily persuaded that it is necessary to make five crossings (back and forth) for everyone to have crossed the bridge. Let us then note that after each crossing, we can sum up the situation (what we call the *state*) by the list of people that are on the original side of the bridge. The goal is therefore that of finding an optimal crossing order (what we call a *policy*) to go from state (a, b, c, d) to state $()$.

After the first crossing, two people remain on the original side of the bridge: either (a, b) after 10 minutes (the crossing time of c and d), or (a, c) after 10 minutes, or (a, d) after five minutes, or (b, c) after 10 minutes, or (b, d) after five minutes, or (c, d) after two minutes. After the return crossing of one person, only four states can appear:

– (a, b, c) are on the original side of the bridge;

– (a, b, d) are on the original side of the bridge;

– (a, c, d) are on the original side of the bridge;

– (b, c, d) are on the original side of the bridge.

State (a, b, d) is reached in three cases:

– *case 1.* (b, d) stayed behind (five minutes) and a came back (one minute);

– *case 2.* (a, d) stayed behind (five minutes) and b came back (two minutes);

– *case 3.* (a, b) stayed behind (10 minutes) and d came back (10 minutes).

In the first case, it took $5 + 1 = 6$ minutes to get to state (a, b, d). In the second case, it took $5 + 2 = 7$ minutes, and in the last case, it took $10 + 10 = 20$ minutes. Now, whatever happens next, the decisions made to get to state (a, b, d) will have no influence on the time needed to go from state (a, b, d) to state $()$, where all the individuals have crossed the bridge. Therefore, cases 2 and 3 can be ignored in the follow-up of the search.

In the same way, we can easily show that at best we need 11 minutes to get to state (a, b, c), three minutes to get to state (a, c, d) and four minutes to get to state (b, c, d). In summary, here are the relevant results to retain after the first two crossings (two people cross the bridge on the outward journey and one person comes back on the return journey):

– at best 11 minutes are needed to reach state (a, b, c);

– at best 6 minutes are needed to reach state (a, b, d);

– at best 3 minutes are needed to reach state (a, c, d);

– at best 4 minutes are needed to reach state (b, c, d).

After the third crossing, (a), (b), (c) or (d) remains on the original side of the bridge. State (a) comes about in three cases:

– *case 1.* we were in state (a, b, c) (11 minutes) and (b, c) have crossed (5 minutes);

– *case 2.* we were in state (a, b, d) (6 minutes) and (b, d) have crossed (10 minutes);

– *case 3.* we were in state (a, c, d) (3 minutes) and (c, d) have crossed (10 minutes).

Therefore, at best $10 + 3 = 13$ minutes are needed to reach state (a) after the third crossing. Repeating the same reasoning, we find:

– at best 13 minutes are needed to reach state (a);

– at best 14 minutes are needed to reach state (b);

– at best 13 minutes are needed to reach state (c);

– at best 8 minutes are needed to reach state (d).

One person comes back again, therefore (a, b), (a, c), (a, d), (b, c), (b, d) or (c, d) remain on the original side of the bridge. State (a, b) comes about in two cases:

– *case 1.* (a) stayed behind (13 minutes) and b came back (2 minutes);

– *case 2.* (b) stayed behind (14 minutes) and a came back (1 minute).

Therefore, at best $13 + 2 = 14 + 1 = 15$ minutes are needed to reach state (a, b) after the fourth crossing. In the same way, we show that:

– at best 15 minutes are needed to reach state (a, b);

– at best 14 minutes are needed to reach state (a, c);

– at best 9 minutes are needed to reach state (a, d);

– at best 15 minutes are needed to reach state (b, c);

– at best 10 minutes are needed to reach state (b, d);

– at best 13 minutes are needed to reach state (c, d).

The two people remaining cross the bridge for the last time and the crossing is finished. For example, if after four crossings we are in state (a, b), $15 + 2 = 17$ minutes will have been needed in total to reach state $()$. We can easily check that this is the best we can do. We thus show that at best 17 minutes are needed for the bridge crossing.

This solution approach is summed up in Table 4.1. Each column corresponds to a forward or backward crossing, and it contains all the possible states after this crossing. The best time for reaching a given state is shown in bold. For example, the line a: $10 + 1 = 11$ at the top of column 2 signifies that a came back on the second crossing and that after this crossing 11 minutes have gone by. We cannot do any better to reach state (a, b, c). We can put together an optimal policy by going back in time from the optimal duration of 17 minutes:

– a, b cross the bridge on the fifth crossing and so we were in state (a, b) after the fourth crossing;

– a comes back on the fourth crossing and so we were in state (b) after the third crossing;

– c, d crossed the bridge on the third crossing and so we were in state (b, c, d) after the second crossing;

– b comes back after the second crossing and so we were in state (c, d) after the first crossing;

– a, b cross the bridge on the first crossing.

The dynamic programming solution approach requires 42 additions and 27 comparisons to find an optimal policy. There are $6 \times 2 \times 3 \times 3 = 108$ possible policies and an exhaustive enumeration approach would therefore have required $4 \times 108 = 432$

1 (outward)		*2* (inward)		*3* (outward)	
(a,b)	c,d: **10**	(a,b,c)	a: **10+1=11**	(a)	b,c: 11+5=16
			b: 10+2=12		b,d: 6+10=16
(a,c)	b,d: **10**		c: 10+5=15		c,d: **3+10=13**
		(a,b,d)	a: **5+1=6**	(b)	a,c: 11+5=16
(a,d)	b,c: **5**		b: 5+2=7		a,d: 6+10=16
			d: 10+10=20		c,d: **4+10=14**
(b,c)	a,d: **10**	(a,c,d)	a: **2+1=3**	(c)	a,b: **11+2=13**
			c: 5+5=10		a,d: **3+10=13**
(b,d)	a,c: **5**		d: 10+10=20		b,d: 4+10=14
		(b,c,d)	b: **2+2=4**	(d)	a,b: **6+2=8**
(c,d)	a,b: **2**		c: 5+5=10		a,c: **3+5=8**
			d: 10+10=20		b,c: 4+5=9
4 (inward)		*5* (outward)			
(a,b)	a: **14+1=15**	$()$	a,b: **15+2=17**		
	b: **13+2=15**				
(a,c)	a: **13+1=14**		a,c: 14+5=19		
	c: 13+5=18				
(a,d)	a: **8+1=9**		a,d: 9+10=19		
	d: 13+10=23				
(b,c)	b: **13+2=15**		b,c: 15+5=20		
	c: 14+5=19				
(b,d)	b: **8+2=10**		b,d: 10+10=20		
	d: 14+10=24				
(c,d)	c: **8+5=13**		c,d: 13+10=23		
	d: 13+10=23				

Table 4.1. *Solution of the bridge-crossing problem*

additions and 107 comparisons. Thus there is a substantial gain in efficiency by solving the problem using dynamic programming. In the following section, we describe a general formalism that allows us to apply this method, and we give conditions that guarantee its accuracy. Many combinatorial problems can be solved by this method, by breaking them down into a sequence of problems of the same type.

4.3. Formalization

4.3.1. *State space, decision set, transition function*

Let $n \in \mathbb{N}$ be an integer. We assume that the problem that we wish to solve can be broken down into $n+1$ periods $0, \cdots, n$. We consider for each period i, a set S_i representing all states at the end of the period i. We further assume that:

– $S_i \cap S_j = \emptyset$ (if $i \neq j$), which is not a restrictive hypothesis (if this is not the case, we can always redefine states by adding the index of the corresponding period).

– there is a unique initial state s_0 and a unique final state s_n. Here, once more, this hypothesis is not restrictive since if this is not the case, we can always add an artificial initial and/or final state (by increasing the number of periods).

Finally, we will denote $S = \bigcup_{i=0}^{n} S_i$ as the set of states.

For example, in the bridge-crossing problem, let us remember that a period corresponds to a crossing, and that a state corresponds to a list of people situated on the original side of the bridge (the initial state being (a, b, c, d) and the final state $()$).

The aim of the problem is to "take the right decisions", that is to take a decision at each period such that at the end of the process these decisions are optimal for certain criteria. Let X be a set representing all the possible decisions. Not all decisions can be taken in each state: more exactly, for each state there is a corresponding set (included in X) representing the possible decisions in this state. Let us thus consider a set $A \subset S \times X$: $(s, x) \in A$ signifies that the decision x can be taken in state s.

Finally, a decision taken at the period i makes us pass from one state to another. We consider a transition function $t : A \to S$ where $t(s, x)$ indicates the state in which we find ourselves after having taken the decision x in state s. For coherence, we must add the following natural hypothesis:

$$\left.\begin{array}{l} s \in S_i \\ (s, x) \in A \end{array}\right\} \Longrightarrow t(s, x) \in S_{i+1}$$

This hypothesis simply means that a decision makes us pass from a state at the end of period i to a state at the end of period $i + 1$.

In the example in section 4.2, if we are in state (a, b, c) after the second crossing, there are three possible decisions: either b and c cross over and we find ourselves in state (a), or a and c cross over and we find ourselves in state (b), or a and b cross over and we find ourselves in state (c).

Let us note that a convenient representation of the set of states and transitions can be made with the help of a *state space graph* defined as follows:

– the set of vertices is S;

– the set of successors of a vertex s is $\Gamma^+(s) = \{t(s, x) : (s, x) \in A\}$.

The state space graph for the bridge-crossing problem is shown in Figure 4.1.

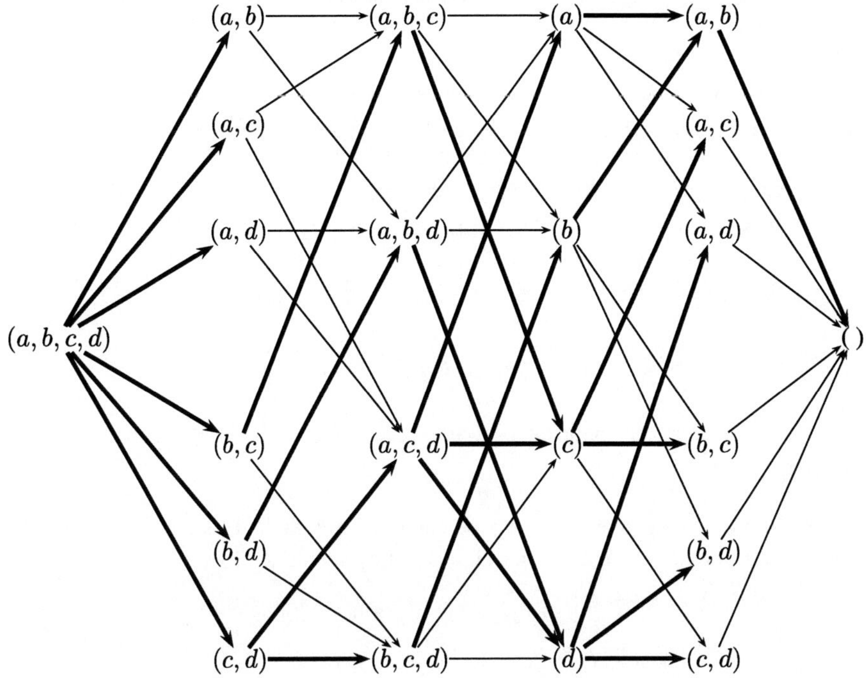

Figure 4.1. *State space graph for the bridge-crossing problem*

4.3.2. *Feasible policies, comparison relationships and objectives*

In this section, we introduce the idea of a *feasible policy*, and that of a *policy value* that allows us to define a choice criterion.

DEFINITION 4.1.– *We call any sequence of decisions a policy. Let* $s_i \in S_i$ *and* $s_j \in S_j$ *(where* $j > i$*). A feasible policy from* s_i *to* s_j *is a sequence of* $(j-i)$ *decisions* $(x_{i+1}, x_{i+2}, \cdots, x_j) \in X^{j-i}$ *such that in going from state* s_i*, these decisions lead us to state* s_j*. In other words, this sequence is such that for* $k \in \{i, i+1, \cdots, j-1\}$*:*

$$(s_k, x_{k+1}) \in A, \text{ and } s_{k+1} = t(s_k, x_{k+1})$$

A policy is said to be feasible from the end of period i *to the end of period* j *if and only if there are* $s_i \in S_i$ *and* $s_j \in S_j$ *such that this policy is feasible from* s_i *to* s_j*.*

We will express by $\Delta(s_i, s_j)$ the set of feasible policies from state s_i to state s_j, by $\Delta_{i,j}$ the feasible policies from period i to period j, and by Δ the set of all feasible policies. We assume that $\Delta(s_0, s_n) \neq \emptyset$, that is a feasible policy always exists from s_0 to s_n.

Since the solution method is based on a comparison of the solutions appearing in our framework as the feasible policies from s_0 to s_n, we assume that a valuation of the feasible policies is given, that is to say a function $v : \Delta \rightarrow Z$, where Z is a space of valuations endowed with an order $\leqslant$ that we assume to be complete (see section 4.7.1 for the case of a partial order). Let us remember that a complete order $\leqslant$ on the set Z is a binary relation fulfilling the following properties:

– transitivity, that is for each $(z_1, z_2, z_3) \in Z^3$, if $z_1 \leqslant z_2$ and $z_2 \leqslant z_3$, then $z_1 \leqslant z_3$;

– antisymmetry, that is for each $(z_1, z_2) \in Z^2$, if $z_1 \leqslant z_2$ and $z_2 \leqslant z_1$, then $z_1 = z_2$;

– completeness, that is for each $(z_1, z_2) \in Z^2$, $z_1 \leqslant z_2$ or $z_2 \leqslant z_1$.

We now have at our disposal a binary relation that allows us to compare feasible policies.

The aim of the problem can now be easily formulated. It consists of establishing an optimal policy $\delta^* \in \Delta(s_0, s_n)$, that is a policy such that:

$$\forall \delta \in \Delta(s_0, s_n), v(\delta^*) \leqslant v(\delta)$$

In the state space graph in Figure 4.1, a path made up of arcs in bold constitutes an optimal policy between its two extremities.

COMMENT.– Symmetrically, in the case where valuations represent profits, and where we are looking from a maximization angle, we obviously need to establish a policy $\delta^* \in \Delta(s_0, s_n)$ such that:

$$\forall \delta \in \Delta(s_0, s_n), v(\delta^*) \geqslant v(\delta)$$

Without loss of generality, we nevertheless assume in what follows that we seek a minimal valuation policy.

The exactness of the solution method (as we will see in section 4.5) relies on a fairly natural property of this comparison relationip between feasible policies, which we call *monotonicity*.

DEFINITION 4.2.– *The valuation function v is said to be:*

– *right monotonic if and only if for each $s_i \in S_i$, each $s_j \in S_j$, each $(\delta, \delta') \in \Delta(s_i, s_j)^2$ and for each $x \in X$ such that $(s_j, x) \in A$:*

$$v(\delta) \leqslant v(\delta') \Longrightarrow v(\delta, x) \leqslant v(\delta', x)$$

– *left monotonic if and only if for each $s_i \in S_i$, each $s_j \in S_j$, each $(\delta, \delta') \in \Delta(s_i, s_j)^2$ and for each $x \in X$ such that there exists s_{i-1} with $(s_{i-1}, x) \in A$ and $t(s_{i-1}, x) = s_i$:*

$$v(\delta) \leqslant v(\delta') \Longrightarrow v(x, \delta) \leqslant v(x, \delta')$$

– *monotonic if and only if it is both left and right monotonic.*

The monotonicity assumption reflects the idea that the order of the policies is not changed if we extend the policies identically.

4.4. Some other examples

We have seen how the introductory example fits well into this formal framework. Nevertheless, dynamic programming is a method that is applied to a fairly large group of problems (see for example [BEL 54, DRE 77]). Here we will give some classical examples that allow us to get an idea of the diversity of the possible applications.

In this part, we simply describe various problems and their modeling in the previously explained framework. In section 4.6, we will suggest a solution to these examples using dynamic programming.

4.4.1. *Stock management*

This first example is one of the most classic of the problems solved using dynamic programming. This is due, in part, to the fact that expressing it in the formalism of dynamic programming is direct and very natural.

Let us consider a company that produces (and sells) a certain piece of merchandise M. Production requires a particular component P that the company buys from a supplier. The company owns a warehouse where it can store up to five components P. At the beginning of each month it buys a certain quantity of components P to enable it to satisfy the demand for the month. The purchase price of the components P fluctuates from month to month but is supposed to be known in advance. Furthermore, the company has at its disposal the demand for the product M for each of the periods that we have under consideration. These data (purchase price of P and demand for M to be satisfied) are presented in Table 4.2.

	Month 1	Month 2	Month 3	Month 4	Month 5
Unit purchase price p_i	3	2	2	4	3
Demand d_i	1	4	2	3	1

Table 4.2. *Demand and purchase price*

Given that one component P is needed for the production of an object M, the aim of the problem is to establish an optimal purchase strategy for the company: more exactly, it is about deciding what quantity of P to buy at the beginning of each month in a way that satisfies demand while minimizing the total cost. We will further assume that the initial stock is zero and that we want zero final stock.

Let us express this problem in the formalism presented in section 4.3:

– *Periods*: the breakdown into periods is obvious. The ith month corresponds to the period i, and we add the initial period 0.

– *State space*: the states of period i correspond to the stock at the end of the period. There is therefore a state $s_0 \in S_0$ that corresponds to zero stock, 6 states in each of the periods 1 to 4 that correspond to a stock of between 0 and 5, and a state for period 5 that corresponds to a final zero stock.

– *Decision set*: the decisions to be taken are for the quantity of components P to buy at the beginning of the period. If we have a stock of k_{i-1} and the end of the period $i-1$, we can buy at most $l_i = 5 - k_{i-1}$ components so as not to exceed the storage capacity. Furthermore, at least d_i components are needed in order to satisfy demand. This therefore shows us the possible decisions in each state, that is the set A: given a stock of k_{i-1} in period $i-1$, $(k_{i-1}, l_i) \in A \Leftrightarrow k_{i-1} + l_i \in \{d_i, d_{i+1}, \cdots, 5\}$.

– *Transition function*: this corresponds to the calculation of the stock k_i at the end of the period i as a function of the stock k_{i-1} at the end of the period $i-1$ and the quantity l_i bought at the beginning of the period i, that is $k_i = k_{i-1} + l_i - d_i$.

– *Policy valuation*: the valuation is also very natural. The goal is to minimize the sum of purchase prices of the raw material. Let us therefore consider a feasible policy where we buy l_i components P in period i, l_{i+1} in period $i+1$, $\cdots$, l_j in period j. We define the value of this policy as the cost occasioned by these purchases: $v(l_i, \cdots, l_j) = \sum_{m=i}^{j} p_m l_m$. This valuation (for real values) is evidently monotonic.

COMMENT.– We will see in section 4.7.2 that we can equally consider the case where demand is random and where we seek to minimize the expected cost.

4.4.2. *Shortest path bottleneck in a graph*

This classical problem in graph theory is modeled (and can therefore be solved) as a dynamic programming problem. Let us consider the graph in Figure 4.2.

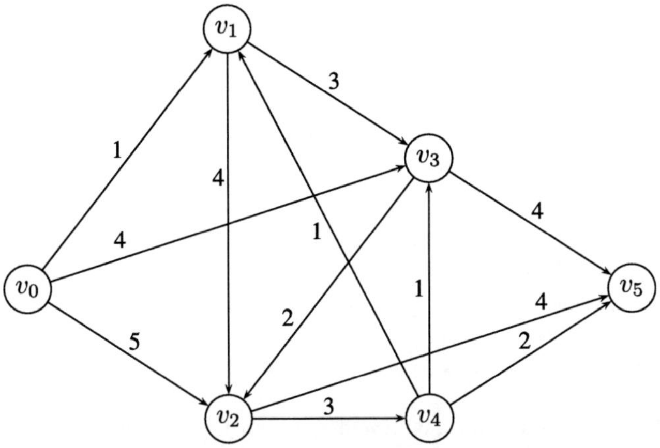

Figure 4.2. *Example of bottleneck shortest path*

Our objective is to find an optimal path from vertex v_0 to vertex v_5, taking *the greatest of the valuations of the edges that make it up* as the value of a path (*bottleneck-type* problem). This type of problem comes up, for example, in the choice of an itinerary for a convoy transporting high-risk products. The arcs of the graph correspond to a level of qualitative risk. It is therefore about finding the itinerary that minimizes the maximum risk along the way. We will express as $\Gamma^+(v_i)$ the set of successors of v_i, that is the vertices v_j for which an arc (v_i, v_j) exists, and we will express as $d(v_i, v_j)$ the length of the arc (v_i, v_j) (with, by convention, $d(v_i, v_i) = -\infty$).

We can easily show, without making any supplementary hypotheses about the lengths of the arcs, that there exists a shortest path *bottleneck* following at most five arcs (more generally $n-1$ where n is the number of vertices of the graph, since any loop cannot strictly decrease the value of a path). It is this property that will show us a possible modeling in the dynamic programming context: we will calculate for each k between 1 and 5 the value of a shortest path comprising at most k arcs going from vertex v_0 to each of the other vertices. The optimal value of a path comprising at most five arcs between v_0 and v_5 is the one we seek. More exactly, this leads to the following modeling:

– *Periods*: they go from 0 to 5 and correspond to the maximum number of arcs in the paths under consideration.

– *State spaces*: a state $(v_j, i) \in S_i$ corresponds to the paths going from v_0 to v_j following at most i arcs. For period 0 there is only one state $(v_0, 0)$ that corresponds to vertex v_0 (the only state that can be reached using 0 arcs). Similarly, there is only one state $(v_5, 5)$ at period 5 corresponding to the final vertex. For the other periods i, there are six states $(v_j, i), j = 0, \ldots, 5$.

– *Decision set*: a decision here consists of staying put or moving towards a successor; the set of possible decisions in state (v_j, i) is therefore $\Gamma^+(v_j) \cup \{v_j\}$, for $i = 0, \ldots, 4$ (where v_k refers to the decision to move towards v_k). For period 5, only the decision v_5 is likely to be acceptable because $S_5 = (v_5, 5)$.

– *Transition function*: this follows on from the definition of our decisions: $t((v_j, i), v_k) = (v_k, i + 1)$.

– *Policy valuation*: when we are in state (v_j, i), the valuation must correspond to the optimal path following at most i arcs between v_0 and v_j. We can therefore define the value of the policy $(v_{k_i}, v_{k_{i+1}}, \ldots, v_{k_j})$ as the value of the path taken:

$$v((v_{k_i}, v_{k_{i+1}}, \ldots, v_{k_j})) = \max\{d(v_{k_l}, v_{k_{l+1}}) : l = i, \ldots, j - 1\}$$

The valuation function here is monotonic. For right monotonicity, for example, it is enough to notice that given a path δ arriving at state (v_k, j) and a decision v_l feasible in state (v_k, j), $v(\delta, v_l) = \max(v(\delta), d(v_k, v_l))$. The addition of the decision v_l does not therefore change the order of preference.

COMMENT.– We have assumed here that the value of a path was the maximum valuation of its edges. We can perfectly well adapt the modeling to other transport problems. All that is needed is to change the valuation of the policies. If, for example, we consider that the value of a path is the sum of the valuations of the arcs making it up, we can then write $v((v_{k_i}, v_{k_{i+1}} \ldots, v_{k_j})) = \sum_{l=i}^{j-1} d(v_{k_l}, v_{k_{l+1}})$; the valuation is also monotonic.

4.4.3. *Knapsack problem*

Let us consider a set of objects $\{1, \ldots, n\}$. A profit p_i as well as a weight w_i is associated with each object. We also have at our disposal a knapsack whose total weight must not exceed b. The aim of the problem is to establish which objects to take in the knapsack in such a way as to maximize the profit (while satisfying the maximum weight constraint). In other words, it is about solving the following binary variable linear program:

$$\begin{cases} \max\ z = \sum_{i=1}^{n} p_i x_i \\ s.c. \left| \begin{array}{l} \sum_{i=1}^{n} w_i x_i \leqslant b \\ x_i \in \{0,1\},\ i = 1, \ldots, n \end{array} \right. \end{cases}$$

To model this problem in the dynamic programming framework, we will solve the following family of problems: for each $i \in \{1, \dots, n\}$ and for each $j \in \{0, \dots, b\}$, how do we choose objects among the first i in such a way that their total weight is at most j and the associated profit is maximum? The solution of the initial problem will therefore be the policy obtained for the problem where $i = n$ and $j = b$.

Let us consider the following instance: we have five objects of respective weights 6, 5, 3, 2 and 1, the associated profits being respectively 20, 19, 9, 5 and 2, and a knapsack whose weight must not exceed 9:

– *Periods*: these go from 0 to 5 and correspond to the subset of the objects under consideration.

– *State space*: a state $(i, j) \in S_i$ (with $i \in \{1, \dots, 5\}$) corresponds to the subsets of $\{1, \dots, i\}$ of a weight less than or equal to j. For periods i between 1 and 4, there are 10 states (i, j), $j \in \{0, \dots, 9\}$. For period 0, there is only one state $(0, 0)$; for period 5, we consider state $(5, 9)$.

– *Decision set*: a decision at period $i \in \{1, \dots, 4\}$ here consists of either adding or not adding the object i to the knapsack (provided that the total weight does not exceed b). There are at most two possible decisions in each state, which we will express as 0 (do not add the object) and 1 (add the object). However, in the last period, given that we wish to end up at state $(5, 9)$, there are only two states where we can make a decision: state $(4, 9)$, where we make the decision 0 not to add the object, and state $(4, 8)$, where we make the decision 1 to add the object to the knapsack.

– *Transition function*: this expresses the influence of the decision on the weight of the knapsack. Thus, if we are in state (i, j), and if we add the object $i + 1$ (decision 1), this decision sends us to state $(i + 1, j + w_{i+1})$; on the other hand, the decision 0, which does not change the weight of the knapsack, sends us to state $(i + 1, j)$.

– *Policy valuation*: this simply shows the profit associated with the elements put in the knapsack:

$$v(x_i, \dots, x_j) = \sum_{\{k : x_k = 1\}} p_k$$

The valuation function is obviously monotonic.

4.5. Solution

Having seen the formal framework and some examples of problems belonging to it, we can now consider an effective solution, that is the method of dynamic programming. As we have seen in the introduction and the examples following it, the aim of the method is to solve not one, but a whole family of problems by establishing recurrence relationships between the optimal solutions of these problems. This will allow us to enumerate only a (relatively limited) subset of solutions, which is significantly more efficient than using the brute force method consisting of enumerating all the solutions.

More exactly, the problems that are under consideration are broken down into $n+1$ periods, for which we have a set of possible states. The central point of this method will be to calculate recursively an optimal policy to get to each of the states of each period i. The hypothesis of monotonicity will allow us to calculate these policies easily and thus to reach the solution of the problem. This is the method known as the forward procedure, which we set out and prove in section 4.5.1.

Next we will set out the second dynamic programming method, known as the backward procedure. Here we start not from the initial state, but from the final state. We will illustrate this method using the example in section 4.2.

To start with, we will clarify the links between the monotonicity hypothesis, on which we base the solution method, and the principle of optimality set out by Bellman, which we often consider as underpinning the method of dynamic programming.

4.5.1. *Forward procedure*

Let us consider a problem for which the valuation function is right monotonic. The forward procedure can be expressed as follows:

1) define $S'_0 = \{s_0\}$ and $\delta^*(s_0) = ()$ (the empty policy);

2) for i from 1 to n:

2a) compute $S'_i = \{t(s_{i-1}, x) : s_{i-1} \in S'_{i-1} \text{ and } (s_{i-1}, x) \in A\}$

2b) for each $s_i \in S'_i$, establish:

$$\delta^*(s_i) \in \arg\min\{v(\delta^*(s_{i-1}), x) : (s_{i-1}, x) \in A \text{ and } t(s_{i-1}, x) = s_i\}$$

3) return $\delta^*(s_n)$

S'_i represents the set of states that we can obtain starting from the initial state s_0 (it is therefore the set of states $s_i \in S_i$ such that $\Delta(s_0, s_i) \neq \emptyset$). For each of these states, we calculate $\delta^*(s_i)$ which represents an optimal policy for going from the initial state to s_i.

THEOREM 4.1.– *If the valuation function is right monotonic, then $\delta^*(s_n)$ is an optimal policy for going from the initial state to the final state.*

Proof. We will show through recurrence on i that for each state $s_i \in S'_i$, $\delta^*(s_i)$ is an optimal policy for going from s_0 to s_i, that is:

$$\forall \delta \in \Delta(s_0, s_i),\ v(\delta^*(s_i)) \leqslant v(\delta)$$

This is obvious for $i = 0$. Assume that it is true for all states belonging to S'_{i-1}. Let $\delta = (x_1, x_2, \ldots, x_i) \in \Delta(s_0, s_i)$. Let us denote by s_{i-1} the state obtained by applying the policy $(x_1, x_2, \ldots, x_{i-1})$ from s_0. So by the recurrence hypothesis, $v(\delta^*(s_{i-1})) \leqslant v((x_1, \ldots, x_{i-1}))$. Thus, by right monotonicity:

$$v(\delta^*(s_{i-1}), x_i) \leqslant v(\delta) \quad [4.1]$$

Furthermore, from the definition of $\delta^*(s_i)$:

$$v(\delta^*(s_i)) \leqslant v(\delta^*(s_{i-1}), x_i) \quad [4.2]$$

We deduce from inequalities [4.1] and [4.2] that $v(\delta^*(s_i)) \leqslant v(\delta)$. We can therefore immediately conclude that $\delta^*(s_n)$ is an optimal policy for going from the initial state to the final state. ∎

4.5.2. *Backward procedure*

Let us go back to the example of the bridge-crossing problem. It is possible to apply "backwards" the approach previously used. Indeed, let us consider the final state () where everyone has crossed the bridge. At the end of the preceding period, we can easily infer that there are six possible states: (a, b), (a, c), (a, d), (b, c), (b, d) and (c, d). Coming from state (a, b), 2 minutes were needed to reach state (). Similarly, coming from states (a, c), (a, d), (b, c), (b, d) or (c, d), 5, 10, 5, 10 or 10 minutes, respectively, were needed to reach state ().

In the same way, four states are possible at the end of the preceding period: (a), (b), (c) and (d). Let us take the example of state (a). Leaving from this state, we can reach state (a, b) in 2 minutes, state (a, c) in 5 minutes or state (a, d) in 10 minutes. Therefore, to go from (a) to (), at best $\min\{2+2, 5+5, 10+10\}$ minutes are needed. Thus an optimal policy for going from (a) to () consists of bringing b back in period 4.

Using this method to calculate for each state an optimal policy for going from this state to the final state, we end up with an optimal policy from the initial state to the final state.

More generally, we now introduce the solution method known as the backward procedure. It consists of following a similar logic to that of the forward procedure, but starting from the final state. We now need left monotonicity for the evaluation function.

1) define $S''_n = \{s_n\}$ and $\delta^*(s_n) = ()$ (the empty policy);
2) for i from $n - 1$ to 0:

2a) compute $S_i'' = \{s_i : \exists x \in X,\ (s_i, x) \in A \text{ and } t(s_i, x) \in S_{i+1}''\}$

2b) for each $s_i \in S_i''$, compute:

$$\delta^*(s_i) \in \arg\min\{v(x, \delta^*(s_{i+1})) : (s_i, x) \in A \text{ and } t(s_i, x) = s_{i+1}\}$$

3) return $\delta^*(s_0)$.

This time, S_i'' represents the set of states of period i from which we can reach the final state s_n. For each of these states, we calculate $\delta^*(s_i)$ which represents an optimal policy for getting to the final state leaving from state s_i. We have the following theorem.

THEOREM 4.2.– *If the valuation function is left monotonic, then $\delta^*(s_0)$ is an optimal policy for going from the initial state to the final state.*

Proof. This is the same as the previous one and we will proceed to show, through descending recurrence on i, that for each state $s_i \in S_i''$, $\delta^*(s_i)$ is an optimal policy for going from s_i to s_n.

This is obvious for $i = n$. Let us assume that it is true for all states belonging to S_{i+1}''. Let $\delta = (x_{i+1}, x_{i+2}, \ldots, x_n) \in \Delta(s_i, s_n)$, and $s_{i+1} = t(s_i, x_{i+1})$. Therefore, using the recurrence hypothesis, $v(\delta^*(s_{i+1})) \leqslant v((x_{i+2}, \ldots, x_n))$, and by left monotonicity $v(x_{i+1}, \delta^*(s_{i+1})) \leqslant v(\delta)$. The definition of $\delta^*(s_i)$ allows us to assert that $v(\delta^*(s_i)) \leqslant v(x_{i+1}, \delta^*(s_{i+1}))$, from which we obtain: $v(\delta^*(s_i)) \leqslant v(\delta)$. ∎

COMMENT.– The backward procedure method does not necessarily lead us to cover the same subset of vertices of the state space graph as that of the forward procedure. Indeed, the states reachable from the initial state are not necessarily the same as those from which we can reach the final state (the reader may go back to the stock management example to convince himself of this).

4.5.3. *Principles of optimality and monotonicity*

To justify the use of dynamic programming, we often refer to the fact that Bellman's principle of optimality holds. Here is how it is set out in [BEL 57] (what we call the *first form* of the principle of optimality): "An optimal policy has the property that whatever the initial state and the initial decision are, the remaining decisions must constitute an optimal policy with regard to the state resulting from the first decision."

This principle therefore stipulates that, in a process where we must make n decisions, the restriction of an optimal policy to the last $(n-1)$ decisions (and so, through recurrence to the last k, for $k \in \{1, \ldots, n-1\}$) must be optimal.

This principle is similar to our left monotonicity hypothesis, the basis of the proof of the backward procedure. Furthermore, we could clearly go back to the proof of the method, using the principle of optimality as a hypothesis, which would constitute a sufficient condition for the backward procedure to be applicable.

We often express the principle of optimality in the following way: "any subpolicy of an optimal policy is optimal" (what we call the *second form* of the principle of optimality). This form is (slightly) more restrictive than the first, and so comes closer to monotonicity; in this case we can apply both the forward and backward procedures.

Nevertheless, note that the monotonicity hypothesis allows us to consider problems for which Bellman's principle of optimality does not hold (see for example [MIN 83]).

Let us go back to the example (see section 4.4.2) of the shortest path *bottleneck* problem in a graph, where the value of a path is the greatest valuation of the edges that make it up. As we have seen, the policy valuation function is monotonic. Let us now consider the graph in Figure 4.3, in which we seek an optimal path between v_1 and v_5.

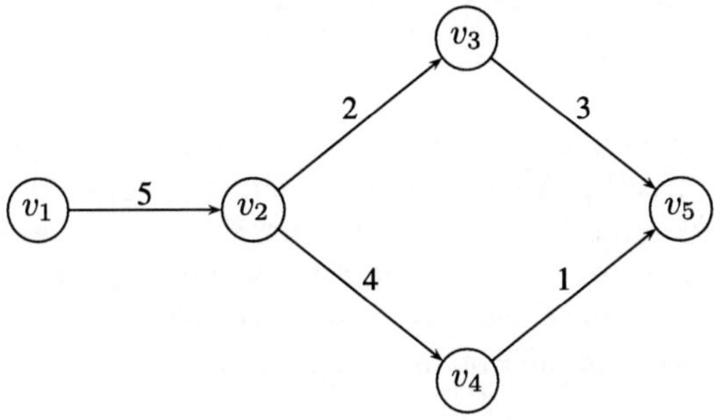

Figure 4.3. *Violation of the principle of optimality*

We can easily see that path (v_1, v_2, v_4, v_5), of value 5, is optimal. On the other hand, subpath (v_2, v_4, v_5), of value 4, is not optimal for going from v_2 to v_5. The other path (that is to say (v_2, v_3, v_5)) is indeed of value 3.

COMMENT.– Strictly speaking, the principle of optimality does not imply monotonicity. To realize this, it is enough to note that it only concerns optimal policies, while monotonicity concerns all policies.

Let us now imagine that we modify the monotonicity hypothesis slightly by introducing the notion of strict monotonicity.

DEFINITION 4.3.– *The valuation function v is said to be:*

– *strictly right monotonic if and only if for each* $s_i \in S_i$, *each* $s_j \in S_j$, *each* $(\delta, \delta') \in \Delta(s_i, s_j)^2$ *and for each* $x \in X$ *such that* $(s_j, x) \in A$:

$$\begin{cases} v(\delta) < v(\delta') \Longrightarrow v(\delta, x) < v(\delta', x) \\ v(\delta) = v(\delta') \Longrightarrow v(\delta, x) = v(\delta', x) \end{cases}$$

– *strictly left monotonic if and only if for each* $s_i \in S_i$, *each* $s_j \in S_j$, *each* $(\delta, \delta') \in \Delta(s_i, s_j)^2$ *and for each* $x \in X$ *such that there exists* s_{i-1} *with* $(s_{i-1}, x) \in A$ *and* $t(s_{i-1}, x) = s_i$:

$$\begin{cases} v(\delta) < v(\delta') \Longrightarrow v(x, \delta) < v(x, \delta') \\ v(\delta) = v(\delta') \Longrightarrow v(x, \delta) = v(x, \delta') \end{cases}$$

– *strictly monotonic if and only if it is both strictly right and left monotonic.*

If we assume that the valuation function is strictly monotonic (or strictly left monotonic), the principle of optimality in its second form (or, respectively, in its first form) is then automatically verified. Indeed, let us assume for example that v is strictly left monotonic, and let us consider an optimal policy $\delta^* = (x_1, x_2, \ldots, x_n)$. Let δ be a policy from $s_1 = t(s_0, x_1)$ to s_n. The hypothesis of strict monotonicity immediately rules out the possibility of $v(\delta) < v(x_2, \ldots, x_n)$, therefore $(x_2, \ldots, x_n)$ is optimal from s_1 to s_n.

The strict nature of monotonicity, if it leads to Bellman's principle of optimality, is only of any interest if we seek not just one, but all of the optimal policies. Indeed, by slightly modifying the forward and backward procedures in such a way as to keep all the partial optimal policies, the strict monotonicity hypothesis guarantees that all optimal policies are obtained. However, general monotonicity does not guarantee this, as we can see from the example in Figure 4.3.

4.6. Solution of the examples

4.6.1. *Stock management*

In this problem, the goal is to establish how many components P must be bought at the start of each month in such a way as to minimize the total cost (while satisfying demand). The approach using dynamic programming here consists of solving the following family of problems: for each period i and for each stock at the end of this period, establish an optimal policy to get to this level of stock. We solve the example from section 4.4.1 in Figure 4.4. In this table, the index of a column corresponds to the period under consideration and the index of a row corresponds to the level of stock reached at the end of this period. In the associated cell are the value of an optimal policy up to this state and the vector of decisions defining such an optimal policy.

	Month 1	Month 2	Month 3	Month 4	Month 5
Unit purchase price p_i	3	2	2	4	3
Demand d_i	1	4	2	3	1

Stock \ Period	0	1	2	3	4	5
0	**0** ()	**3** $(\mathbf{1})$	**11** $(\mathbf{1,4})$	15 $(1,4,2)$	**21** $(\mathbf{1,4,5,0})$	**24** $(\mathbf{1,4,5,0,1})$
1		6 (2)	13 $(1,5)$	17 $(1,4,3)$	25 $(1,4,5,1)$	
2		9 (3)		19 $(1,4,4)$	29 $(1,4,5,2)$	
3		12 (4)		**21** $(\mathbf{1,4,5})$		
4		15 (5)				
5						

Figure 4.4. *Solution of the stock problem*

4.6.2. *Shortest path* bottleneck

In this problem, the goal is to establish a bottleneck shortest path for going from a vertex v_0 to a vertex v_n. The dynamic programming approach here consists of solving the following family of problems: for each $i \leqslant n$ and for each vertex of the graph, establish an optimal path comprising at most i arcs from v_0 to this vertex. We solve the example from section 4.4.2 in Figure 4.5. In the table, the index of a column corresponds to the maximum number of arcs in the paths under consideration and the index of a row corresponds to the vertex of the graph reached. As before, the associated cell contains the value of an optimal policy up to this state and the vector of decisions defining such an optimal policy.

4.6.3. *Knapsack*

The solution of this problem using dynamic programming consists of establishing for each $i \in \{1, \ldots, 5\}$ and for each $j \in \{0, \ldots, 9\}$ an optimal solution to the following problem: establish a subset of objects among $\{1, \ldots, i\}$ of a weight less than or equal to j and of maximum profit.

The table in Figure 4.6 shows the solution of the example from section 4.4.3. The cell at the intersection of the column $i \in \{1, \ldots, 5\}$ with the row $j \in \{0, \ldots, 9\}$

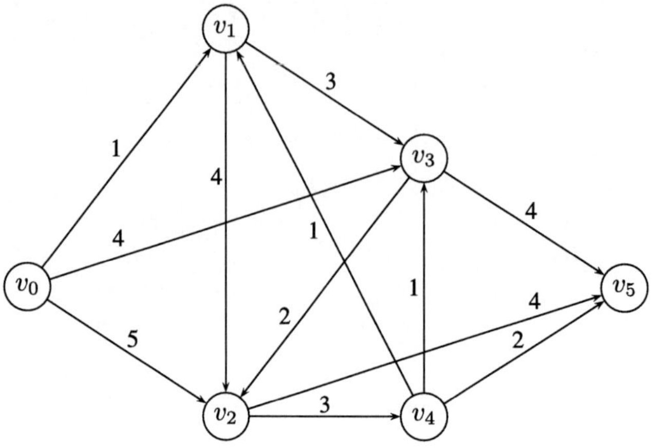

Vertex \ Period	0	1	2	3	4	5
v_0	$-\infty$ 0	$-\infty$ (v_0)	$-\infty$ (v_0, v_0)	$-\infty$ (v_0, v_0, v_0)	$-\infty$ (v_0, v_0, v_0, v_0)	
v_1		**1** $(\mathbf{v_1})$	1 (v_1, v_1)	1 (v_1, v_1, v_1)	1 (v_1, v_1, v_1, v_1)	
v_2		5 (v_2)	4 (v_1, v_2)	**3** $(\mathbf{v_1, v_3, v_2})$	3 (v_1, v_3, v_2, v_2)	
v_3		4 (v_3)	**3** $(\mathbf{v_1, v_3})$	3 (v_1, v_3, v_3)	3 (v_1, v_3, v_3, v_3)	
v_4			5 (v_2, v_4)	4 (v_1, v_2, v_4)	**3** $(\mathbf{v_1, v_3, v_2, v_4})$	
v_5			4 (v_3, v_5)	4 (v_3, v_5, v_5)	4 (v_3, v_5, v_5, v_5)	**3** $(\mathbf{v_1, v_3, v_2, v_4, v_5})$

Figure 4.5. *Solution of the bottleneck shortest path problem*

shows the value of an optimal solution for the corresponding problem, as well as the vector of decisions defining such a solution.

4.7. A few extensions

Having looked at some classical examples of solving problems that were directly expressed in the framework used here, in this part we will introduce some extensions that allow us to solve other problems using the dynamic programming method. In section 4.7.1, we extend the method to cases where the set of values Z comes with an order assumed to be only partial (which is notably true for multicriteria optimization).

	Object 1	Object 2	Object 3	Object 4	Object 5
Weights	6	5	3	2	1
Profits	20	19	9	5	2

Weight \ Period	0	1	2	3	4	5
0	**0** ()	**0** **(0)**	0 (0, 0)	0 (0, 0, 0)	0 (0, 0, 0, 0)	
1		0 (0)	0 (0, 0)	0 (0, 0, 0)	0 (0, 0, 0, 0)	
2		0 (0)	0 (0, 0)	0 (0, 0, 0)	5 (0, 0, 0, 1)	
3		0 (0)	0 (0, 0)	9 (0, 0, 1)	9 (0, 0, 1, 0)	
4		0 (0)	0 (0, 0)	9 (0, 0, 1)	9 (0, 0, 1, 0)	
5		0 (0)	**19** **(0, 1)**	19 (0, 1, 0)	19 (0, 1, 0, 0)	
6		20 (1)	20 (1, 0)	20 (1, 0, 0)	20 (1, 0, 0, 0)	
7		20 (1)	20 (1, 0)	20 (1, 0, 0)	24 (0, 1, 0, 1)	
8		20 (1)	20 (1, 0)	**28** **(0, 1, 1)**	**28** **(0, 1, 1, 0)**	
9		20 (1)	20 (1, 0)	29 (1, 0, 1)	29 (1, 0, 1, 0)	**30** **(0, 1, 1, 0, 1)**

Figure 4.6. *Solution of the knapsack problem*

Next, in section 4.7.2, we look at dynamic programming with uncertainty. Finally, in section 4.7.3, we present generalized dynamic programming, which has been introduced to deal with problems where the valuation function is not monotonic.

4.7.1. *Partial order and multicriteria optimization*

4.7.1.1. *New formulation of the problem*

We assume here that the policy valuation function v uses values from a set Z that has a partial order $\preccurlyeq$. This means that two policies are not necessarily comparable. We therefore introduce the notion of a non-dominated policy: a policy δ_0 is said to be non-dominated if and only if there does not exist a policy δ such that $v(\delta) \prec v(\delta_0)$.

The aim of the problem is therefore changed: it is no longer about finding a policy $\delta^* \in \Delta(s_0, s_n)$ better than all the others (because often such a policy will not exist), but rather about finding the set of non-dominated policies[1].

Let Δ_0 be a set of feasible policies. We express by $ND(\Delta_0)$ a subset of Δ_0 minimal for inclusion, such that:

$$\forall \delta \in \Delta_0, \exists \delta^* \in ND(\Delta_0) : v(\delta^*) \preccurlyeq v(\delta)$$

Solving the problem thus consists of finding a set $ND(\Delta(s_0, s_n))$.

4.7.1.2. *Solution*

The forward and backward solution methods are still applicable under the right or left monotonicity hypotheses of the valuation function. They must, however, be adapted to the new problem set by establishing, for each state s_i of each period i, the set $\Delta^*(s_i)$ of non-dominated policies from s_0 to s_i (or from s_i to s_n).

The forward procedure is expressed as follows:

1) define $S'_0 = \{s_0\}$ and $\Delta^*(s_0) = \{()\}$ (the set containing one element: the empty policy);

2) for i from 1 to n:

2a) compute $S'_i = \{t(s_{i-1}, x) : \ s_{i-1} \in S'_{i-1}$ and $(s_{i-1}, x) \in A\}$

2b) for each $s_i \in S'_i$, compute a set:

$$\Delta^*(s_i) = ND(\{(\delta, x) : \delta \in \Delta^*(s_{i-1}) \text{ and } t(s_{i-1}, x) = s_i\})$$

3) return $\Delta^*(s_n)$.

This forward procedure method is valid under the right monotonicity hypothesis:

THEOREM 4.3.– *If the valuation function is right monotonic then* $\Delta^*(s_n) = ND(\Delta(s_0, s_n))$.

The proof, similar to the one in the traditional framework, has been omitted.

4.7.1.3. *Examples*

In multicriteria problems [TRZ 94], we have q valuation functions $v_1, \dots, v_q$ from Δ to $\mathbb{N}$ (the *criteria*), *which we assume to be monotonic*.

1. In reality, if two non-dominated policies have the same valuation, we will only keep one of them.

Thus each feasible policy corresponds to a point in the *criteria space*. The partial order most used between cost vectors is weak *Pareto-dominance* $\preccurlyeq_P$:

$$\forall x, y \in \mathbb{N}^q,\ x \preccurlyeq_P y \iff \forall i \in \{1, \ldots, q\},\ x_i \leqslant y_i$$

Suppose that we have modeled a multicriteria problem in the form of the state space graph in Figure 4.7. At the moment of execution of the forward procedure mentioned in the previous section, the sets $\Delta^*(v_i)$ retained are (denoting by v_j the decision leading to state v_j):

- $\Delta^*(v_1) = \{(v_1)\}$
- $\Delta^*(v_2) = \{(v_2)\}$
- $\Delta^*(v_3) = \{(v_1, v_3)\}$
- $\Delta^*(v_4) = \{(v_1, v_4), (v_2, v_4)\}$
- $\Delta^*(v_5) = \{(v_2, v_5)\}$
- $\Delta^*(v_6) = \{(v_2, v_4, v_6), (v_2, v_5, v_6)\}$

For $\Delta^*(v_4)$ for example, the two possible policies (v_1, v_4) and (v_2, v_4), of respective values $(5, 3)$ and $(3, 4)$, must be retained because they are both non-dominated.

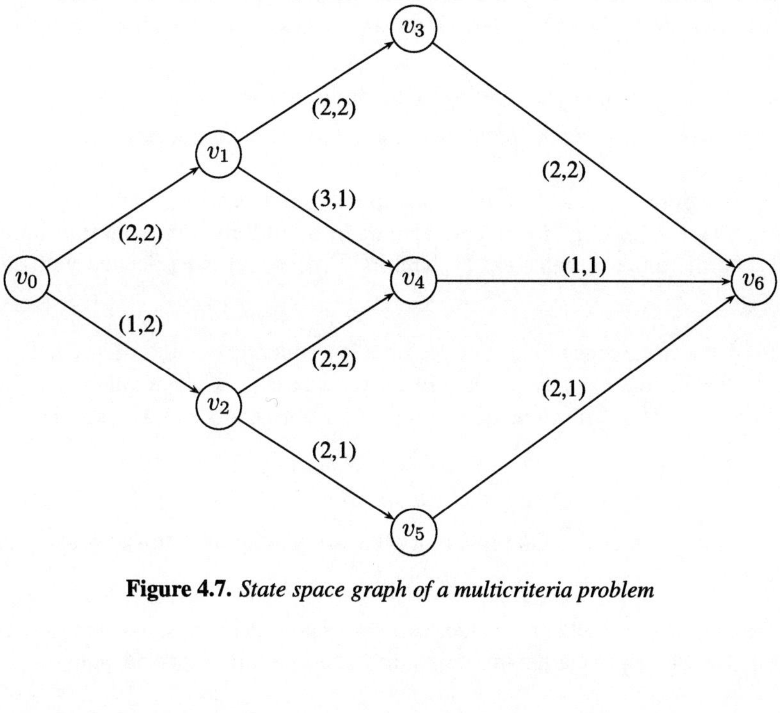

Figure 4.7. *State space graph of a multicriteria problem*

4.7.2. *Dynamic programming with variables*

In most real world problems, we have to take into account uncertainty. The process is no longer affected just by the choices of the decision maker, but also by an uncertain quantity linked to exterior parameters: we therefore have to modify the deterministic framework studied up until now. In a stock management problem, for example, it is rare for demand to be deterministic. Statistical data nevertheless generally allow us to model the uncertainty in the form of a random variable corresponding to demand. We then evaluate the sequence of choices with regard to a given decision criterion (for example the expected value of the solution).

4.7.2.1. *Sequential decision problems under uncertainty*

In this new framework, a period is divided into two successive phases: a decision, then the intervention of chance that determines the state in which we will be at the end of this period. We consider that, given a state and a decision, the chance is given by a set of states that may be reached and by a probability for each one of them being reached. Thus, it comes down to considering a transition function that is no longer deterministic, but random.

A policy here no longer consists of a sequence of decisions, because such a sequence would have no meaning in the presence of uncertainty. Now it is about giving, for each state $s \in S$, the decision $x(s)$ that we would make (if the preceding decisions and chance led us to it). It is convenient to represent the sequential decision problem process by a special graph, called a *decision tree*, where we distinguish two types of vertex:

– vertices $s_i \in S_i$ corresponding to the different states;

– vertices $h_i \in H_i$ corresponding to the intervention of uncertainty.

In this graph, an arc (s_i, h_i) corresponds to a decision, and the set of arcs $\{(h_i, s_{i+1}) : s_{i+1} \in \Gamma^+(h_i)\}$ corresponds to the different random transitions possible. Furthermore, a transition probability $P(h_i, s_{i+1})$ is associated with each arc (h_i, s_{i+1}).

In such a context, the final state possible is frequently not unique: we have a set of final states S_n that we may reach. Moreover, we assign a scalar value $v(s_n)$ to each final state $s_n \in S_n$. The value of a policy δ thus corresponds to the expected value:

$$E(\delta) = \sum_{s_n \in S_n} P_\delta(s_n) v(s_n)$$

where $P_\delta(s_n)$ represents the probability that we end up in state s_n by using the policy δ.

Depending on whether the values associated with the final states represent costs or profits, we seek a policy that minimizes or maximizes the expected value.

4.7.2.2. *Solution*

The value of an optimal policy is obtained by going back from the final states towards the initial state according to the following recurrence relation (without loss of generality we assume here that we want to minimize the expected value):

$$\begin{aligned} v(s_i) &= \min\{v(h_i) : h_i \in \Gamma^+(s_i)\} \\ v(h_i) &= \textstyle\sum_{s_{i+1} \in \Gamma^+(h_i)} P(h_i, s_{i+1}) v(s_{i+1}) \end{aligned}$$

We associate a label with each vertex of the decision graph corresponding to the value of an optimal policy from this vertex to the final state. An optimal decision $x^*(s_i)$ in a state s_i corresponds to an arc (s_i, h_i) for which the label associated with the vertex h_i is minimal. The set of these optimal decisions makes up an optimal policy.

4.7.2.3. *Example*

Let us consider an insurance contract that we can decide to take out for a year 1 and/or a year 2. Obviously we do not know whether or not we will need to make a claim following an accident. Furthermore, the probability of having an accident during year 2 is affected by whether or not we have already had an accident in year 1. The decision tree corresponding to this problem is represented in Figure 4.8. Vertex d_1 concerns the choice of taking out the contract (branch 1) or not (branch 0) in year 1. Vertices h_1 and h_2 concern the occurrence (branch 1) or not (branch 0) of an event necessitating making a claim on this contract ($P(0) = 2/5$ and $P(1) = 3/5$). The vertices d_2, d_3, d_4 and d_5 concern the choice of taking out the contract or not in year 2. Lastly, vertices h_3, h_4, h_5, h_6, h_7, h_8, h_9 and h_{10} concern the occurrence or not of an event necessitating making a claim on this contract in the second year. If we have already had an accident in year 1, the probability of such an event happening is $P(0) = 3/4$ (and therefore $P(1) = 1/4$). On the other hand, if we have not had an accident in year 1, the probabilities remain $P(0) = 2/5$ and $P(1) = 3/5$. In each leaf of the decision tree is the total cost generated. Taking out a contract costs 1, while having an accident that is not covered costs 2 (as opposed to 0 if we are insured). The reader can check without difficulty that such parameters lead us to associate the value $\min\{6/5, 1\} = 1$ with the vertex d_2, 5/2 with the vertex d_3, 2 with the vertex d_4 and 3/2 with the vertex d_5. As a result, the label of vertex h_1 is worth $1 \times 2/5 + 5/2 \times 3/5 = 19/10$, and the label of vertex h_2 is worth $17/10$. The optimal policy therefore consists of taking out the contract for the first year, and only taking out the contract for the second year if we have not had an accident in the first year.

4.7.3. *Generalized dynamic programming*

Here we look at a problem similar to the one mentioned in an article by Sniedovich [SNI 81], concerning Kao's procedure [KAO 78] for a traveling salesman problem

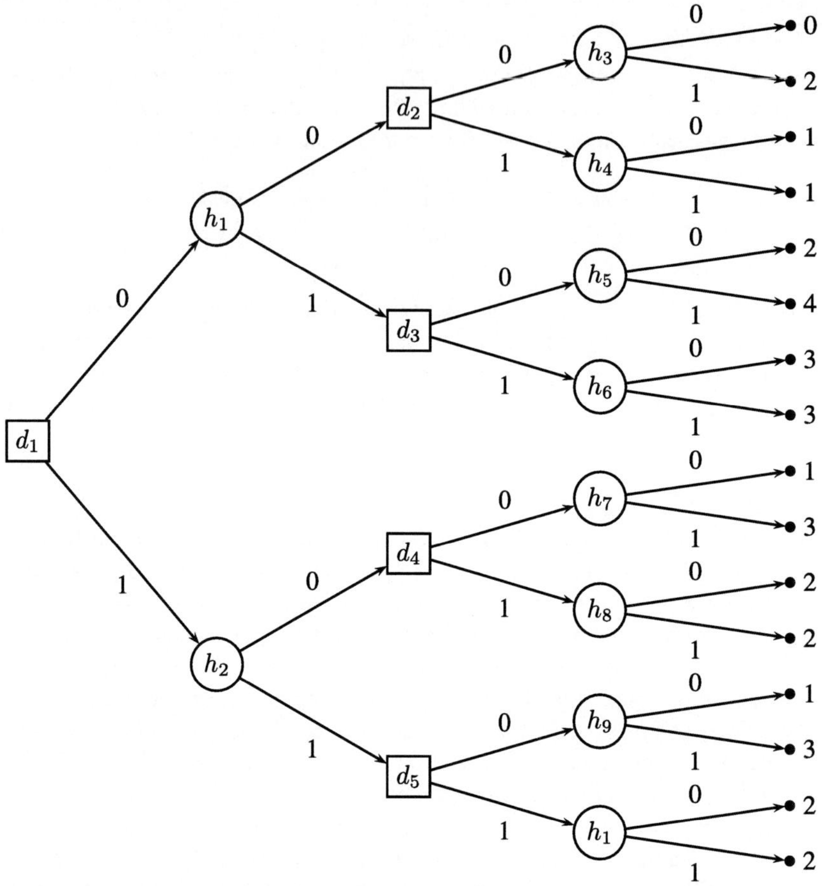

Figure 4.8. *Decision tree for the insurance contract problem*

with stochastic costs. In this context, we study a stochastic shortest path problem where the probability of lateness is associated with each arc. Let us consider the graph in Figure 4.9. We leave from the vertex s at 12:00 and we assume that we take buses at each vertex of the graph, in order to reach the vertex t where we have a train to take that leaves at 16:00. At each vertex of the graph, and every hour on the hour, a bus leaves for the next vertex with a journey time that is usually less than one hour. Nevertheless, fluctuations in traffic can increase the journey time beyond an hour, and are modeled by the probability of this event occurring. We seek the most suitable path for arriving at the destination before 16:00. Let us note that this timetable is kept to if and only if at most one bus is late. Let $R(P)$ be the random variable representing the number of buses that are late along path P. We seek therefore to maximize the probability $P(R(P) \leqslant 1)$. At the vertex t, we have $P(R(s,1,3,t) \leqslant 1) < P(R(s,2,3,t) \leqslant$

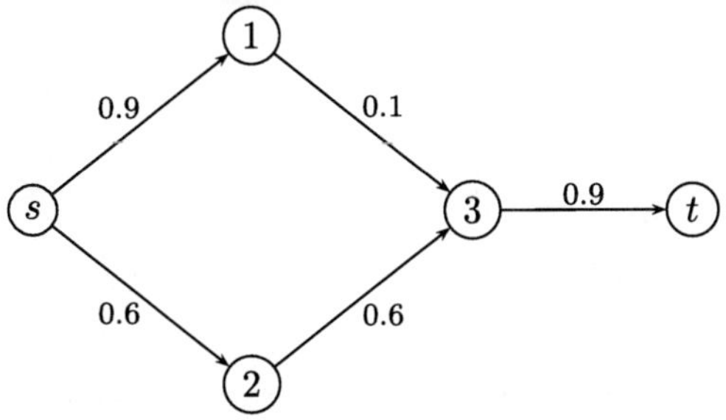

Figure 4.9. *An instance of the stochastic shortest path problem*

1). Indeed, adding together the probabilities that all the buses are on time, with the probabilities that only one bus is late, we obtain:

$$\begin{cases} P(R(s,1,3,t) \leqslant 1) &= 0,009 + 0,9 \times 0,09 + 0,1 \times 0,01 + 0,9 \times 0,09 \\ &= 0,172 \\ P(R(s,2,3,t) \leqslant 1) &= 0,016 + 0,6 \times 0,04 + 0,6 \times 0,04 + 0,9 \times 0,16 \\ &= 0,208 \end{cases}$$

and so the optimal solution is path $(s, 2, 3, t)$. Unfortunately, if we take the probability $P(R(P) \leqslant 1)$ as the value of path P, application of the dynamic programming method leads to an error because the criterion is not monotonic. Indeed, at vertex 3 we have:

$$\begin{cases} P(R(s,1,3) \leqslant 1) &= 1 - 0,09 &= 0,91 \\ P(R(s,2,3) \leqslant 1) &= 1 - 0,36 &= 0,64 \end{cases}$$

As a consequence, path $(s, 2, 3)$ would be discarded at the execution of the method.

Generalized dynamic programming has been developed by Carraway *et al.* [CAR 89, CAR 90] to deal with this type of situation. It is based on a weak optimality principle which is described as follows: *an optimal policy is made up of subpolicies that may be part of an optimal policy.*

More exactly, it is about adapting the principle of optimality to the case where the policy valuation function is not monotonic, by eliminating the partial policies that we know it is not possible to complete to give an optimal policy. For this, we use for each state s the partial pre-order $\preccurlyeq_s$, such that for each $(\delta_1, \delta_2) \in \Delta(s_0, s)^2$:

$$\delta_1 \preccurlyeq_s \delta_2 \Leftrightarrow \exists \delta' \in \Delta(s, s_n) : \forall \delta \in \Delta(s, s_n), v(\delta_1, \delta') \leqslant v(\delta_2, \delta)$$

The generalized dynamic programming method then consists of only retaining, for each state s, the policies $\delta \in \Delta(s_0, s)$ for which there is not a feasible policy δ' from

s_0 to s such that $\delta' \prec_s \delta$. By proceeding in this way, we end up in the last period with a set of feasible policies from s_0 to s_n containing all the optimal policies.

In practice, it may be difficult to construct the entire pre-order $\preccurlyeq_s$; it is nevertheless often possible to establish a sufficient condition so that $\delta \prec_s \delta'$. This allows us to perform pruning and thus a noticeable gain in efficiency relative to an exhaustive method.

In the stochastic shortest path example, such an approach would consist of noting that a partial path P_1 cannot lead to an optimal path from the moment that it is strictly dominated by another path P_2 at the same vertex, in the sense of stochastic dominance $\preccurlyeq_S$, defined as follows:

$$P_2 \preccurlyeq_S P_1 \iff \forall k \in \{0,1,2,3\},\ P(R(P_2) \leqslant k) \geqslant P(R(P_1) \leqslant k)$$

From that moment, a path is discarded by the algorithm only if it fulfills this condition. In the graph in Figure 4.9, path $(s,2,3)$ is not discarded at vertex 3 because $P(R(s,2,3) \leqslant 0) > P(R(s,1,3) \leqslant 0)$.

By evaluating partial policies in relation to their impact on the global problem, the potential errors tied to the local view of traditional dynamic programming are avoided. In this way, the exactness of the solution procedure is guaranteed.

4.8. Conclusion

In this chapter, we have introduced a formal framework for dynamic programming, an implicit enumeration method in common use in combinatorial optimization. This approach offers the advantage of being applicable to a wide variety of problems, of which the sequential aspect is not necessarily foremost. Two solution procedures are put forward in the literature – a forward procedure and a backward procedure – for which we have given sufficient conditions of validity (right monotonicity, left monotonicity, monotonicity). We have then illustrated these procedures using examples that are typical of dynamic programming (stock management, shortest path, knapsack). Lastly, a few classical extensions of our framework (multicriteria or non-monotonic valuation function, sequential decisions under uncertainty) have been studied.

4.9. Bibliography

[BEL 54] BELLMAN R., "Some applications of the theory of dynamic programming – a review", *Journal of the Operational Research Society of America*, vol. 2, num. 3, p. 275–288, 1954.

[BEL 57] BELLMAN R., *Dynamic Programming*, Princeton University Press, Princeton, 1957.

[BEL 84] BELLMAN R., *Eye of the Hurricane*, World Scientific Publishing Company, Singapore, 1984.

[CAR 89] CARRAWAY R., MORIN T. , MOSKOWITZ H., "Generalized dynamic programming for stochastic combinatorial optimization", *Operations Research*, vol. 37, num. 5, p. 819–829, 1989.

[CAR 90] CARRAWAY R., MORIN T., MOSKOWITZ H., "Generalized dynamic programming for multicriteria optimization", *European Journal of Operational Research*, vol. 44, p. 95–104, 1990.

[DRE 77] DREYFUS S., LAW A., *The Art and Theory of Dynamic Programming*, Academic Press, Orlando, 1977.

[DRE 02] DREYFUS S., "Richard Bellman on the birth of dynamic programming", *Operations Research*, vol. 50, num. 1, p. 48–51, 2002.

[KAO 78] KAO E., "A preference order dynamic program for a stochastic traveling salesman problem", *Operations Research*, vol. 26, num. 6, p. 1033–1045, 1978.

[KAR 67] KARP R., HELD M., "Finite-state processes and dynamic programming", *SIAM Journal on Applied Mathematics*, vol. 15, num. 3, p. 693–718, 1967.

[LAU 79] LAURIÈRE J., *Éléments de programmation dynamique*, Gauthier-Villars, Paris, 1979.

[MIN 83] MINOUX M., *Programmation mathématique, théorie et algorithmes*, vol. 2 of *Collection Technique et Scientifique des Télécommunications*, Bordas, Paris, 1983.

[MIT 64] MITTEN L., "Composition principles for synthesis of optimal multistage processes", *Operations Research*, vol. 12, num. 4, p. 610–619, 1964.

[MIT 74] MITTEN L., "Preference order dynamic programming", *Management Science*, vol. 21, num. 1, p. 43–46, 1974.

[MOR 82] MORIN T., "Monotonicity and the principle of optimality", *Journal of Mathematical Analysis and Applications*, vol. 86, p. 665–674, 1982.

[SNI 81] SNIEDOVICH M., "Analysis of a preference order traveling salesman problem", *Operations Research*, vol. 29, num. 6, p. 1234–1237, 1981.

[TRZ 94] TRZASKALIK T., "Multiple criteria discrete dynamic programming", *Mathematics Today*, vol. XII-A, p. 173–199, 1994.

PART III

Elements from Mathematical Programming

Chapter 5

Mixed Integer Linear Programming Models for Combinatorial Optimization Problems

5.1. Introduction

A combinatorial optimization problem can be typically formulated as a mixed integer linear programming (MILP) model by first translating each elementary decision into a variable and then by specifying the objective function and the related constraints. To obtain a *linear* programming model, the objective is also required to be a *linear* function of the variables and the constraints are required to be *linear (in)equalities* involving the variables. While modeling can sometimes be considered to be an "art" which cannot be reduced to a standard set of procedures, it is of prime interest to devise at least some general rules to help with the formulation process. Note that, even though nearly all books on operations research and combinatorial optimization deal broadly with MILP models, only a few references (see for instance [PLA 02, WIL 90]) discuss some sort of systematic approach in the MILP modeling of combinatorial optimization problems. The purpose of this chapter is to provide an insight into MILP modeling of combinatorial optimization problems. First, introductory MILP models are recalled together with general modeling techniques; then more or less standard MILP formulations of several combinatorial optimization problems are discussed.

5.1.1. *Preliminaries*

A general MILP model is composed of three fundamental elements:

Chapter written by Frédérico DELLA CROCE.

– Decision variables $(x_1, x_2, ..., x_j, ..., x_n)$: the unknowns of the problem, namely those entities such that *given any (feasible) solution of the problem, if we know their value, then we can immediately verify if the solution is feasible and compute the cost function value.* Identification of the decision variables is the primary issue in ILP modeling and the determination of a solution for an ILP model consists of assigning a value to the decision variables.

– Constraints: the mathematical relations that must express the requirements of the considered problem. When the variables set is properly defined, then all the requirements can be naturally formulated by means of equalities or inequalities on the given variables.

– Objective function: the function f of the decision variables $x_1, x_2, ..., x_j, ..., x_n$ to be minimized (maximized).

As we are dealing with MILP models, both the objective function and the constraints are restricted to *linear* expressions of the variables.

The general formulation of a MILP model is:

$$\min\,(\max)\;\; f(x_1, ..., x_n) = c_1x_1 + c_2x_2 + ... + c_jx_j + ... + c_nx_n$$

$$a_{11}x_1 + a_{12}x_2 + ... + a_{1j}x_j + ... + a_{1n}x_n \gtreqless b_1$$

$$...$$

$$a_{i1}x_1 + a_{i2}x_2 + ... + a_{ij}x_j + ... + a_{in}x_n \gtreqless b_i \qquad [5.1]$$

$$...$$

$$a_{m1}x_1 + a_{m2}x_2 + ... + a_{mj}x_j + ... + a_{mn}x_n \gtreqless b_m$$

$$x_j \geqslant 0, \quad j = 1, ..., n$$

with $x_1, ..., x_j, ..., x_n$, decision variables, $c_1, ..., c_j, ..., c_n$, cost (profit) coefficients in the objective function for a minimization (maximization) problem, $b_1, ..., b_i, ..., b_m$, so-called right-end sides in the constraints set, $a_{11}, ..., a_{ij}, ..., a_{mn}$, variables coefficients in the constraints set.

The decision variables are typically split into three main categories:

1) positive real variables ($x_j \geqslant 0$);

2) positive integer variables ($x_j \geqslant 0$, integer);

3) binary variables ($x_j \in \{0, 1\}$).

Note that we can immediately handle problems where a given variable x_j is free ($-\infty \leqslant x_j \leqslant +\infty$) or non-positive ($x_j \leqslant 0$). To do this, if x_j is free, it is sufficient to set $x_j = u_j - v_j$ with $u_j \geqslant 0$, $v_j \geqslant 0$ and substitute x_j with $u_j - v_j$. Analogously, if $x_j \leqslant 0$, it is sufficient to set $x_j = -y_j$ with $y_j \geqslant 0$ and substitute x_j with $-y_j$.

In the following we provide a miscellany of basic MILP formulations of well-known combinatorial optimization problems.

5.1.2. *The knapsack problem*

In the knapsack problem, a hiker must decide which items to include in his/her knapsack on a forthcoming trip. He/she must choose from among a set N of n items, where item i has weight w_i (say in kilograms) and utility u_i. The objective is to maximize the hiker's trip utility subject to the limitation that he/she can carry at most W kg. Let x_i be a 0–1 variable such that $x_i = 1$ if item i is included in the knapsack, $x_i = 0$ if it is not.

The standard MILP formulation of this model is as follows:

$$\max \sum_{i \in N} u_i x_i \qquad [5.2]$$

subject to

$$\sum_{i \in N} w_i x_i \leqslant W \qquad [5.3]$$

$$x_i \in \{0, 1\} \quad \forall i \in N \qquad [5.4]$$

where the objective function [5.2] maximizes the hiker's trip utility, constraint [5.3] forbids exceeding the weight capacity W, and constraints [5.4] impose the restriction that the x_i variables must be binary.

5.1.3. *The bin-packing problem*

In the bin-packing problem, there is a given set N of n items with sizes $\{s_1, ..., s_n\}$ to be packed into bins such that the sum of the sizes of all items assigned to each bin does not exceed the bin capacity, W (where $s_j \leqslant W \quad \forall j \in N$, or else the problem is not feasible), and the objective is to use as few bins as possible. Let u_i ($i = 1, ..., n$) be a 0–1 variable such that $u_i = 1$ if bin i is used, $u_i = 0$ if it is not. Let x_{ij} ($i = 1, ..., n$, $j \in N$) be a 0–1 variable such that $x_{ij} = 1$ if item j is assigned to bin i, $x_{ij} = 0$ if it is not.

The standard MILP formulation of this model is as follows:

$$\min \sum_{i=1}^{n} u_i \qquad [5.5]$$

subject to

$$\sum_{i=1}^{n} x_{ij} = 1 \quad \forall j \in N \qquad [5.6]$$

$$\sum_{j \in N} s_j x_{ij} \leqslant W u_i \quad \forall i = 1, ..., n \qquad [5.7]$$

$$u_i, x_{ij} \in \{0, 1\} \quad \forall i = 1, ..., n, \ \forall j \in N \qquad [5.8]$$

where the objective function [5.5] minimizes the number of bins used, constraints [5.6] impose the restriction that each item is assigned to exactly one bin, constraints [5.7] impose the restriction that the weight capacity W for each bin is not excedded and constraints [5.8] impose the restriction that the variables u_i and x_{ij} must be binary.

5.1.4. *The set covering/set partitioning problem*

Consider a $0 - -1$ $m \times n$ matrix $A = (a_{ij})$. Let $M = (1, ..., m)$ be the set of rows of A and correspondingly let $N = (1, ..., n)$ be the set of columns of A. Let c_j ($j = 1, ..., n$) denote the cost of column j, where we assume $c_j > 0, \forall j$. We say that a column $j \in N$ covers a row $i \in M$ if $a_{ij} = 1$. In the set covering problem we search for a minimum cost subset $S \subseteq N$ of columns such that each row $i \in M$ is covered by at least one column $j \in N$. We are dealing with a set partitioning problem if the requirement is that each row $i \in M$ is covered by exactly one column $j \in N$. Let x_j be a 0–1 variable such that $x_j = 1$ if column j is selected ($j \in S$), $x_j = 0$ if it is not.

The standard MILP formulation of the set covering problem is as follows:

$$\min \sum_{j \in N} c_j x_j \qquad [5.9]$$

subject to

$$\sum_{j \in N} a_{ij} x_j \geqslant 1 \quad \forall i \in M \qquad [5.10]$$

$$x_j \in \{0, 1\} \quad \forall j \in N \qquad [5.11]$$

where the objective function [5.9] minimizes the weighted sum of selected columns, constraints [5.10] impose the restriction that each row is covered by at least one column and constraints [5.11] impose the restriction that the x_j variables must be binary.

To deal with the set partitioning problem, it is sufficient to substitute inequalities [5.10] with equalities as follows:

$$\sum_{j \in N} a_{ij} x_j = 1 \quad \forall i \in M. \qquad [5.12]$$

5.1.5. *The minimum cost flow problem*

In the minimum cost flow problem we want to determine a least cost shipment of a commodity through a network in order to satisfy demands at certain nodes from available supplies at other nodes.

Consider a directed network $G(N, A)$ defined by a set N of n nodes and a set A of m directed arcs. Each arc (i, j) is associated with a cost c_{ij} denoting the cost per unit flow on that arc where the flow cost is assumed to vary linearly with the amount of flow. Each node $i \in N$ is associated with an integer value b_i representing its supply/demand. If $b_i > 0$, i is a supply node; if $b_i < 0$, i is a demand node; finally, if $b_i = 0$, i is a transhipment node. We assume that the network is balanced, that is $\sum_{i \in N} b_i = 0$.

Let $x_{ij} \geqslant 0$ denote the flow on arc $(i, j) \in A$. A standard MILP formulation of the minimum cost flow problem is as follows:

$$\min \sum_{(i,j)\in A} c_{ij}x_{ij} \qquad [5.13]$$

subject to

$$\sum_{k:(j,k)\in A} x_{jk} - \sum_{i:(i,j)\in A} x_{ij} = b_j \quad \forall j \in N \qquad [5.14]$$

$$x_{ij} \geqslant 0 \quad \forall (i,j) \in A \qquad [5.15]$$

where the objective function [5.13] minimizes the total flow cost, constraints [5.14], the so-called *flow balance constraints*, impose the restriction that, for each node, the flow emanating from the node minus the flow entering the node must be equal to the supply/demand of that node, and constraints [5.15] indicate that the variables x_{ij} are positive.

Note that, as the constraints coefficients matrix is totally unimodular, if b_j are integers then an optimal solution of the problem in which all x_{ij} are integers exists.

5.1.6. *The maximum flow problem*

In the maximum flow problem we still consider a directed network $G(N, A)$ defined by a set N of n nodes and a set A of m directed arcs. We want to determine the maximum amount of flow from a specified source node s to another specified sink node t. No cost is associated with the flow, but each arc $(i, j) \in A$ is associated with a capacity u_{ij} that denotes the maximum amount that can flow on the arc. Let $x_{ij} \geqslant 0$ denote, as before, the flow on arc $(i, j) \in A$ and let v denote the flow sent from s to t. A standard MILP formulation of the maximum flow problem is as follows:

$$\max v \qquad [5.16]$$

subject to

$$\sum_{k:(j,k)\in A} x_{jk} - \sum_{i:(i,j)\in A} x_{ij} = \begin{cases} v, & j = s \\ 0, & j \neq s,t \\ -v, & j = t \end{cases} \qquad [5.17]$$

$$0 \leqslant x_{ij} \leqslant u_{ij} \quad \forall (i,j) \in A \qquad [5.18]$$

where the objective function [5.16] maximizes the flow v sent from s to t (with $v = \sum_{k:(s,k)\in A} x_{sk} = \sum_{i:(i,t)\in A} x_{it}$), constraints [5.17] represent the *flow balance constraints* and constraints [5.18] indicate that the flow on arc $(i,j) \in A$ is positive and not greater than the arc capacity u_{ij}. Note that, as for the minimum cost flow problem, the constraints coefficients matrix is totally unimodular, so, if u_{ij} are integers then an optimal solution of the problem in which all x_{ij}s and v are integers exists.

5.1.7. *The transportation problem*

The transportation problem is a special case of the minimum cost flow problem where the node set N is partitioned into two subsets N_1 and N_2; each node $i \in N_1$ is a supply node providing a positive amount $a_i > 0$ of goods to be shipped and each node $j \in N_2$ is a demand node requiring a positive amount $b_j > 0$ of the same goods. Correspondingly, for each arc $(i,j) \in A$, we have $i \in N_1$ and $j \in N_2$. Also, we still assume that the network is balanced, that is $\sum_{i\in N_1} a_i = \sum_{j\in N_2} b_j$.

By denoting with $x_{ij} \geqslant 0$ the amount of goods sent on arc $(i,j) \in A$ from node $i \in N_1$ to node $j \in N_2$, we get the following standard MILP formulation of the transportation problem:

$$\min \sum_{(i,j)\in A} c_{ij} x_{ij} \qquad [5.19]$$

subject to

$$\sum_{j\in N_2} x_{ij} = a_i \quad \forall i \in N_1 \qquad [5.20]$$

$$\sum_{i\in N_1} x_{ij} = b_j \quad \forall j \in N_2 \qquad [5.21]$$

$$x_{ij} \geqslant 0 \quad \forall (i,j) \in A \qquad [5.22]$$

where the objective function [5.19] minimizes the total transportation cost, constraints [5.20] impose the restriction that for each node $i \in N_1$ the total amount of shipped goods is equal to a_i, constraints [5.21] impose the restriction that for each node $j \in N_2$ the total amount of received goods is equal to b_j, and constraints [5.22] indicate that the variables x_{ij} are positive.

Note that, as the transportation problem is a special case of the minimum cost flow problem, if all a_is and b_js are integers, then an optimal solution of the problem in which all x_{ij}s are integers exists.

5.1.8. *The assignment problem*

In the assignment problem, two equally sized sets N_1 and N_2 (with $|N_1| = |N_2|$) and a collection of pairs $A \subseteq N_1 \times N_2$ representing possible assignments are given. Each pair $(i, j) \in A$ with $i \in N_1$ and $j \in N_2$ is associated with a cost c_{ij}. In the assignment problem we search for a minimum cost assignment where each element $i \in N_1$ is assigned to exactly one element $j \in N_2$. Let x_{ij} be a 0–1 variable such that $x_{ij} = 1$ if element $i \in N_1$ is assigned to element $j \in N_2$, $x_{ij} = 0$ if it is not.

The standard formulation of the assignment problem is as follows:

$$\min \sum_{(i,j) \in A} c_{ij} x_{ij} \qquad [5.23]$$

subject to

$$\sum_{j \in N_2} x_{ij} = 1 \quad \forall i \in N_1 \qquad [5.24]$$

$$\sum_{i \in N_1} x_{ij} = 1 \quad \forall j \in N_2 \qquad [5.25]$$

$$x_{ij} \in \{0, 1\} \quad \forall (i, j) \in A \qquad [5.26]$$

where the objective function [5.23] minimizes the total assignment cost, constraints [5.24] impose the restriction that each element $i \in N_1$ is assigned to exactly one element $j \in N_2$, constraints [5.25] impose the restriction that each element $j \in N_2$ is assigned to exactly one element $i \in N_1$, and constraints [5.26] impose the restriction that the variables x_{ij} must be binary.

Note that the assignment problem can be viewed as a special case of the transportation problem with $|N_1| = |N_2|$ and $a_i = b_j = 1,\ \forall i \in N_1,\ \forall j \in N_2$. Hence, the integrality constraints of the variables can be dropped. Furthermore, as constraints [5.24] imply that all x_{ij} variables are $\leqslant 1$, [5.26] can be substituted by constraints:

$$x_{ij} \geqslant 0 \quad \forall (i,j) \in A \tag{5.27}$$

5.1.9. *The shortest path problem*

In the shortest path problem, we search for the minimum distance path between two nodes s and t of a given network. This problem can be modeled as a special case of the minimum cost flow problem where node s is a supply node with unitary supply ($b_s = 1$), node t is a demand node with unitary demand ($b_t = -1$), and all other nodes are transhipment nodes ($b_k = 0, \ \forall k \in N, \ k \neq s,t$). The model is then:

$$\min \sum_{(i,j)\in A} c_{ij} x_{ij} \tag{5.28}$$

subject to

$$\sum_{k:(j,k)\in A} x_{jk} - \sum_{i:(i,j)\in A} x_{ij} = \begin{cases} 1, & j = s \\ 0, & j \neq s,t \\ -1, & j = t \end{cases} \tag{5.29}$$

$$x_{ij} \geqslant 0 \quad \forall (i,j) \in A \tag{5.30}$$

Indeed, as all b_is are integers, the solution to this special minimum cost flow problem is an integer and will send exactly one unit of flow from node s to node t along the shortest path.

5.2. General modeling techniques

In the following we introduce several modeling tricks that enable generation of linear models in the presence of special types of non-linear expressions or in the presence of logic conditions between real/integer variables and/or between binary variables.

5.2.1. *Min-max, max-min, min-abs models*

Consider a mathematical programming model where the constraints are expressed by linear expressions of the decision variables, but the objective function is as follows:

$$\min \max\{e_1, e_2, ..., e_n\}$$

where $e_1, e_2, ..., e_n$ are linear expressions of the variables. We have a so-called *min-max model* that is non-linear but can be transformed into an equivalent linear model. We need to introduce a fictitious variable y in the objective function to be minimized. We then have as the objective function:

$$\min y$$

Furthermore, we need to add to the constraints set the following inequalities:

$$y \geqslant e_1, \;\; y \geqslant e_2, \;\; ..., \;\; y \geqslant e_n$$

The generated model is linear and is equivalent to the corresponding original model: indeed, the value of y corresponds exactly to the value $\max\{e_1, e_2, ..., e_n\}$, as it is constrained by the introduced inequalities to be not less than the largest of the e_i terms and is minimized by the objective function. Hence, this implies that in the optimal solution at least one of the constraints $y \geqslant e_i$ is saturated, namely $\exists i : y = e_i$.

Analogously, for a problem of the type:

$$\max \min\{e_1, e_2, ..., e_n\}$$

where $e_1, e_2, ..., e_n$ are linear expressions of the variables, we can generate an equivalent linear model. In this case, the objective function (having introduced a fictitious variable y) becomes:

$$\max y$$

and the original constraints set must be integrated with the following inequalities:

$$y \leqslant e_1, \;\; y \leqslant e_2, \;\; ..., \;\; y \leqslant e_n$$

Finally, consider a so-called *min-abs model* with an objective function of the type:

$$\min |e|$$

where e is a linear expression of the variables. Here we want to minimize the absolute value (which is non-linear) of a linear expression. As $|e| = \max\{e, -e\}$, the *min-abs model* can be written as a *min-max model* and correspondingly transformed into a linear model. Notice that the same can also be done for expressions of the type $\min |e_1| + |e_2| + ... + |e_n|$ provided that $e_1, ..., e_n$ are linear expressions of the decision variables. However, it is not possible to derive equivalent linear formulations in the presence of objective functions such as $\min\{\min\{e_1, e_2, ..., e_n\}\}$, $\max\{\max\{e_1, e_2, ..., e_n\}\}$, or $\max\{|e_1| + |e_2| + ... + |e_n|\}$.

5.2.2. *Handling logic conditions*

In many cases the real/integer decision variables of the problem are subject to further logic conditions. These relations can be either specific requirements on the value of a single variable or may provide logic conditions on two or more variables. It is possible to model these relations by introducing fictitious binary variables, linking these variables to the original decision variables, and finally adding linear constraints on the binary variables. Some examples of this technique are given below.

5.2.2.1. *One-to-many conditions*

Assume that a given positive variable x_j is restricted to having only k values $p_1,\ p_2, ...,\ p_k$. This condition cannot be expressed linearly if only variable x_j is considered. Besides, by introducing k binary variables $y_1,\ y_2, ...,\ y_k$, the condition can be expressed by means of the equations:

$$\sum_{i=1}^{k} y_i = 1 \qquad [5.31]$$

$$x_j = \sum_{i=1}^{k} p_i y_i \qquad [5.32]$$

where equation [5.31] imposes the restriction that only one of the variables y_i is equal to 1 and, correspondingly, equation [5.32] imposes the restriction that variable x_j can have only one of the values $p_1,\ p_2, ...,\ p_k$.

5.2.2.2. *Fixed charge constraints*

Consider the so-called *fixed charge* case. Here we have a given variable x_j which cannot exceed a given threshold value M. This variable (typically corresponding to an activity) is associated in the objective function not only with a unitary cost coefficient c_j but also with a *fixed charge* f_j that must be considered only if x_j has the value > 0. To model this requirement, it is sufficient to introduce a fictitious binary variable y_j so that the global cost of the activity becomes:

$$c_j x_j + f_j y_j \qquad [5.33]$$

and add the constraint:

$$x_j \leqslant M y_j. \qquad [5.34]$$

Note that, as variable y_j can only have values $\{0, 1\}$, then either we have $x_j = y_j = 0$ or $y_j = 1$, which only implies that x_j is limited so as not to exceed M.

5.2.2.3. *Big-M constraints formulation*

The linear formulation of the *fixed charge* constraint seen above is a special case of the so-called *big-M* constraints formulation. Assume that a given (real or integer) positive variable x_j, which cannot exceed a given threshold value M, is restricted to having either value 0 or a value not less than a constant K. To model this requirement, it is sufficient to introduce a fictitious binary variable y_j and add the following constraints:

$$x_j \leqslant M y_j \qquad [5.35]$$

$$x_j \geqslant K y_j. \qquad [5.36]$$

Note that, as the variable y_j can only have values $\{0, 1\}$, then either we have $x_j = 0$ when $y_j = 0$ from [5.35], or we have $K \leqslant x_j \leqslant M$ when $y_j = 1$ from both [5.35] and [5.36].

The *big-M* constraints formulation can also be used to model logic conditions on real/integer variables as the following example on two variables $x_i \geqslant 0$ and $x_j \geqslant 0$

shows. Suppose that x_i and x_j, where x_i (x_j) cannot exceed a given threshold value M_i (M_j), represent two activities that are incompatible. This implies the following logic condition:

IF $x_i > 0$ THEN $x_j = 0$ AND (VICE VERSA) IF $x_j > 0$ THEN $x_i = 0$.

To model this condition, it is sufficient to introduce two binary variables $y_i, y_j \in \{0, 1\}$ and add the following constraints:

$$x_i \leqslant My_i \quad [5.37]$$

$$x_j \leqslant My_j \quad [5.38]$$

$$y_i + y_j \leqslant 1. \quad [5.39]$$

Constraints [5.37] and [5.38] are introduced to link the real variables to the corresponding binary ones, while constraint [5.39] induces the logic condition on the real variables, namely it forbids the x_i, x_j variables to be contemporaneously > 0.

5.2.2.4. *Either–or constraints*

Consider the case where a choice must be made between two constraints so that one must hold, but not both. Namely, consider an *either–or* constraint of the following type:

EITHER

$$f(x_1, ..., x_n) \geqslant k \quad [5.40]$$

OR

$$f(x_1, ..., x_n) \leqslant \gamma \quad [5.41]$$

where we know that $f(x_1, ..., x_n)$ is bounded above by some large coefficient M. To model this constraint, it is sufficient to introduce a $0--1$ variable y_j, such that $y_j = 1$ ($y_j = 0$) if $f(x_1, ..., x_n) \geqslant k$ ($f(x_1, ..., x_n) \leqslant \gamma$) and add the constraints:

LOGIC CONDITION	LINEAR CONSTRAINT
$y_1 \rightarrow y_2$	$y_1 \leqslant y_2$
$y_1 \rightarrow \overline{y_2}$	$y_1 \leqslant 1 - y_2$
$(y_1 \bigcap y_2) \rightarrow y_3$	$y_1 + y_2 - 1 \leqslant y_3$
$(y_1 \bigcup y_2) \rightarrow y_3$	$y_1 + y_2 \leqslant 2y_3$
$\bigcap_{(i \in I)} y_i \rightarrow \omega$	$\sum_{i \in I} y_i - \lvert I \rvert + 1 \leqslant \omega$
$\bigcup_{(i \in I)} y_i \rightarrow \omega$	$\sum_{i \in I} y_i \leqslant \lvert I \rvert \cdot \omega$

Table 5.1. *Correspondence between logic conditions and linear constraints on* $0 - -1$ *variables*

$$f(x_1, ..., x_n) \leqslant \gamma + My_j \qquad [5.42]$$

$$f(x_1, ..., x_n) \geqslant k - M(1 - y_j) \qquad [5.43]$$

where, either $y_j = 0$, constraint [5.42] holds and reduces to constraint [5.41], while constraint [5.43] becomes redundant, or $y_j = 1$, constraint [5.43] holds and reduces to constraint [5.40], while constraint [5.42] becomes redundant.

5.2.2.5. *Linearization of quadratic* $0 - -1$ *variables*

Consider a mathematical programming model involving quadratic terms of the form $x_i x_j$ with $x_i, x_j \in \{0, 1\}$. This can be transformed into an equivalent linear model by introducing a new variable $y_{ij} \in \{0, 1\}$ such that $y_{ij} = x_i x_j$. As x_i and x_j are $0 - -1$ variables, this corresponds to $y_{ij} = \min\{x_i, x_j\}$ and can be made explicit by means of the following three linear constraints:

$$x_i \geqslant y_{ij}$$

$$x_j \geqslant y_{ij}$$

$$x_i + x_j \leqslant 1 + y_{ij}$$

To conclude this section, Table 5.1 depicts several examples of logic conditions and the corresponding linear constraints on binary variables to be added in a MILP model (the left column indicates the logic condition and the right column the corresponding linear constraint with $y_i \in \{0,1\}, \forall i$). For an insightful work on how to give tight representations (with respect to the convex hull of the considered formulation) of logical conditions by means of linear constraints, see [YAN 99].

5.3. More advanced MILP models

In this section we discuss fairly standard MILP formulations of several known combinatorial optimization problems spanning locations, graphs, networks, and scheduling problems.

5.3.1. *Location models*

5.3.1.1. *The p-median problem*

In the p-median problem we consider a set N of n potential facilities and a set M of m clients. We want to open at most p facilities (medians) and assign each client to each nearest open facility so as to minimize the sum of the distances between the clients and their nearest open facility.

Let y_j be a 0–1 variable such that $y_j = 1$ if node j is a median (it is open as a facility), $y_j = 0$ if it is not. Let x_{ij} be a 0–1 variable such that $x_{ij} = 1$ if client i is allocated to facility j (and correspondingly facility j is median), $x_{ij} = 0$ if it is not. Let d_{ij} be the cost of allocating client i to facility j. A standard MILP formulation of the p-median problem (see [DAS 95]) is as follows:

$$\min \sum_{i=1}^{m} \sum_{j=1}^{n} d_{ij} x_{ij} \qquad [5.44]$$

subject to

$$\sum_{j=1}^{n} x_{ij} = 1 \quad i = 1, ..., m \qquad [5.45]$$

$$\sum_{j=1}^{n} y_j \leqslant p \qquad [5.46]$$

$$x_{ij} \leqslant y_j \quad i = 1, ..., m, \; j = 1, ..., n \qquad [5.47]$$

$$x_{ij}, y_j \in \{0, 1\} \quad i = 1, ..., m, \; j = 1, ..., n \qquad [5.48]$$

where the objective function [5.44] minimizes the total distance, constraints [5.45] impose the restriction that each client is allocated to a facility, constraint [5.46] requires there to be at most p medians, constraints [5.47] forbid the assignment of a client to a closed facility, and constraints [5.48] impose the restriction that the variables x_{ij} and y_j are binary.

5.3.1.2. *The p-center problem*

In the p-center problem, we still consider (as for the p-median problem) a set N of n potential facilities and a set M of m clients, where we want to open at most p facilities and assign each client to each nearest open facility. However, the objective function in this case is to minimize the radius, namely the maximum distance between a client and the nearest facility.

Let y_j be a 0–1 variable such that $y_j = 1$ if node j is open as a facility, $y_j = 0$ if it is not. Let x_{ij} be a 0–1 variable such that $x_{ij} = 1$ if client i is allocated to facility j, $x_{ij} = 0$ if he is not. Let d_{ij} be the cost of allocating client i to facility j. Let z be the radius that we want to minimize. A standard MILP formulation of the p-center problem (see [DAS 95]) is as follows:

$$\min z \qquad [5.49]$$

subject to

$$\sum_{j=1}^{n} x_{ij} = 1 \quad i = 1, ..., m \qquad [5.50]$$

$$\sum_{j=1}^{n} y_j \leqslant p \qquad [5.51]$$

$$x_{ij} \leqslant y_j \quad i = 1, ..., m, \; j = 1, ..., n \qquad [5.52]$$

$$\sum_{j=1}^{n} d_{ij}x_{ij} \leqslant z \quad i = 1, ..., m \qquad [5.53]$$

$$x_{ij}, y_j \in \{0,1\} \quad i = 1, ..., m, \ j = 1, ..., n \qquad [5.54]$$

where the objective function [5.49] minimizes the value of the radius, while constraints [5.50] impose the restriction that each client is allocated to a facility, constraint [5.51] requires that there are at most p open facilities, constraints [5.52] forbid the assignment of a client to a closed facility, constraints [5.53] force z to be not less than the distance from any client to the assigned facility, and constraints [5.54] impose the restriction that the variables x_{ij} and y_j are binary.

A different formulation of the p-center problem is provided in [ELL 04]. Let us arrange the distance matrix (that is the matrix of the d_{ij} values) in such a way that $D^{\min} = D^0 < D^1 < ... < D^K = D^{\max}$ are the sorted different values in that matrix. Let y_j always be a 0–1 variable such that $y_j = 1$ if node j is open as a facility, $y_j = 0$ if it is not. Let z_k ($k = 1, ..., K$) be a 0–1 variable such that $z_k = 0$ only if it is possible to choose p facilities and cover all the clients within the radius D^{k-1}. Note that in an optimal solution $z_k = 0$ implies $z_{k+1} = ... = z_K = 0$ and the objective function value is strictly less than D^k. Analogously, $z_k = 1$ implies $z_{k-1} = ... = z_1 = 1$. The formulation is as follows:

$$\min D^0 + \sum_{k=1}^{K}(D^k - D^{k-1})z_k \qquad [5.55]$$

subject to

$$\sum_{j=1}^{n} y_j \geqslant 1 \qquad [5.56]$$

$$\sum_{j=1}^{n} y_j \leqslant p \qquad [5.57]$$

$$z_k + \sum_{j:d_{ij}<D^k} y_j \geqslant 1 \quad i = 1, ..., n, \ k = 1, ..., K \qquad [5.58]$$

$$y_j, z_k \in \{0,1\} \quad j = 1, ..., n, \; k = 1, ..., K. \qquad [5.59]$$

Here the objective function [5.55] still minimizes the value of the radius (note that all the cost coefficients are positive). Constraint [5.56] (which is redundant in the previous formulation) discards solutions with no open facilities. Constraint [5.57] requires there to be at most p open facilities. Constraints [5.58] indicate that for a given k, $z_k = 0$ if and only if all clients can be served at a distance strictly less than D^k. Hence, when $z_k = 1$, the value $D^k - D^{k-1}$ is added to the radius in the objective function [5.55]. Finally, constraints [5.59] impose the restriction that the variables y_j and z_k are binary.

5.3.2. *Graphs and network models*

5.3.2.1. *The maximum clique problem*

Consider a graph $G = (V, E)$ composed of a set V of n vertices and a set E of m edges. When two vertices are connected by an edge, they are denoted as adjacent. A clique in G is a subgraph where the vertices are all pairwise adjacent. The maximum clique problem consists of finding a clique of maximum cardinality.

Let x_j ($j \in V$) be a 0–1 variable such that $x_j = 1$ if vertex j belongs to the clique, $x_j = 0$ if it does not.

The standard MILP formulation of this model is as follows:

$$\max \sum_{j \in V} x_j \qquad [5.60]$$

subject to

$$x_i + x_j \leqslant 1 \quad \forall i, j : (i, j) \notin E \qquad [5.61]$$

$$x_j \in \{0,1\} \quad \forall j \in V \qquad [5.62]$$

where the objective function [5.60] maximizes the cardinality of the clique, constraints [5.61] impose the restriction that for all pairs of non-adjacent vertices at most one of them can belong to the clique, and constraints [5.62] impose the restriction that the variables x_j must be binary.

With the above formulation, the set of constraints [5.61] has cardinality $|\overline{E}|$. Let $\overline{E_j}$ indicate the set of vertices non-adjacent to vertex j. A more compact reformulation (involving just $|V|$ constraints, though widening the solutions space of the corresponding linear programming relaxation, see [CRO 94]) of this constraints set is as follows:

$$\sum_{i\in\overline{E_j}} x_i + |\overline{E_j}|x_j \leqslant |\overline{E_j}| \quad \forall j \in V \qquad [5.63]$$

where constraints [5.63] indicate that for each vertex j, if j is in the clique ($x_j = 1$), then all vertices non-adjacent to j cannot belong to the clique.

5.3.2.2. *The graph coloring problem*

Consider a graph $G = (V, E)$ with $|V| = n$. A coloring of G is an assignment of colors to the vertices such that two adjacent vertices have different colors. A k-coloring of G is a coloring that uses k colors. The chromatic number of G is the smallest number of colors needed to color G and it is denoted by $\chi(G)$. The graph coloring problem consists of finding $\chi(G)$. Let u_i ($i = 1, ..., n$) be a 0–1 variable such that $u_i = 1$ if color i is used (as there are n vertices, at most n colors will be used), $u_i = 0$ if it is not. Let x_{ij} ($i = 1, ..., n$, $j \in N$) be a 0–1 variable such that $x_{ij} = 1$ if vertex j is colored with color i, $x_{ij} = 0$ if it is not.

The standard MILP formulation (see, for example, [COL 02]) of this model is as follows:

$$\min \sum_{i=1}^{n} u_i \qquad [5.64]$$

subject to

$$\sum_{i=1}^{n} x_{ij} = 1 \quad \forall j \in N \qquad [5.65]$$

$$x_{ij} + x_{kj} \leqslant u_j \quad \forall i \in V,\ k \in V,\ (i,k) \in E,\quad j = 1, ..., n \qquad [5.66]$$

$$u_i, x_{ij} \in \{0, 1\} \quad \forall i = 1, ..., n,\ \forall j \in N \qquad [5.67]$$

where the objective function [5.64] minimizes the number of colors, constraints [5.65] impose the restriction that each vertex is colored, constraints [5.66] impose the restriction that two adjacent vertices cannot have the same color and that $u_j = 1$ when some vertex has color j, and constraints [5.67] impose the restriction that the variables u_i and x_{ij} must be binary.

5.3.2.3. *The shortest spanning tree problem*

The shortest spanning tree problem can be expressed in the following way. A graph $G = (V, E)$ has a set N of n vertices and a set E of m edges, where each edge $(i, j) \in E$ has a cost (distance) c_{ij}. We search for a spanning tree (a connected acyclic graph) $T = (V, E')$ in the graph G such that the cost of the edges of T, that is $\sum_{(i,j)\in E'}$, is minimum. Note that $|E'| = n - 1$, that is a spanning tree in a graph with n vertices contains exactly $n - 1$ edges. In a tree each of the vertices can be considered, without loss of generality, as being the root vertex 1: then, given vertex 1, the edges of a tree can always be split into levels (at most $n - 1$), where the edges containing vertex 1 are assigned to the first level, the edges not containing vertex 1 but containing vertices adjacent to vertex 1 are assigned to the second level and so on. Let x_{ij}^k be a 0–1 variable such that $x_{ij}^k = 1$ if edge (i, j) belongs to the spanning tree and is assigned to level k, $x_{ij}^k = 0$ if it does not. The following MILP formulation holds for the minimum spanning tree problem:

$$\min \sum_{k=1}^{n-1} \sum_{(i,j)\in E} c_{ij} x_{ij}^k \qquad [5.68]$$

subject to

$$\sum_{k=1}^{n-1} x_{ij}^k \leqslant 1 \quad \forall (i, j) \in E \qquad [5.69]$$

$$\sum_{j>1, j\in V} x_{1j}^1 \geqslant 1 \qquad [5.70]$$

$$\sum_{k=1}^{n-1} \Big(\sum_{i<j} x_{ij}^k + \sum_{l>j} x_{jl}^k\Big) \geqslant 1 \quad \forall j \neq 1 \in V \qquad [5.71]$$

$$\sum_{k=1}^{n-1} \sum_{(i,j)\in E} x_{ij}^k = n - 1 \qquad [5.72]$$

$$\sum_{h<i} x_{hi}^{k-1} + \sum_{l>i,\ l\neq j} x_{il}^{k-1} + \sum_{h<j\ h\neq i} x_{hj}^{k-1} + \sum_{l>j} x_{jl}^{k-1} \geqslant x_{ij}^{k} \quad \forall k = 2, ..., n \quad \forall (i,j) \in E \qquad [5.73]$$

$$x_{ij}^{k} \in \{0,1\} \quad \forall k = 1, ..., n-1, \quad \forall (i,j) \in E \qquad [5.74]$$

where the objective function [5.68] minimizes the total cost (distance) of the tree; constraints [5.69] impose the restriction that each edge can be assigned to at most one level of the tree; constraint [5.70] imposes the restriction that vertex 1 is the root vertex, that is at least one edge containing vertex 1 must be assigned to the first level; constraints [5.71] impose the restriction that for each vertex j at least one edge containing vertex j belongs to the tree and is assigned to some level $1 \leqslant k \leqslant n-1$; constraint [5.72] imposes the restriction that the tree is composed of exactly $n-1$ edges; constraints [5.73] impose the restriction that for each edge (i,j) and for each level $k \geqslant 2$, if edge (i,j) is assigned to level k, then at least one edge containing either vertex i or vertex j must be assigned to level $k-1$; finally, constraints [5.74] impose the restriction that the variables x_{ij}^{k} must be binary.

5.3.2.4. *The traveling salesman problem*

The traveling salesman problem can be expressed in the following way. Given a directed graph $G = (N, A)$ with a set N of n nodes and a set A of m arcs, where each arc $(i,j) \in A$ has a cost (distance) c_{ij}, we search for the minimum cost (distance) Hamiltonian circuit on graph G. Note that a circuit is Hamiltonian if and only if all nodes are visited exactly once by the circuit. Let A_S denote the subset of arcs linking the nodes belonging to subset $S \subset N$. Let x_{ij} be a 0–1 variable such that $x_{ij} = 1$ if arc (i,j) belongs to the circuit, $x_{ij} = 0$ if it does not. The standard MILP formulation (see [DAN 54]) of the traveling salesman problem is as follows:

$$\min \sum_{i,j\in N,\ i\neq j} c_{ij} x_{ij} \qquad [5.75]$$

subject to

$$\sum_{i\in N,\ i\neq j} x_{ij} = 1 \quad \forall j \in N \qquad [5.76]$$

$$\sum_{j\in N,\ j\neq i} x_{ij} = 1 \quad \forall i \in N \qquad [5.77]$$

$$\sum_{(i,j)\in A_S} x_{ij} \leqslant |S| - 1, \quad \forall S : S \subseteq N,\ 2 \leqslant |S| \leqslant n-2 \qquad [5.78]$$

$$x_{ij} \in \{0,1\} \quad \forall i,j \in N, \quad i \neq j \qquad [5.79]$$

where the objective function [5.75] minimizes the total cost (distance) of the circuit, constraints [5.76] impose the restriction that exactly one arc enters each node, constraints [5.77] impose the restriction that exactly one arc exits from each node, constraints [5.78] are the so-called subtour elimination constraints, namely they impose the restriction that no subtour can exist, and constraints [5.79] impose the restriction that the variables x_{ij}^k must be binary.

The above formulation presents a limited number of variables, just $O(m)$, but an exponential number of subtour elimination constraints [5.78]. A formulation with a polynomial number of variables and constraints is proposed in [MAC 03]. Here, another formulation with $O(mn)$ variables and $O(n^2)$ constraints based on positional variables is given. Let x_{ij}^k be a 0–1 variable such that $x_{ij}^k = 1$ if arc (i,j) is the k-th arc of the circuit, $x_{ij}^k = 0$ if it is not. Note that any node can be arbitrarily selected to be the first node of the circuit (and correspondingly one of its outgoing arcs will be the first arc of the circuit). A MILP formulation of the traveling salesman problem is as follows:

$$\min \sum_{k=1}^{n} \sum_{i,j\in N,\ i\neq j} c_{ij} x_{ij}^k \qquad [5.80]$$

subject to

$$\sum_{k=1}^{n} \sum_{i\in N,\, i\neq j} x_{ij}^k = 1 \quad \forall j \in N \qquad [5.81]$$

$$\sum_{k=1}^{n} \sum_{j\in N,\, j\neq i} x_{ij}^k = 1 \quad \forall i \in N \qquad [5.82]$$

$$\sum_{i,j\in N,\ i\neq j} x_{ij}^k = 1 \quad \forall k = 1, ..., n \qquad [5.83]$$

$$\sum_{i \in N,\, i \neq j} x_{ij}^k = \sum_{l \in N,\, l \neq j} x_{jl}^{k+1} \quad \forall j \in N, \quad \forall k = 1, ..., n-1 \qquad [5.84]$$

$$\sum_{i \in N,\, i \neq j} x_{ij}^n = \sum_{l \in N,\, l \neq j} x_{jl}^1 \quad \forall j \in N \qquad [5.85]$$

$$x_{ij}^k \in \{0, 1\} \quad \forall k = 1, ..., n, \ \forall i, j \in N, \ i \neq j \qquad [5.86]$$

where the objective function [5.80] minimizes the total cost (distance) of the circuit; constraints [5.81] impose the restriction that exactly one arc enters each node at exactly one position; constraints [5.82] impose the restriction that exactly one arc exits from each node in exactly one position; constraints [5.83] impose the restriction that exactly one arc is assigned to each position in the circuit; constraints [5.84] impose the restriction that for each node j, if it has an ingoing arc at position $k < n$ in the circuit, then it must also have an outgoing arc at position $k + 1$ in the circuit (*vice versa*, if it does not have an ingoing arc at position k in the circuit, then it cannot have an outgoing arc at position $k + 1$ in the circuit); constraints [5.85] impose the restriction that for each node j, if it has an ingoing arc at the last position in the circuit, then it must also have an outgoing arc at the first position in the circuit (*vice versa*, if it does not have an ingoing arc at the last position in the circuit, then it cannot have an outgoing arc at the first position in the circuit); finally, constraints [5.86] impose the restriction that the variables x_{ij}^k must be binary.

5.3.2.5. *The multicommodity network flow problem*

The minimum cost multicommodity network flow problem generalizes the minimum cost flow problem to multiple commodities. Here we want to determine a least cost shipment of several commodities through a network in order to satisfy demands at certain nodes from available supplies at other nodes, where the commodities share the capacity of the same arcs and all the flows are integer. Consider a network $G = (N, A)$ composed of a set N of n nodes, and a set A of m arcs, where the direct arc (i, j) connects node i to node j. Let K be a set of k commodities where each commodity must generally be shipped from multiple sources to multiple destinations, and where sources and destinations are nodes of the considered network. A parameter b_i^k is associated with each node i and each commodity k. If $b_i^k > 0$, node i is the source node for commodity k with a supply of b_i^k units of flow. If $b_i^k < 0$, node i is the destination node for commodity k with a demand of $-b_i^k$ units of flow. If $b_i^k = 0$, node i is a transhipment node for commodity k. Let c_{ij}^k denote the unit flow cost of commodity k on arc (i, j). We assume $c_{ij}^k \geqslant 0 \ \forall (i, j) \in A, \forall k \in K$. Each arc (i, j) has a capacity $u_{ij} \geqslant 0$ that restricts the total flow of all commodities on that arc. Let $x_{ij}^k \geqslant 0$ denote the flow of commodity k on arc (i, j). The so-called node–arc formulation (see, for instance, [AHU 93]) of the minimum cost multicommodity network flow problem is as follows:

$$\min \sum_{k \in K,\ (i,j) \in A} c_{ij}^k x_{ij}^k \qquad [5.87]$$

subject to

$$\sum_{l \in N} x_{jl}^k - \sum_{i \in N} x_{ij}^k = b_j^k \quad \forall j \in N, \quad \forall k \in K \qquad [5.88]$$

$$\sum_{k \in K} x_{ij}^k \leqslant u_{ij} \quad \forall (i,j) \in A \qquad [5.89]$$

$$x_{ij}^k \ integer \geqslant 0 \quad \forall (i,j) \in A, \ \forall k \in K \qquad [5.90]$$

where the objective function [5.87] minimizes the total multicommodity flow cost; constraints [5.88] represent the ordinary mass balance constraints, that is the requirement that, for each commodity, the difference between the flow that enters a node and the flow that leaves that node is equal to the supply/demand of that node; constraints [5.89] are the so-called bundle constraints, that is they state that the total flow on each arc must be less than or equal to its capacity; finally, constraints [5.90] impose the restriction that the variables x_{ij}^k must be positive integers.

A different formulation of the same problem (the so-called arc–path formulation, see [AHU 93]), but with an exponential number of variables, can be obtained in the following way. Let P_k denote the collection of all directed paths p between all source–destination pairs of commodity k. Also, let the set of all sources and destinations of commodity k be represented by Q_k. Let δ_{ij}^{pk} be a 0–1 arc–path constant where $\delta_{ij}^{pk} = 1$ if arc (i,j) belongs to path p of commodity k, $\delta_{ij}^{pk} = 0$ if it does not. Let γ_i^{pk} be a node–path constant where $\gamma_i^{pk} = 1$ if node i is the source of path p of commodity k, $\gamma_i^{pk} = -1$ if node i is the destination of path p of commodity k, $\gamma_i^{pk} = 0$ if node i is neither the source nor the destination of path p of commodity k. Let $c_p^k = \sum_{(i,j) \in A} c_{ij}^k \delta_{ij}^{pk}$ denote the unit flow cost of path p of commodity k. Let y_p^k denote the integer flow on path p of commodity k. The arc–path formulation of the minimum cost multicommodity network flow problem is as follows:

$$\min \sum_{k \in K,\ p \in P_k} c_p^k y_p^k \qquad [5.91]$$

subject to

$$\sum_{p \in P_k} \gamma_{ij}^{pk} y_p^k = b_i^k \quad \forall i \in Q^k, \quad \forall k \in K \qquad [5.92]$$

$$\sum_{k \in K,\ p \in P_k} \delta_{ij}^{pk} y_p^k \leqslant u_{ij} \quad \forall (i,j) \in A \qquad [5.93]$$

$$y_p^k \ \mathit{integer} \geqslant 0 \quad \forall k \in K \quad \forall p \in P_k \qquad [5.94]$$

where the objective function [5.91] minimizes the total multicommodity flow cost, constraints [5.92] represent the ordinary mass balance constraints, constraints [5.93] are the bundle constraints, and constraints [5.94] impose the restriction that the variables y_p^k must be positive integers.

5.3.3. *Machine scheduling models*

5.3.3.1. *The job shop problem*

In the job shop problem a set of m machines and a set of n jobs are given. Each job $j \in J$ consists of a set of m operations to be executed on distinct machines, ordered according to a given permutation $\sigma_j = (\sigma_j^1, \sigma_j^2, ..., \sigma_j^m)$ of the machines: job j must be executed first on machine σ_j^1, then on σ_j^2, and so on. Let $p_j(i)$ be the processing time of job j on machine i. Each job cannot start before its release date r_j and pre-emption is not allowed, that is once an operation has started execution, it must be executed for its entire duration without interruption. Each resource can execute at most one operation at a time. For each job, the first operation starts after the job release date and the operations are processed according to the specified order. Let $C_j(i)$ denote the completion time of job j on machine i. Note that the final completion time for job j corresponds to $C_j(\sigma_j^m)$, namely the completion time for job j on its last visited machine σ_j^m. Let $C_{\max}$ denote the maximum of the jobs' final completion times. Let $x_{kl}(i)$ be a 0–1 variable such that $x_{kl}(i) = 1$ if job k is processed before job l on machine i, $x_{kl}(i) = 0$ if it is not, i.e. when job l is processed before job k on machine i. Let M be some large constant such that $C_j(i) \leqslant M \ \forall j \in J, \forall i = 1, ..., m$ is always verified. The standard MILP model (see [BAK 74]) based on a big-M formulation (see section 5.2.2.3) of the job shop problem is as follows:

$$\min C_{\max} \qquad [5.95]$$

subject to

$$C_j(\sigma_j^m) \leqslant C_{\max} \quad \forall j \in J \qquad [5.96]$$

$$C_j(\sigma_j^1) \geqslant r_j + p_j(\sigma_j^1) \quad \forall j \in J \qquad [5.97]$$

$$C_j(\sigma_j^{t+1}) \geqslant C_j(\sigma_j^t) + p_j(\sigma_j^{t+1}) \quad \forall j \in J, \quad \forall t = 1, ..., m-1 \qquad [5.98]$$

$$\begin{cases} C_k(i) \geqslant C_l(i) + p_k(i) - Mx_{kl}(i) \quad \forall k, l \in J : k < l, \quad \forall i = 1, ..., m \\ C_l(i) \geqslant C_k(i) + p_l(i) - M[1 - x_{kl}(i)] \quad \forall k, l \in J : k < l, \quad \forall i = 1, ..., m \end{cases} \qquad [5.99]$$

$$C_{\max}, C_j(i) \geqslant 0 \quad \forall j \in J, \ \forall i = 1, ..., m \qquad [5.100]$$

$$x_{kl}(i) \in \{0, 1\} \quad \forall k, l \in J : k < l, \ \forall i = 1, ..., m \qquad [5.101]$$

where the objective function [5.95] minimizes the maximum of the jobs' completion times, constraints [5.96] ensure that the maximum of the jobs' completion times is not less than the final completion time of each job j, constraints [5.97] impose the restriction that each job cannot start processing before its release date, constraints [5.98] impose the precedence conditions between the operations for each job, constraints [5.99] impose the restriction that for each machine i either job k precedes job l or job l precedes job k, constraints [5.100] impose the restriction that $C_{\max}$ and the variables $C_j(i)$ must be positive, and constraints [5.101] impose the restriction that the variables $x_{kl}(i)$ must be binary.

The continuous relaxation of this big-M formulation does not allow tightening of the lower bounds because the links in constraints [5.99] between the decision variables $x_{kl}(i)$ and the secondary variables $C_j(i)$ are quite weak. To get a tighter, though more complex, formulation (requiring $O(m^2n^2)$ variables and $O(m^3n^3)$ constraints), we use positional completion time variables (see [LAS 92] where this type of formulation was introduced for single machine models) related to each job operation. Given n jobs and m machines, we then have $\tau = mn$ operations. We know that for any solution of the job shop problem, any given operation i is processed in some position k if there are $k-1$ operations completing not later than operation i and the remaining $\tau - k$ operations do not complete before operation i. Let $y_{i,k}$ be a 0–1 variable such that $y_{i,k} = 1$ if operation i is processed in position k, $y_{i,k} = 0$ if it is not. Let p_i denote the processing time of operation i. Let M_i denote the set of operations to be executed on the same machine on which operation i is processed. Let s_i denote the operation following i for the same job. Let C_k be the completion time of the k-th operation. The MILP model based on positional completion times of the job shop problem is as follows:

$$\min C_\tau \qquad [5.102]$$

subject to

$$\sum_{k=1}^{\tau} y_{i,k} = 1 \quad \forall i = 1, ..., \tau \qquad [5.103]$$

$$\sum_{i=1}^{\tau} y_{i,k} = 1 \quad \forall k = 1, ..., \tau \qquad [5.104]$$

$$C_k \geqslant C_{k-1} \quad \forall k = 2, ..., \tau \qquad [5.105]$$

$$C_k \geqslant C_l + p_i[y_{i,k} + \sum_{j \in \{M_i \cup s_i\}} y_{j,l} - 1] \quad \forall k = 2, ..., \tau, \ \forall l = 1, ..., k-1, \ \forall i = 1, ..., \tau \qquad [5.106]$$

$$\sum_{j=1}^{k-1} y_{i,j} \geqslant y_{s_i,k} \quad \forall k = 2, ..., \tau, \quad \forall i : s_i \neq \{\} \qquad [5.107]$$

$$C_k \geqslant 0 \quad \forall j \in J, \ \forall k = 1, ..., \tau \qquad [5.108]$$

$$y_{i,k} \in \{0,1\} \quad \forall i = 1, ..., \tau \quad \forall k = 1, ..., \tau \qquad [5.109]$$

where the objective function [5.102] minimizes the maximum of the jobs' completion times, that is the completion time of the last operation; constraints [5.103] impose the restriction that each operation occupies exactly one position; constraints [5.104] impose the restriction that each position is occupied by exactly one operation; constraints [5.105] impose the restriction that, given an operation in some position k, it cannot be completed before the operation in position $k-1$; constraints [5.106] impose the restriction that, given an operation i in some position k, the previous operation of the same job, and all operations processed on the same machine as operation i, must be completed before the start of operation i if they are placed in a position $l < k$; constraints [5.107] impose the restriction that, given two consecutive operations i, s_i of the same job with s_i placed in position k, operation i must be placed in a position $j < k$; finally, constraints [5.108] impose the restriction that the variables C_k must be positive, and constraints [5.109] impose the restriction that the variables $y_{i,k}$ must be binary.

5.3.3.2. *The single machine problem: time-indexed formulation*

In the non-pre-emptive single-machine problem, a set J of n jobs has to be scheduled, where each job j has a processing time p_j, a release time r_j, and the machine can handle no more than one job at a time. The objective function is the minimization of the weighted sum of the start times. Assuming that all p_js are integers, a time-indexed formulation [AKK 00] based on time discretization, i.e. time is divided into periods, where period t starts at time $t-1$ and ends at time t, can be devised. The planning horizon is denoted by T, which means that all jobs have to be completed by time T. It is sufficient, for instance, to assume that $T \geqslant \max_j r_j + \sum_j p_j$. Let c_{jt} be the cost of job j if it is started in period t and let x_{jt} be a 0–1 variable such that $x_{jt} = 1$ if job j is started at the beginning of period t, $x_{jt} = 0$ if it is not. The time-indexed formulation is as follows:

$$\min \sum_{j=1}^{n} \sum_{t=1}^{T-p_j+1} c_{jt} x_{jt} \qquad [5.110]$$

subject to

$$\sum_{t=1}^{T-p_j+1} x_{jt} = 1 \quad \forall j = 1, ..., n \qquad [5.111]$$

$$\sum_{j=1}^{n} \sum_{s=t-p_j+1}^{t} x_{js} \leqslant 1 \quad \forall t = 1, ..., T \qquad [5.112]$$

$$x_{jt} = 0 \quad \forall j = 1, ..., n \quad \forall t = 1, ..., r_j \qquad [5.113]$$

$$x_{jt} \in \{0, 1\} \quad \forall j = 1, ..., n \quad \forall t = 1, ..., T - p_j + 1 \qquad [5.114]$$

where the objective function [5.110] minimizes the weighted sum of the start times, constraints [5.111] impose the restriction that each job is processed between time 0 and time T, constraints [5.112] impose the restriction that the machine can handle no more than one job at a time, constraints [5.113] impose the restriction that each job cannot start processing before its release time, and constraints [5.114] impose the restriction that the variables x_{jt} must be binary. This formulation can be used to model several single-machine scheduling problems using an appropriate choice of the objective coefficients and possibly restriction of the set of variables. It is, however, pseudo-polynomial in the number of variables.

5.3.3.3. *The single machine problem with tardy jobs cost function*

In this case we always have a non-pre-emptive single-machine problem, where a set J of n jobs has to be scheduled: each job j has a processing time p_j, release time 0, a due date d_j, a deadline $\tilde{d}_j$ and a weight w_j, and the machine can handle no more than one job at a time. Let C_j denote the completion time of job j. We say that a job j is tardy (early) if its completion time C_j is such that $C_j > d_j$ ($C_j \leqslant d_j$). Let x_j be a 0–1 variable such that $x_j = 1$ if job j is early, $x_{jt} = 0$ if it is not, namely if job j is tardy. Consider as the objective function the minimization of the weighted number of tardy jobs. Notice that the presence of deadlines imposes the restriction for each job j that $C_j \leqslant \tilde{d}_j$. Assume that the jobs are indexed in non-decreasing order of due dates, that is $d_1 \leqslant d_2 \leqslant ... \leqslant d_n$. A peculiarity of this problem is that for any feasible solution, it is sufficient to know for each job j whether it is early or tardy in order to get the corresponding schedule. Indeed, if we know the values x_j $\forall j$, then the related schedule can be obtained by sequencing the jobs in non-decreasing order of $d_j x_j + \tilde{d}_j(1 - x_j)$. Based on the above remark, the following formulation [BAP 02] with n variables and $2n$ constraints holds for the considered problem:

$$\max \sum_{j=1}^{n} w_j x_j \qquad [5.115]$$

subject to

$$\sum_{i:\ \tilde{d}_i \leqslant d_j} p_i + \sum_{k:\ d_k \leqslant d_j \cap \ \tilde{d}_k > d_j} p_k x_k \leqslant d_j \quad \forall j = 1, ..., n \qquad [5.116]$$

$$\sum_{i:\ \tilde{d}_i \leqslant \tilde{d}_j} p_i + \sum_{k:\ d_k \leqslant \tilde{d}_j \cap \ \tilde{d}_k > \tilde{d}_j} p_k x_k \leqslant \tilde{d}_j \quad \forall j = 1, ..., n \qquad [5.117]$$

$$x_j \in \{0, 1\} \quad \forall j = 1, ..., n \qquad [5.118]$$

where the objective function [5.115] maximizes the weighted number of early jobs, which is equivalent, apart from a constant factor, to minimizing the weighted number of tardy jobs; constraints [5.116] impose the restriction that for each due date d_j all jobs i with deadlines not greater than d_j ($\tilde{d}_i \leqslant d_j$) plus all early jobs k with deadlines greater than d_j ($\tilde{d}_k > d_j$) but due date not greater than d_j ($d_k \leqslant d_j$) must be completed by d_j; constraints [5.117] impose the restriction that for each deadline $\tilde{d}_j$, all jobs i with deadlines not greater than $\tilde{d}_j$ ($\tilde{d}_i \leqslant \tilde{d}_j$) plus all early jobs k with deadlines greater than $\tilde{d}_j$ ($\tilde{d}_k > \tilde{d}_j$) but due date not greater than $\tilde{d}_j$ ($d_k \leqslant \tilde{d}_j$) must be completed by $\tilde{d}_j$; finally, constraints [5.118] impose the restriction that the variables x_j must be binary.

5.4. Conclusions

In this chapter we have discussed basic and advanced MILP models for known combinatorial optimization problems and we have outlined some general techniques for MILP modeling.

Typically, a MILP model is considered "good" if its continuous linear programming relaxation is sufficiently tight, that is the optimal solution value of the continuous linear programming relaxation is sufficiently close to the optimal solution value of the considered problem.

We wish to point out that, to derive a "good" MILP model, it is strongly recommended that all the structural properties of the considered problem be taken into account (and this is why modeling is sometimes considered to be an "art"). Indeed, the more we know about a problem, the better will be the corresponding MILP model, where the general modeling techniques presented here, together with a reasonable background on known MILP models, are often a sufficient tool to derive such a model for problems having strong structural properties.

5.5. Bibliography

[AHU 93] R.K. AHUJA, T.L. MAGNANTI, J.B. ORLIN, *Network Flows: Theory, Algorithms, and Applications*, Prentice Hall, New Jersey, 1993.

[AKK 00] J.M. VAN DEN AKKER, C.A.J. HURKENS, M.W.P. SAVELSBERGH, "Time-indexed formulations for machine scheduling problems: column generation", *INFORMS Journal on Computing*, 12, 2000, 111–124.

[BAK 74] K.R. BAKER, *Introduction to Sequencing and Scheduling*, Wiley, New York, 1974.

[BAP 02] Ph. BAPTISTE, personal communication, 2002.

[COL 02] P. COLL, J. MARENCO, I. MENDEZ DIAZ, P. ZABALA, "Facets of the graph coloring polytope", *Annals of Operations Research*, 116, 2002, 79–90.

[DAN 54] G.B. DANTZIG, R. FULKERSON, S.M. JOHNSON, "Solution of a large-scale traveling salesman problem", *Operations Research*, 2, 1954, 393–410.

[DAS 95] M.S. DASKIN, *Network and Discrete Location*, Wiley, New York, 1995.

[CRO 94] F. DELLA CROCE, R. TADEI, "A multi-KP modeling for the maximum clique problem", *European Journal of Operational Research*, 73, 1994, 555–561.

[ELL 04] S. ELLOUMI, M. LABBÉ, Y. POCHET, "A new formulation and resolution method for the p-center problem", *INFORMS Journal on Computing*, 16, 2004, 84–94.

[LAS 92] J.B. LASSERRE, M. QUEYRANNE, "Generic scheduling polyhedra and a new mixed-integer formulation for single-machine scheduling", in E. BALAS, G. CORNUÉJOLS, R. KANNAN eds., *Integer Programming and Combinatorial Optimization, Proceedings of the Second IPCO-Conference*, University Printing and Publications, Carnegie Mellon University, Pittsburgh, 1992, 136–149.

[MAC 03] N. MACULAN, G. PLATEAU, A. LISSER, "Integer linear models with a polynomial number of variables and constraints for some classical combinatorial problems", *Pesquisa Operacional*, 23, 1, 2003, 161–168.

[PLA 02] F. PLASTRIA, "Formulating logical implications in combinatorial optimisation", *European Journal of Operational Research*, 140, 2002, 338–353.

[WIL 90] H. P. WILLIAMS, *Model Building in Mathematical Programming*, Wiley, Chichester, 1990.

[YAN 99] H. YAN, J.N. HOOKER, "Tight representation of logical constraints as cardinality rules", *Mathematical Programming*, 85, 1999, 363–378.

Chapter 6

Simplex Algorithms for Linear Programming

6.1. Introduction

The purpose of this chapter is to introduce the main simplex algorithms for linear programming, namely the primal simplex algorithm, the dual simplex algorithm, and the primal–dual simplex algorithm, and related issues. It is largely based on Luenberger's excellent textbook [LUE 84], as is the notation used in most cases. Some extensions of reduced costs and pseudo-costs of the simplex solution in integer linear programming are also presented.

6.2. Primal and dual programs

In a linear programming (LP) problem in *standard form*, a matrix $A \in \mathbb{R}^{m \times n}$, a vector $\overline{b} \in \mathbb{R}^m$, a cost vector $\overline{c} \in \mathbb{R}^n$, and a variable vector $\overline{x} \in \mathbb{R}^n$ are given; the problem requires us to

$$\text{minimize } z = \overline{c}^T \overline{x} \qquad [6.1]$$

subject to

$$A\overline{x} = \overline{b} \qquad [6.2]$$

$$\overline{x} \geqslant \overline{0} \qquad [6.3]$$

Chapter written by Frédérico DELLA CROCE and Andrea GROSSO.

The set $\mathcal{S} = \{\overline{x}\colon A\overline{x} = b,\ \overline{x} \geqslant 0\}$ is called the *feasible region* of the problem. By introducing Lagrange multipliers $\overline{\lambda} \in \mathbb{R}^m$ and $\overline{\mu} \in \mathbb{R}^n_+$ for the constraints sets [6.2] and [6.3], respectively, the lagrangian function of the problem is written as

$$\mathcal{L}(\overline{\lambda}, \overline{\mu}, \overline{x}) = \overline{c}^T\overline{x} + \overline{\lambda}^T\left(\overline{b} - A\overline{x}\right) - \overline{\mu}^T\overline{x} = \overline{\lambda}^T\overline{b} + \left(\overline{c}^T - \overline{\lambda}^T A - \overline{\mu}^T\right)\overline{x} \qquad [6.4]$$

When $\overline{x}$ lies in the feasible region $\mathcal{S}$, $\mathcal{L}(\overline{\lambda}, \overline{\mu}, \overline{x}) = \overline{c}^T\overline{x} - \overline{\mu}^T\overline{x} \leqslant \overline{c}^T\overline{x}$, hence minimizing $\mathcal{L}$ over $\overline{x} \in \mathbb{R}^n$ gives a lower bound on the optimal value of problem [6.1]–[6.3] (depending on $\overline{\lambda}$ and $\overline{\mu}$):

$$\begin{aligned} w &= \min\left\{\mathcal{L}(\overline{\lambda}, \overline{\mu}, \overline{x})\colon \overline{x} \in \mathbb{R}^n\right\} = \\ &= \overline{\lambda}^T\overline{b} + \min\left\{\left(\overline{c}^T - \overline{\lambda}^T A - \mu^T\right)\overline{x},\ \overline{x} \in \mathbb{R}^n\right\} \end{aligned} \qquad [6.5]$$

The lower bound w is well defined only for

$$\overline{c}^T - \overline{\lambda}^T A - \overline{\mu}^T = 0 \qquad (\overline{\lambda} \in \mathbb{R}^m, \overline{\mu} \in \mathbb{R}^n_+)$$

or equivalently

$$\overline{\lambda}^T A \leqslant \overline{c}^T \qquad (\overline{\lambda} \in \mathbb{R}^m)$$

otherwise the $\min\{\cdot\}$ argument would be unbounded from below – say $w = -\infty$. For $\overline{\lambda}$, $\overline{\mu}$ such that $w > -\infty$, the lower bound value is $w = \overline{\lambda}^T\overline{b}$. The tightest lower bound is determined by solving the so-called *dual* linear program associated with the *primal* [6.1]–[6.3]:

$$\text{maximize } w = \overline{\lambda}^T\overline{b} \qquad [6.6]$$

subject to

$$\overline{\lambda}^T A \leqslant \overline{c}^T \qquad \overline{\lambda} \in \mathbb{R}^m \qquad [6.7]$$

A dual-feasible region can be introduced for the dual problem:

$$\mathcal{D} = \left\{\overline{\lambda} \in \mathbb{R}^m : \overline{\lambda}^T A \leqslant \overline{c}^T\right\}$$

6.2.1. *Optimality conditions and strong duality*

At an optimal point $\overline{x}^*$ for the primal problem, the general Karush–Kuhn–Tucker necessary conditions (see [NOC 99]) require that there must exist multipliers $\overline{\lambda}^*$, $\overline{\mu}^*$ such that

$$\overline{0} = \nabla_{\overline{x}}\mathcal{L} = \overline{c}^T - \overline{\lambda}^{*T} A - \overline{\mu}^{*T} \qquad [6.8]$$

$$A\overline{x}^* - \overline{b} = 0 \qquad [6.9]$$

$$\overline{\mu}^{*T}\overline{x}^* = 0 \qquad [6.10]$$

Note that condition [6.8] stands for $\overline{\lambda} \in \mathcal{D}$, and condition [6.9] stands for $\overline{x} \in \mathcal{S}$. Condition [6.10] is also known as complementary slackness.

Conditions [6.8]–[6.10] are easily seen to also be sufficient for optimality: multiplying condition [6.8] by $\overline{x}^*$ yields the scalar equation

$$\left(\overline{c}^T - \overline{\lambda}^{*T} A - \overline{\mu}^{*T}\right) \overline{x}^* = 0$$

and, taking into account conditions [6.9] and [6.10],

$$\overline{c}^T \overline{x}^* = \overline{\lambda}^{*T} A\overline{x} + \overline{\mu}^{*T} \overline{x} = \overline{\lambda}^{*T} \overline{b}$$

The equality with the lower bound $w = \overline{c}^T \overline{x}^* = \overline{\lambda}^{*T} \overline{b}$ certifies optimality.

The necessary and sufficient optimality conditions are often stated as the following classical theorem.

THEOREM 6.1.– *The strong duality theorem of LP. Given a primal problem* [6.1]–[6.3] *and a primal-feasible solution* $\overline{x}^*$, *and the associated dual problem* [6.6]–[6.7] *with a dual-feasible solution* $\overline{\lambda}^*$, *the following conditions are equivalent*

(i) $\overline{x}^*$ *is primal-optimal, and* $\overline{c}^T \overline{x}^* = \overline{\lambda}^{*T} \overline{b}$;

(ii) $\overline{\lambda}^*$ *is dual-optimal, and* $\overline{\lambda}^{*T} \overline{b} = \overline{c}^T \overline{x}^*$;

(iii) $(\overline{\lambda}^{*T} A - \overline{c}^T)\overline{x}^* = 0$ *(complementary slackness).*

6.2.2. *Symmetry of the duality relation*

Problem [6.6]–[6.7] itself is a linear program, whose dual is exactly problem [6.1]–[6.3]. The dual problem [6.6]–[6.7] can be rewritten in standard form as follows, with variables $\overline{\lambda}_+, \overline{\lambda}_- \in \mathbb{R}^m$, $\overline{\mu} \in \mathbb{R}^n$:

$$\text{minimize } w = -(\overline{b}^T \quad -\overline{b}^T \quad \overline{0}) \begin{pmatrix} \overline{\lambda}_+ \\ \overline{\lambda}_- \\ \overline{\mu} \end{pmatrix}$$

subject to

$$(A^T \quad -A^T \quad I) \begin{pmatrix} \overline{\lambda}_+ \\ \overline{\lambda}_- \\ \overline{\mu} \end{pmatrix} = \overline{c} \qquad (I \in \mathbb{R}^{n \times n})$$

$$\overline{\lambda}_+ \geqslant \overline{0}, \overline{\lambda}_- \geqslant \overline{0}, \overline{\mu} \geqslant \overline{0}$$

The dual of this problem, written with variables $\overline{y} = (y_1, \ldots, y_n)^T$, is

$$\text{maximize } z = \overline{y}^T\overline{c} = \overline{c}^T\overline{y} \qquad [6.11]$$

subject to

$$A\overline{y} \leqslant -\overline{b} \qquad [6.12]$$

$$-A\overline{y} \leqslant \overline{b} \qquad [6.13]$$

$$\overline{y} \leqslant 0. \qquad [6.14]$$

By noting that constraints [6.12] and [6.13] are equivalent to $A\overline{y} = -\overline{b}$, and substituting $\overline{x} = -\overline{y}$, problem [6.11]–[6.14] immediately reduces to the original primal [6.1]–[6.3]. Hence primal–dual pairs of problems are often regarded as being substantially the same problem, since the "dual of the dual" is the primal problem.

6.2.3. *Weak duality*

From [6.5] it turns out that every dual solution $\overline{\lambda}$ provides a lower bound on the optimal value of the primal problem. A slightly less obvious result is that, if the primal objective function $z = \overline{c}^T\overline{x}$ is unbounded from below, the dual necessarily has no feasible solution.

THEOREM 6.2.– *Weak duality theorem*

(i) $\overline{x} \in \mathcal{S}, \overline{\lambda} \in \mathcal{D} \implies \overline{c}^T\overline{x} \geqslant \overline{\lambda}^T\overline{b}$.

(ii) $\inf\left\{z = \overline{c}^T\overline{x} \colon \overline{x} \in \mathcal{S}\right\} = -\infty \implies \mathcal{D} = \emptyset$.

(iii) $\sup\left\{w = \overline{\lambda}^T\overline{b} \colon \overline{\lambda} \in \mathcal{D}\right\} = +\infty \implies \mathcal{S} = \emptyset$.

The theorem still leaves open the possibility that both the primal and the dual problems have empty feasible regions. This can actually happen, as the reader can easily verify with the following example.

primal	dual
minimize $z = -2x_1 + x_2$	maximize $w = \lambda_1 - 2\lambda_2$
subject to	subject to
$x_1 - x_2 = 1$	$\lambda_1 - \lambda_2 \leqslant -2$
$-x_1 + x_2 = -2$	$-\lambda_1 + \lambda_2 \leqslant 1$
$x_1, x_2 \geqslant 0.$	

6.2.4. *Economic interpretation of duality*

Consider the following problem. A foundry produces various types of steel, and the steel-making process must be supplied with sufficient quantities of chromium, molybdenum and manganese. Such elements are available on the market in packaged form as "boxes" containing all the three elements in different quantities, and sold at different prices.

Package type	Unit price(euro)	Contents (tons/package) Cr	Mo	Mn
1	10	2	1	2
2	15	3	1	2
3	20	2	5	1

For the current month's production the foundry needs supplies of at least 30, 180 and 130 tons of chromium, molybnenum and manganese, respectively.

The linear program for computing the minimum-cost supplies is the following, where x_1, x_2, x_3 represent the number of boxes of type $1, 2, 3$, respectively, to be purchased. The surplus variables s_1, s_2, s_3 enable us to write $\geqslant$ constraints as linear equalities.

$$\text{minimize } z = 10x_1 + 15x_2 + 20x_3$$

subject to

$$\begin{aligned} 2x_1 + 3x_2 + 2x_3 - s_1 &= 30 && (\geqslant 30 \text{ tons chromium}) \\ x_1 + \ x_2 + 5x_3 - s_2 &= 180 && (\geqslant 180 \text{ tons molybdenum}) \\ 2x_1 + 2x_2 + \ x_3 - s_3 &= 130 && (\geqslant 130 \text{ tons manganese}) \end{aligned}$$

$$x_1, x_2, x_3, s_1, s_2, s_3 \geqslant 0$$

The dual problem is formulated on the variables λ_1, λ_2, λ_3 as follows:

$$\text{maximize } w = 30\lambda_1 + 180\lambda_2 + 130\lambda_3$$

subject to

$$\begin{aligned} 2\lambda_1 + \lambda_2 + 2\lambda_3 &\leqslant 10 \\ 3\lambda_1 + \lambda_2 + 2\lambda_3 &\leqslant 15 \\ 2\lambda_1 + 5\lambda_2 + \lambda_3 &\leqslant 20 \\ -\lambda_1 \leqslant 0, -\lambda_2 &\leqslant 0, -\lambda_3 \leqslant 0 \end{aligned}$$

The dual problem models the choice of a retailer that wants to sell out-of-the-box chromium, molybdenum and manganese to the foundry, at prices of λ_1, λ_2 and λ_3 euros/ton for the three components. The seller aims to maximize his own profit w from selling the whole stock to the foundry.

The dual constraints reflect the need for the retailer to be competitive with the market prices for packaged components. The first dual constraint represents the total cost incurred by the foundry for buying the contents $(2 + 1 + 2$ tons) of a type 1 box from the retailer; if the condition

$$2\lambda_1 + \lambda_2 + 2\lambda_3 \leqslant 10$$

was violated, it would be more convenient for the foundry to buy a box on the market. Similarly, the other constraints will force the retailer to be competitive with matket prices for type-2 and type-3 boxes.

The LP duality theorem formally states that the foundry management and the dual retailer can actually sign a fair agreement, where the foundry can get the required supplies at minimum cost z^* and the retailer can get the maximum profit $w^* = z^*$.

The optimal dual variables $\overline{\lambda}^*$ also give insights into the sensitivity of the optimal primal cost/dual profit in response to "small" changes in the foundry's requirements: for small changes $30 + \Delta_{\text{Cr}}$, $180 + \Delta_{\text{Mo}}$, $130 + \Delta_{\text{Mn}}$ in the required quantities of chromium, molybdenum and manganese the foundry could expect a rise in the total cost of

$$\lambda_1^*\Delta_{\text{Cr}} + \lambda_2^*\Delta_{\text{Mo}} + \lambda_3^*\Delta_{\text{Mn}}$$

λ_1^*, λ_2^* and λ_3^* specify the costs of one ton of chromium, molybdenum or manganeses (perceived by the foundry) in the steel-making process.

6.3. The primal simplex method

6.3.1. *Basic solutions*

Consider the linear programming problem [6.1]–[6.3] in standard form, which we rewrite here as

$$\text{minimize } \overline{c}^T\overline{x}$$

$$A\overline{x} = \overline{b}$$

$$\overline{x} \geqslant \overline{0}$$

Let us select from A a set of m linearly independent columns (such a set exists if the rank of A is m). This set constitutes a non-singular $m \times m$ matrix B. Let D be the matrix of the remaining columns of A. Let us associate with B the related variables vector $\overline{x}_B$. Similarly let us associate with D the related variables vector $\overline{x}_D$.

Correspondingly, the constraints set [6.2]–[6.3] becomes

$$B\overline{x}_B + D\overline{x}_D = \overline{b}$$

and gives the unique solution $\overline{x}_B = B^{-1}\overline{b}$ if $\overline{x}_D = \overline{0}$.

DEFINITION 6.1.– *Basic solution* $\overline{x} = (\overline{x}_B, \overline{x}_D = \overline{0})$ *denotes the* basic solution *of* [6.2]–[6.3], *where* $\overline{x}_B$ *is the set of* basic variables *and* $\overline{x}_D$ *is the set of* non-basic variables.

Such a basic solution is also defined as *degenerate* if at least one basic variable x_j exists such that $x_j = 0$. In general, the constraints set [6.2]–[6.3] may have no basic solutions. To avoid this we make the following assumptions on matrix A, namely:

1) $m < n$;

2) the m rows of A are linearly independent.

Basic solutions are fundamental in linear programming because of the following result (for the proof, see [LUE 84]).

THEOREM 6.3.– *Fundamental LP theorem. Consider the LP problem* [6.1]–[6.3] *in standard form with A being an $m \times n$ matrix (of rank m).*

I. If there is a feasible solution, there is a basic solution.

II. If there is an optimal feasible solution, there is a an optimal basic feasible solution.

The importance of this theorem is that it reduces the task of solving a linear programming problem to that of searching for the best among the basic feasible solutions. Furthermore, the total number of these solutions is finite, namely

$$\begin{pmatrix} n \\ m \end{pmatrix} = \frac{n!}{m!\,(n-m)!}$$

while there are in general infinite feasible solutions.

From [6.3], theorem 6.1 can also be expressed as follows.

THEOREM 6.4.– *The strong LP duality theorem (reformulated). Consider a primal problem* [6.1]–[6.3] *and a primal optimal basic feasible solution* $\overline{x} = (\overline{x}^*_B, \overline{x}_D = \overline{0})$. *Let the matrix cost A be partitioned as $A = [B, D]$ with the related cost coefficients vector $\overline{c}^T$ partitioned as $\overline{c} = [\overline{c}_B, \overline{c}_D]$. Then, $\overline{\lambda}^{*T} = \overline{c}_B^T B^{-1}$ is an optimal feasible solution to the associated dual problem* [6.6]–[6.7].

6.3.2. *Canonical form and reduced costs*

Consider the constraints set

$$B\overline{x}_B + D\overline{x}_D = \overline{b}:$$

By premultiplying each term by B^{-1}, we get

$$B^{-1}B\overline{x}_B + B^{-1}D\overline{x}_D = B^{-1}\overline{b}$$

that is

$$\overline{x}_B = B^{-1}\overline{b} - B^{-1}D\overline{x}_D$$

This is the so-called *canonical form* of the constraints set with respect to the basic variables set $\overline{x}_B$. Indeed, whenever a basic solution is considered with $\overline{x}_D = 0$, we immediately derive the value of the basic variables as $\overline{x}_B = B^{-1}\overline{b}$.

Now, consider the objective function $z = \overline{c}^T\overline{x} = \overline{c}_B^T\overline{x}_B + \overline{c}_D^T\overline{x}_D$. It can be expressed as

$$\begin{aligned} z &= \overline{c}_B^T\overline{x}_B + \overline{c}_D^T\overline{x}_D = \overline{c}_B^T(B^{-1}\overline{b} - B^{-1}D\overline{x}_D) + \overline{c}_D^T\overline{x}_D = && [6.15] \\ &= \overline{c}_B^T B^{-1}\overline{b} + (\overline{c}_D^T - \overline{c}_B^T B^{-1}D)\overline{x}_D = && [6.16] \\ &= z_0 + (\overline{c}_D^T - \overline{c}_B^T B^{-1}D)\overline{x}_D && [6.17] \end{aligned}$$

where we denote the value of the current basic feasible solution by $z_0 = \overline{c}_B^T B^{-1}\overline{b}$.

Given the current basic solution in canonical form, from the above expression [6.15], an immediate optimality condition can be derived:

$\overline{x}_B$ is an optimal basic solution if

$\overline{c}_D^T - \overline{c}_B^T B^{-1}D \geqslant 0$, that is (as we are dealing with positive variables) if

$c_j - \overline{c}_B^T B^{-1}\overline{a}_j \geqslant 0 \ \forall j \in B$, namely if

$r_j \geqslant 0 \, \forall j \in B$, where r_j denotes the reduced cost of variable x_j.

If, however, non-basic variables x_j with $r_j < 0$ exist, the solution may be non-optimal as it should be possible to find a better solution by assigning to variable x_j a value strictly greater than 0. As we search for basic solutions, in order to improve the solution value, we may consider swapping a basic variable x_i with a non-basic variable x_j so as to obtain a new improved basic solution including x_j as part of the basis. This swap is often called a *pivoting operation.*

The pivoting operation, by swapping a basic variable with a non-basic one, does not necessarily preserve the non-negativity (feasibility) of the solution. Given the non-basic variable that enters the basis, the following conditions must hold for the basic variable selected to leave the basis. Let $\overline{x} = [x_1, ..., x_m, 0, ..., 0]$ be a basic feasible solution. Let $\overline{e}_i$ be the unit vector with only the i-th non-zero component. The corresponding constraints set in canonical form presents a system of equations of the following type.

$$x_i + \overline{e}_i B^{-1}\overline{a}_{m+1}x_{m+1} + ... + \overline{e}_i B^{-1}\overline{a}_q x_q + ...+ \overline{e}_i B^{-1}\overline{a}_n x_n = \overline{e}_i B^{-1}\overline{b} \quad [6.18]$$

If only x_q varies (as it becomes part of the basic solution) among the non-basic variables, equations [6.18] can be expressed as

$$x_i + \overline{e}_i B^{-1}\overline{a}_q x_q = \overline{e}_i B^{-1}\overline{b}$$

To preserve feasibility, we must have

$$x_i = \overline{e}_i B^{-1}\overline{b} - \overline{e}_i B^{-1}\overline{a}_q x_q \geqslant 0 \, \forall i$$

namely

$$x_q \leqslant \frac{\overline{e}_i B^{-1}\overline{b}}{\overline{e}_i B^{-1}\overline{a}_q} \, \forall i : \overline{e}_i B^{-1}\overline{a}_q > 0$$

Hence, a new improved basic feasible solution can be obtained for

$$x_q = \min_i \{ \frac{\overline{e}_i B^{-1}\overline{b}}{\overline{e}_i B^{-1}\overline{a}_q} : \overline{e}_i B^{-1}\overline{a}_q > 0 \} \quad [6.19]$$

and the variable x_i that leaves the basis is the one corresponding to the minimizing index in [6.19].

If the minimum in [6.19] is achieved by more than a single index i, then any of the variables can be regarded as the one that leaves the basis, but after the pivoting operations the next basic solution will be degenerate. Besides, if $\overline{e}_i B^{-1} \overline{a}_q \leqslant 0 \; \forall i$, then no new basic feasible solution can obtained. Indeed, for any positive value of x_q each basic variable x_i increases its value by $\overline{e}_i B^{-1} \overline{a}_q x_q$ and there are feasible (non-basic) solutions where all basic variables and x_q may have arbitrarily large values. As the objective function can be expressed as $z = z_0 + r_q x_q$, the problem is unbounded (recalling that r_q is negative).

We are now ready to introduce the primal simplex algorithm:

– Initial conditions: a basic feasible solution.

– Iterative approach:

1) if the reduced costs of the non-basic variables are positive, STOP the solution is optimal;

2) else swap (if possible) a non-basic variable x_j with negative reduced cost with a basic variable in such a way as to result in a new basic feasible solution (if such a swap is not possible, then STOP the problem is unbounded);

3) Go To 1.

More formally, the primal simplex algorithm can be expressed as follows.

1: Initial conditions: a basic feasible solution B with $\overline{x}_B = B^{-1}\overline{b}$ current vector of the basic variables.
2: **loop**
3: Compute the simplex multipliers $\overline{\lambda}^T = \overline{c}_B^T B^{-1}$ and the reduced costs of the non-basic variables: $r_j = c_j - \overline{\lambda}^T \overline{a}_j \; \forall j \notin B$.
4: **if** $r_j \geqslant 0 \; \forall j$ **then**
5: STOP, $\overline{x} = (\overline{x}_B, \overline{x}_D = 0)$ is optimal
6: **else**
7: go to 2.
8: **end if**
9: Select $r_q < 0$. Vector $\overline{a}_q$ enters the basis. Compute $B^{-1}\overline{a}_q$.
10: **if** $\overline{e}_i B^{-1} \overline{a}_q \leqslant 0, \forall i$ **then**
11: the problem is unbounded, STOP.
12: **else**
13: Compute the minimum ratio $\dfrac{\overline{e}_i B^{-1} \overline{b}}{\overline{e}_i B^{-1} \overline{a}_q}$, with $\overline{e}_i B^{-1} \overline{a}_q > 0$ to determine which vector is to leave the basis.
14: **end if**

15: Update B and correspondingly B^{-1} and the new basic solution $\overline{x}_B = B^{-1}\overline{b}$. Return to step 1.

16: **end loop**

6.4. Bland's rule

When degenerate basic solutions are encountered, the simplex algorithm may not converge and fall into an infinite loop. This may occur (even though very rarely) whenever the current value of the variable deputed to exit the basis is zero. Consider the following example:

$$\text{minimize } z = -10x_1 + 57x_2 - 9x_3 - 24x_4$$

subject to

$$\begin{aligned}
\frac{1}{2}x_1 - \frac{11}{2}x_2 - \frac{5}{2}x_3 + 9x_4 + x_6 &= 0\\
\frac{1}{2}x_1 - \frac{3}{2}x_2 - \frac{1}{2}x_3 + x_4 + x_7 &= 0\\
x_1 &= 1\\
x_1, x_2, x_3, x_4, x_5, x_6, x_7 &\geqslant 0
\end{aligned}$$

The initial basis is $B = [x_5, x_6, x_7]$ with $x_B^T = [0, 0, 1]$ and just one non-basic variable with negative reduced cost, namely $r_1 = -10$. Correspondingly (first iteration), x_1 enters the basis: then two variables (x_5 and x_6) are deputed to exit the basis and we select among them x_5 so that the new basis is $B = [x_1, x_6, x_7]$, stil with $\overline{x}_B^T = [0, 0, 1]$. Now (second iteration), there are two non-basic variables with negative reduced costs, namely $r_2 = -53$ and $r_3 = -41$. This time x_2 is selected to enter the basis and correspondingly (just one option) x_6 exits the basis so that the new basis is $B = [x_1, x_2, x_7]$, still with $\overline{x}_B^T = [0, 0, 1]$. Now (third iteration), there is just one non-basic variable with negative reduced cost, namely $r_3 = -\frac{29}{2}$. Correspondingly, x_3 enters the basis: then, two variables (x_1 and x_2) are deputed to exit the basis and we select among them x_1 so that the new basis is $B = [x_3, x_2, x_7]$, still with $\overline{x}_B^T = [0, 0, 1]$. Now (fourth iteration), there are two non-basic variables with negative reduced costs, namely $r_4 = -18$ and $r_5 = -15$. This time x_4 is selected to enter the basis and correspondingly (just one option) x_2 exits the basis so that the new basis is $B = [x_3, x_4, x_7]$, still with $\overline{x}_B^T = [0, 0, 1]$. Now (fifth iteration), there is just one non-basic variable with negative reduced cost, namely $r_5 = -\frac{21}{2}$. Correspondingly, x_5 enters the basis: then, two variables (x_3 and x_4) are deputed to exit the basis and we select among them x_3 so that the new basis is $B = [x_5, x_4, x_7]$, still with $\overline{x}_B^T = [0, 0, 1]$. Finally (sixth iteration), there are two non-basic variables with negative reduced costs, namely $r_1 = -22$ and $r_6 = -24$. We select x_6 to enter the basis: correspondingly the unique variable deputed to exit the basis is x_4 and we are back to the initial basis $B = [x_5, x_6, x_7]$, still with $\overline{x}_B^T = [0, 0, 1]$.

In order to avoid this kind of infinite cycle, the following rule (for its proof refer to [BLA 77]) holds.

THEOREM 6.5.– *Bland's rule*

– *Select as the non-basic variable to enter the basis the lowest indexed favorable variable, namely the variable x_q such that*

$$q = \min\{j : r_j < 0\}$$

– *If there are ties on the basic variable deputed to exit the basis, again select the one with the lowest index, namely the variable x_p such that*

$$p = \min\{i : \overline{e}_i B^{-1}\overline{a}_q > 0 \wedge \frac{\overline{e}_i B^{-1} b}{\overline{e}_i B^{-1}\overline{a}_q} \leqslant \frac{\overline{e}_k B^{-1} b}{\overline{e}_k B^{-1}\overline{a}_q} \;\forall k : \overline{e}_k B^{-1}\overline{a}_q > 0\}$$

By applying Bland's rule, the algorithm converges to the optimal solution (or proves unboundedness) within a finite number of iterations.

If Bland's rule is applied, the first five iterations remain the same, reaching the basis $B = [x_5, x_4, x_7]$. Now, there are two non-basic variables with negative reduced costs, namely $r_1 = -22$ and $r_6 = -24$. Unlike the above case, Bland's rule indicates selection of x_1 as the entering variable: correspondingly (just one option) x_4 exits the basis so that the new basis is $B = [x_5, x_1, x_7]$, still with $\overline{x}_B^T = [0, 0, 1]$. But then, the only possible entering variable is x_3: correspondingly (just one option) x_7 exits the basis so that the new basis is $B = [x_5, x_1, x_3]$ with $\overline{x}_B^T = [2, 1, 1]$. This is actually the optimal solution with $z = -1$ as all reduced costs are now positive.

6.4.1. *Searching for a feasible solution*

As indicated in the first step of the primal simplex algorithm, a basic feasible solution is required to proceed with the algorithm, but actually a basic feasible solution is not always available as the constraints set is not necessarily in canonical form. Also, there are unfeasible problems that we would like to be able to detect. To this purpose, it is possible to construct an artificial minimization problem that either provides a basic feasible solution for, or detects unfeasibility of, the original problem. Consider the original constraints set [6.2]–[6.3]

$$A\overline{x} = \overline{b}$$

$$\overline{x} \geqslant \overline{0}$$

and introduce an artificial minimization problem with respect to the constraints set [6.2]–[6.3] that has the following form

$$\text{minimize } \phi = \sum_{i=1}^{m} \psi_i \qquad [6.20]$$

subject to

$$A\overline{x} + \overline{\psi} = \overline{b} \qquad [6.21]$$

$$\overline{x}, \overline{\psi} \geqslant \overline{0} \qquad [6.22]$$

where $\overline{\psi} = [\psi_i, \ \forall i \in 1...m]$ is the so-called set of *artificial* variables.

Note that the LP problem [6.20]–[6.22] is in canonical form, where ψ constitutes the vector of the initial basic variables. Hence, the LP problem [6.20]–[6.22] can be solved using the primal simplex algorithm presented above to give an optimal basic feasible solution $w \geqslant 0$ (problem [6.20]–[6.22] is not unbounded as all variables in the objective function are positive and have positive coefficients). We have then two possible outcomes:

1) If the optimal solution of problem [6.20]–[6.22] is $\phi = 0$ (namely, $\psi_i = 0 \ \forall i$), then this solution is also a feasible solution with respect to the constraints set [6.2]–[6.3]. Note also that if all ψ_i variables are non-basic, then this solution constitutes a basic feasible solution. Alternatively the solution is degenerate with some ψ_i basic variables with value 0 and it is sufficient to perform several further pivots entering some non-basic x_j variables in place of the ψ_i basic variables.

2) Else, ϕ is strictly greater than 0, but this implies that the constraints set [6.2]–[6.3] has no feasible solution.

6.5. Simplex methods for the dual problem

6.5.1. *The dual simplex method*

Recall that a basis B is called feasible (or *primal-feasible*) if $\overline{x}_B = B^{-1}\overline{b}$ has no negative components. Let $A = (\overline{a}_1, \overline{a}_2, \ldots, \overline{a}_n)$ be columns, $B = (\overline{a}_{B_1}, \ldots, \overline{a}_{B_m})$ be the basis matrix, and $D = (\overline{a}_{D_1}, \ldots, \overline{a}_{D_{n-m}})$. A basis B is also called *dual-feasible* if $\lambda^T = c_B^T B^{-1} \in \mathcal{D}$; this means that $\lambda^T B = c_B^T$, and $\lambda^T D \leqslant c_D^T$, or that the primal reduced costs

$$\overline{r}_D = \overline{c}_D^T - \overline{c}_B^T B^{-1} D = c_D^T - \overline{\lambda}^T D \qquad [6.23]$$

are non-negative. The so-called *dual simplex* method plays an important role when (in some way) a dual-feasible basis is available which is *not* primal-feasible. Interestingly, this means that the basis B is *optimal* – from equation [6.23] – but not feasible for the primal problem, i.e. the vector $\overline{x}_B = B^{-1}\overline{b}$ has negative components. Far from being uncommon, this situation arises, for example, when, for an optimal basis, primal feasibility is destroyed because of a perturbation in $\overline{b}$ or by the introduction of a new constraint; the latter is the classical case of a branch in branch-and-bound algorithms, or introduction of *valid inequalities* in the LP relaxation of integer programs.

It is then computationally advantageous to perform basis changes aimed at optimizing the dual program, while primal feasibility will finally be recovered with the dual optimality. In the dual simplex method the dual basis changes are performed by executing pivot operations on the *primal* tableau, with no need for explicitly stating and solving the dual version.

Change of basis in the dual problem

From a primal point of view, changing basis means making a non-basic variable, x_{D_p}, a basic one, and removing a basic variable x_{B_q} from the basis. From a dual point of view, since the current basis B is dual-feasible, the dual solution $\lambda^T = \overline{c}_B^T B^{-1}$ guarantees $\overline{\lambda}^T A \leqslant \overline{c}^T$ by

$$\lambda^T \overline{a}_j = \overline{c}_j \qquad \text{for } j \in \{B_1, \ldots, B_m\}, \qquad [6.24]$$

$$\lambda^T \overline{a}_j \leqslant \overline{c}_j \qquad \text{for } j \in \{D_1, \ldots, D_{n-m}\} \qquad [6.25]$$

With a basis change, the dual solution must be updated to a new one, $\overline{\lambda} + \varepsilon\overline{\delta}$, with $\varepsilon \geqslant 0$ such that one of the non-basic constraints [6.25] saturates and a basic one [6.24] becomes non-saturated. This requires

$$\overline{\delta}^T \overline{a}_j = 0 \qquad \text{for } j \in \{B_1, \ldots, B_m\} \setminus \{B_q\}$$

and $\overline{\delta}^T a_{D_p} < 0$. Also, the dual objective value must not decrease. Let $\overline{e}_i$ always be the unit vector with only the i-th non-zero component. A suitable choice for δ is

$$\delta^T = -\overline{e}_q^T B^{-1}$$

with q such that $x_{B_q} = (B^{-1}\overline{b})_q < 0$ – such an index definitely esists if the basis is primal-unfeasible. The components δ are the q-th row of the basis inverse matrix. Note that

$$(\overline{\lambda} + \varepsilon\overline{\delta})^T \overline{b} = \overline{\lambda}^T \overline{b} - \varepsilon \underbrace{\overline{e}_q^T B^{-1} \overline{b}}_{= x_{B_i} < 0} \geqslant \overline{\lambda}^T \overline{b}.$$

Dual feasibility is preserved if

$$\varepsilon \leqslant \frac{\overline{c}_j - \overline{\lambda}^T \overline{a}_j}{\overline{\delta}^T \overline{a}_j} = \frac{\overline{c}_j - \overline{\lambda}^T \overline{a}_j}{-\overline{e}_q^T B^{-1} \overline{a}_j} =$$

$$= \frac{r_j}{-\overline{e}_q^T B^{-1} \overline{a}_j} \qquad \text{for all } j \in \{D_1, \ldots, D_{n-m}\} \text{ such that } \overline{e}_q^T B^{-1} \overline{a}_j < 0$$

The dual constraint that saturates at the maximum ε will become basic.

The basis change can be carried out on the primal problem, provided that the leaving and entering variables are chosen according to the following.

Dual pivot choice rule

First choose the leaving variable x_{B_q}, such that

$$x_{B_q} < 0$$

Choose the entering variable x_{D_p} so that

$$\frac{r_{D_p}}{-\overline{e}_q^T B^{-1} \overline{a}_{D_p}} = \min \left\{ \frac{r_j}{-\overline{e}_q^T B^{-1} \overline{a}_j} : j \in \{D_1, \ldots, D_{n-m}\}, \overline{e}_q^T B^{-1} \overline{a}_j < 0 \right\}$$

If no $\overline{e}_q^T B^{-1} \overline{a}_j$ happens to be strictly negative, dual-feasible solutions $\overline{\lambda} + \varepsilon \overline{\delta}$ with arbitrarily large values $(\overline{\lambda} + \varepsilon \overline{\delta})^T \overline{b}$ (unbounded from above, as $\varepsilon \to +\infty$) can be constructed, certifying that the primal problem has no feasible solution.

6.5.2. *The primal–dual simplex algorithm*

The primal–dual simplex algorithm is an algorithm that solves a LP problem by generating a sequence of dual-feasible solutions (not necessarily basic)

$$\overline{\lambda}^1, \overline{\lambda}^2, \ldots, \overline{\lambda}^r$$

such that $\overline{\lambda}^{1\,T} \overline{b} \geqslant \overline{\lambda}^{2\,T} \overline{b} \geqslant \cdots \geqslant \overline{\lambda}^{r\,T} \overline{b}$, and an associated sequence of primal, not necessarily feasible, solutions $\overline{x}^1, \ldots, \overline{x}^r$ such that each primal–dual pair $\langle \overline{x}^k, \overline{\lambda}^k \rangle$ satisfies the complementary slackness condition. The final solutions $\langle \overline{x}^r, \overline{\lambda}^r \rangle$ are guaranteed to be primal and dual optimal.

Restricted primal and its dual

Consider a pair of primal and dual problems like [6.1]–[6.3] and [6.6]–[6.7]. At iteration k, suppose a dual-feasible solution $\overline{\lambda}^k = \overline{\lambda}$ is available – let us drop the k superscript in order to simplify the notation. Form the set $J \subseteq \{1, \ldots, n\}$ of indices for which $\overline{\lambda}^T \overline{a}_j - c_j = 0$, and formulate the so-called *restricted primal* problem:

$$\text{minimize } \zeta = \sum_{i=1}^{m} s_i$$

subject to

$$\sum_{j \in J} y_j \overline{a}_j + s_i = b_i \qquad i = 1, \ldots, m$$

$$y_j \geqslant 0,\ j \in J,\ s_i \geqslant 0,\ i = 1, \ldots, m$$

The restricted primal problem has the following characteristics:

– It always has a feasible solution.

– It has an optimal value $\zeta^* = 0$ if and only if $\{y_j^* : j \in J\}$ exist such that

$$\sum_{j \in J} y_j^* \overline{a}_j = b_i \qquad \text{for all } i = 1, \ldots, m.$$

– If $\{y_j^* : j \in J\}$ exists, the solution $\overline{x}^* = (x_1^*, \ldots, x_n^*)^T$ defined by

$$x_j^* = y_j^*, \quad j \in J,$$

$$x_j^* = 0, \quad j \notin J,$$

is feasible and optimal for the primal problem [6.1]–[6.3]. Optimality is certified by the complementary slackness, since

$$(\overline{\lambda}^T A - \overline{c}^T)\overline{x}^* = \sum_{j \in J} \underbrace{(\overline{\lambda}^T \overline{a}_j - c_j)}_{=0} x_j^* + \sum_{j \notin J} (\overline{\lambda}^T \overline{a}_j - c_j) \underbrace{x_j^*}_{=0} = 0.$$

The *dual of the restricted primal* problem is written, with variables $\overline{u} = (u_1, \ldots, u_m)^T$, as

$$\text{maximize } \eta = \overline{u}^T \overline{b}$$

subject to

$$\overline{u}^T \overline{a}_j \leqslant 0 \qquad \text{for } j \in J$$

$$u_i \leqslant 1,\ i = 1, \ldots, m$$

The main iteration

The primal–dual simplex algorithm first defines and then solves the restricted primal problem. The key step of the algorithm is the following. If the restricted primal has $\zeta^* = 0$, a primal-optimal solution is available. Otherwise, the dual of the restricted primal has an optimal value $\eta^* = \zeta^* > 0$; the dual solution $\overline{\lambda}$ can be updated to a new one

$$\overline{\lambda}^{k+1} = \overline{\lambda} + \varepsilon\overline{u}^*$$

where $\overline{u}^*$ is the optimal solution of the dual of the restricted primal problem. The step size $\varepsilon > 0$ is chosen in order to maintain dual feasibility in $\overline{\lambda}^{k+1}$: ε is required to satisfy

$$\varepsilon \leqslant \frac{c_j - \overline{\lambda}^T \overline{a}_j}{\overline{u}^T \overline{a}_j} \qquad \text{for all } j \in \{1, \ldots, n\} \text{ such that } \overline{u}^T \overline{a}_j > 0$$

Note that

(a) $\overline{\lambda}^{k+1\,T}\overline{b} = \overline{\lambda}^T\overline{b} + \varepsilon\eta^* > \overline{\lambda}^T\overline{b}$; and

(b) $\overline{\lambda}^{k+1\,T}\overline{a}_j = c_j$ still holds for every $j \in J$ with $y_j^* > 0$ – since at optimality such variables will have reduced costs $-\overline{u}^{*\,T}\overline{a}_j = 0$.

Condition (b) ensures that the optimal solution of the restricted primal problem remains feasible in the next iteration, allowing for incremental solution of the new restricted primal. If $\overline{u}^T\overline{a}_j \leqslant 0$ for all indices, the dual problem is unbounded and the primal has no feasible solution.

The overall algorithm works as follows ($\overline{\lambda}^1$ is given).

$k = 1$.
loop
  Compute $J = \left\{ j = 1, \ldots n \colon \overline{\lambda}^k \overline{a}_j = c_j \right\}$ and formulate the restricted primal.
  Solve the restricted primal → get the optimal solution $\{y_j^*\}, \{s_i^*\}$.
  if $\zeta^* = \sum_{i=1}^m s_i^* = 0$ **then**
    The primal optimal solution is

$$x_j^* = y_j^*, \quad j \in J,$$
$$x_j^* = 0 \quad , j \notin J.$$

    return $(x_1^*, \ldots, x_n^*)$.
  else
    Compute the restricted dual solution $\overline{u}^*$, and set

$$\varepsilon = \min \left\{ \frac{c_j - \overline{\lambda}^{k\,T}\overline{a}_j}{\overline{u}^T\overline{a}_j} : j \in \{1, \ldots, n\}, \ \overline{u}^T\overline{a}_j > 0 \right\} \cup \{+\infty\}.$$

if $\varepsilon = +\infty$ **then**
return ⟨unfeasible⟩.
else
Set $\overline{\lambda}^{k+1} = \overline{\lambda}^{k} + \varepsilon \overline{u}^{*}$.
end if
end if
Set $k = k + 1$.
end loop

6.6. Using reduced costs and pseudo-costs for integer programming

In integer linear programming (ILP) the usual problem is

$$\text{minimize } z = \overline{c}^T \overline{x} \tag{6.26}$$

subject to

$$A\overline{x} = \overline{b} \tag{6.27}$$

$$\overline{x} \in \mathbb{Z}_+^n \tag{6.28}$$

Such a problem is known to belong, in general, to the **NP**-hard class, and is usually solved by branch-and-bound (and cut) techniques. The simplex method is a natural candidate for solving the LP relaxation of the nodes in the branch tree – the dual simplex is particularly well suited to incrementally solving the relaxed problem after a branch operation.

From the optimal solution of the LP relaxation some useful information for strengthening the problem formulation can be immediately inferred – see for example [NEM 99] for an in-depth discussion of the links between LP and ILP.

6.6.1. *Using reduced costs for tightening variable bounds*

Assume that the LP relaxation of problem [6.26]–[6.28]

$$\text{minimize } z = \overline{c}^T \overline{x}$$

subject to [6.27] and

$$\overline{x} \geqslant \overline{0}$$

has been optimally solved with optimal basis B. The basic optimal (non-integer, to avoid the trivial case) solution $\overline{x}_B^* = B^{-1}\overline{b}, \overline{x}_D^* = \overline{0}$ is the *only* solution with $\overline{x}_D = \overline{0}$, and the objective function of the LP relaxation can be rewritten as

$$z_{\text{LP}} = B^{-1}\overline{b} + \left(\overline{c}_D^T - \overline{c}_B^T B^{-1} D\right) \overline{x}_D = z_{\text{LP}}^* + \overline{r}_D^T \overline{x}_D$$

Given any integer solution $\overline{x}'$, since $\overline{x}'$ is also an element of the LP feasible region $\mathcal{S}$, the objective function difference $z(\overline{x}') - z(\overline{x}^*)$ is

$$z(\overline{x}') - z_{\text{LP}}(\overline{x}^*) = \underbrace{\overline{r}_D^T}_{\geqslant \overline{0}} (\overline{x}'_D - \underbrace{\overline{x}_D}_{=\overline{0}}) = \overline{r}_D^T \overline{x}'_D \qquad [6.29]$$

By focusing on a single non-basic variable x_j

$$z(\overline{x}') - z(\overline{x}^*) = \overline{r}_D^T \overline{x}'_D \geqslant r_j x'_j$$

If an upper bound UB on the optimal value for the integer problem is known, every solution with $z_{\text{LP}}^* + r_j x'_j \geqslant \text{UB}$ can immediately be fathomed. Hence the upper bounds of the non-basic variables in the ILP problem can be (hopefully) strengthened, without loss of optimality, to

$$x_j \leqslant \left\lfloor \frac{\text{UB} - z_{\text{LP}}^*}{r_j} \right\rfloor, \qquad j \in \{D_1, \ldots, D_{n-m}\}$$

Note that from [6.29] even more complex constraints could be inferred:

$$\sum_{j \in Q} r_j x_j \leqslant \text{UB} - z_{\text{LP}}^* \implies$$

$$\implies \sum_{j \in Q} \lfloor r_j \rfloor x_j \text{UB} \leqslant \lfloor \text{UB} - z_{\text{LP}}^* \rfloor \qquad Q \subseteq \{D_1, \ldots, D_{n-m}\}$$

6.6.2. Pseudo-costs for integer programming

Assume that the LP relaxation of an ILP has been solved with continous optimum $\overline{x}^*$, $z^* = \overline{c}^T \overline{x}^*$, and at least one basic variable $x_i^* = \beta \notin \mathbb{Z}_+$. Then, in the classical branch-and-bound environment a branch has to be performed on variable x_{B_i}, introducing one of the constraints

$$x_{B_i} + s = \lfloor \beta \rfloor, \quad s \geqslant 0, \qquad \text{("down" branch)}$$

$$x_{B_i} - s = \lceil \beta \rceil, \quad s \geqslant 0, \qquad \text{("up" branch)}$$

in the resulting descendant problems. A reasonable estimate of the lower bound degradation can be obtained by computing a partial dual pivot step.

Consider the down-branch. Since x_{B_i} is basic, the i-th constraint can be formulated as

$$x_{B_i} + \sum_{j=D_1}^{D_{n-m}} \alpha_j x_j = \beta$$

where $\alpha_j = \overline{e}_i B^{-1} \overline{a}_j$. Subtracting it from the down-branch constraint yields

$$s - \sum_{j=D_1}^{D_{n-m}} \alpha_j x_j = \lfloor \beta \rfloor - \beta$$

This constraint is primal-unfeasible for the primal-optimal solution of the LP relaxation associated with the down-descendant node. The first dual simplex iteration would then perform a pivot by selecting x_{B_i} as the leaving variable and an entering variable x_k such that

$$\frac{r_k}{\alpha_k} = \min \left\{ \frac{r_j}{\alpha_j} : j = D_1, \dots, D_{n-m},\ \alpha_j > 0 \right\}$$

The objective function would be, after one single dual basis change:

$$z^* \underbrace{-r_k \frac{(\lfloor \beta \rfloor - \beta)}{\alpha_k}}_{\geqslant 0}$$

The quantity

$$d_{B_i} = -r_k \frac{(\lfloor \beta \rfloor - \beta)}{\alpha_k}$$

is also called the down-pseudo-cost (see [DRI 66] for a pioneering paper on this topic) of variable x_{B_i}, and is a valid lower bound on the increase of the optimal value of the LP relaxation at the considered node.

Using similar arguments we can derive the up-pseudo-cost

$$u_{B_i} = -r_k \frac{(\lceil \beta \rceil - \beta)}{\alpha_k}$$

with $\quad -\dfrac{r_k}{\alpha_k} = \min \left\{ -\dfrac{r_j}{\alpha_j} : j = D_1, \dots, D_{n-m},\ \alpha_j < 0 \right\}.$

We note that both reduced costs and pseudo-costs provide cost coefficients indicating potential degradation of the objective function value with respect to the modification of the corresponding variable from the continuous solution. Notice, however, that reduced costs are additive, whilst pseudo-costs are not. A more comperehensive analysis of the use of pseudo-costs in integer programming can be found in [LIN 99].

6.7. Bibliography

[BLA 77] BLAND, R.G., "New finite pivoting rules for the simplex method", *Mathematics of Operations Reserach*, vol. 2, 103–107, 1977.

[DRI 66] DRIEBEEK, N.J., "An algorithm for the solution of mixed integer programming problems", *Management Science*, vol. 12, 576–587, 1966.

[LIN 99] LINDEROTH, J.T., SAVELSBERGH, P., LUENBERGER, D.G., " A computational study of search strategies for mixed integer programming", *INFORMS Journal on Computing*, vol. 11, 173–187, 1999.

[LUE 84] LUENBERGER, D.G., *Linear and Nonlinear Programming*, Addison-Wesley, 1984.

[NEM 99] NEMHAUSER, G.L., WOLSEY, L.A., *Integer and Combinatorial Optimization*, Wiley, 1999.

[NOC 99] NOCEDAL, J., WRIGHT, S.J., *Numerical Optimization*, Springer, 1999.

Chapter 7

A Survey of some Linear Programming Methods

7.1. Introduction

After its early stages in the years 1939–1945, operations research grew from 1947 when G. B. Dantzig [DAN 63] proposed the simplex method. It has become a fundamental tool for operations researchers. Apart from solving real linear programs, it has, in particular, greatly contributed to, and still contributes to, solving **NP**-hard combinatorial optimization problems, defining cuts and evaluating solutions at nodes in trees in methods such as *branch and bound*, *branch and cut*, *branch and cut and price*, putting into place column or constraint generation procedures, etc. In fact, the general principle of the simplex method had already been roughly drawn up by Joseph Fourier in around 1820 [SCH 85]; he proposed methods for solving systems of linear inequalities and a minimization method of $\|A.x - b\|_\infty$ that had many points in common with the method invented by Dantzig.

"To reach the lower point of the vase rapidly, we erect in any point of the horizontal plane, for example at the origin of x and y, a vertical ordinate up to the highest point of intersection with the plane, that is to say that among all the points of intersection that we find on this vertical line, we choose the one the furthest away from the plane of x and y. Let m_1 be this point of intersection placed on the extreme plane. We go down this same plane from the point m_1, to the point m_2 of an edge of the polyhedron, and following this edge, we go down from the point m_2 to the vertex m_3 common to the three extreme planes. From the point m_3, we continue going down following a

Chapter written by Pierre TOLLA.

second edge, to a new vertex m_4 and we continue to apply the same procedure, always following the one of the two edges that leads to the least elevated vertex. In this way we arrive very quickly at the lowest point of the polyhedron."

The simplex method has allowed us to solve a good number of linear programs with up to about 50,000 variables, but it suffers from a few imperfections: as with most of the pivot methods, the numerical quality of the solutions is not guaranteed, even at the price of a more than significant increase in the solution time by using cumbersome procedures of the "*affine scaling*" type, or statistical calculations of the number of significant figures of the solution with procedures of the type "*iterative refinement*"; it does not allows us to solve certain linear programs in polynomial time, especially the example of Klee and Minty [KLE 72]. On the other hand, the size of the problems solved in this way is relatively small. In 1979, G. Khachiyan [KHA 79] proposed a reasonable polynomial algorithm; this, although ineffective at a practical level, was of great use in developing polynomial algorithms.

This is why N. Karmarkar [KAR 84] was the first to propose an interior point method that solved linear programs in polynomial time. These algorithms allow us to solve linear programs of a very large size (several million variables and several hundred thousand constraints) in a very small number of iterations; furthermore, they are for the most part self-correcting iterative methods and guarantee very high quality numerical solutions. Later in this chapter the reader will find an introduction to the basic methods as well as the algorithms most widely used in commercial software.

7.2. Dantzig's simplex method

7.2.1. *Standard linear programming and the main results*

Any real linear program can be put into the form:

$$(LP)\left\{\begin{array}{ll} Minimize\ z = c^t.x & \\ S.c. & \left.\begin{array}{c} A.x = b \\ x \geq 0_{\mathbb{R}^n} \end{array}\right\} (P) \end{array}\right.$$

with $c, x \in \mathbb{R}^n$, $b \in \mathbb{R}^m$, A a matrix (m, n) of full rank and (P) the set of points satisfying the constraints; it is a closed convex polyhedron because it is at the intersection of the closed half-spaces defined by the constraints. It is called the feasible point domain.

DEFINITION 7.1.– *Any point of P that cannot be expressed as a convex linear combination of two points of P is called an extreme point or vertex of P.*

PROPERTY 7.1.– If the linear program (*LP*) has a finite optimal solution, this is reached in at least one vertex of P.

DEFINITION 7.2.– *Let B be a square matrix (m, m) taken from A, N be the matrix of the columns of A that are not in B, x_B and x_N be the subvectors of x corresponding to B and N. That is:*

$$\bar{x} = \begin{pmatrix} \bar{x}_B \\ \bar{x}_N \end{pmatrix}$$

where $\bar{x}_B = B^{-1}.b$ and $\bar{x}_N = 0_{\mathbb{R}^{n-m}}$. If $\bar{x}_B \geq 0_{\mathbb{R}^m}, \bar{x}$ is a vertex of P. We call this a feasible base solution. B is the base matrix, x_B and x_N are the vectors of the base and non-base variables, respectively.

PROPERTY 7.2.– Let $\nabla f(x) = c$ be the gradient of the function f. The gradient of a linear function is a strictly increasing direction of this function. A direction making an acute angle with the gradient of a linear function is a strictly increasing direction of the function.

These two properties are used in the majority of optimization methods. The second, in particular, allows us to find improved feasible points when the direction of the gradient does not allow us, when the starting point is on the edge of the polyhedron P, to obtain a new better quality feasible point.

7.2.2. *Principle of the simplex method*

The principle of the simplex algorithm is a movement on the edge of the polyhedron P, from vertex to vertex, such that the value of the objective function strictly increases (see Figure 7.1). If this function is bounded and the number of vertices of the polyhedron is increased by C_n^m, the optimum is reached in a finite number of iterations.

7.2.3. *Putting the problem into canonical form*

Reusing the notation of the analytical definition of a vertex and assuming that we have a feasible base matrix B, the program (*LP*) is written by expressing c_B and c_N, the subvectors of c corresponding to B and N:

$$\begin{cases} \text{Min } f(x) = c_B^t.x_B + c_N^t.x_N \\ S.c. \;\; B.x_B + N.x_N = b \, (1) \\ x_B \geq 0_{\mathbb{R}^m}; \; x_N \geq 0_{\mathbb{R}^{n-m}} \, (2) \end{cases}$$

Let us transform the linear program into an equivalent problem in which the objective function depends only on the non-base variables: $(1) \Leftrightarrow x_B = B^{-1}.b - B^{-1}.N.x_N$. Therefore $z = c_B^t B^{-1}.b + \left[c_N^t - c_B^t.B^{-1}N\right].x_N$.

DEFINITION 7.3.– *$\Delta_N = c_N^t - c_B^t.B^{-1}.N$ is the vector of the marginal costs of the non-base variables. $\pi = c_B^t.B^{-1}$ is the vector of the simplex multipliers.*

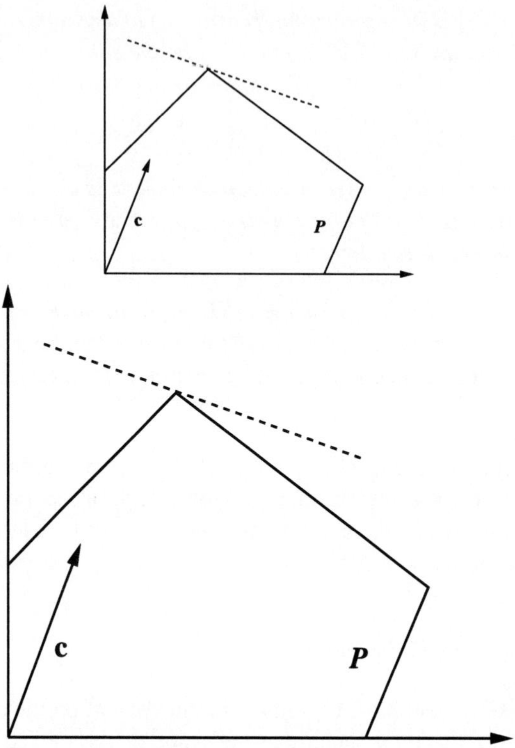

Figure 7.1. *The simplex method*

(*LP*) can therefore be written in the form:

$$(LPC)\left\{\begin{array}{c} \text{Min } f(x) = \Delta_N.x_N + c_B^t.B^{-1}.b \\ S.c. \quad x_B + Y.x_N = B^{-1}b \\ x_B \geq 0_{\mathbb{R}^m};\ x_N \geq 0_{\mathbb{R}^{n-m}} \end{array}\right.$$

where $Y = B^{-1}.N$.

7.2.4. *Stopping criterion, heuristics and pivoting*

At the kth iteration, (*LPC*) is written:

$$(LPC^{(k)})\left\{\begin{array}{c} Min\ f(x) = \ \Delta_{N^{(k)}}.x_{N^{(k)}} + c_{B^{(k)}}^t.\left[B^{(k)}\right]^{-1}.b \\ S.c. \quad x_{B^{(k)}} + Y^{(k)}.x_{N^{(k)}} = \left[B^{(k)}\right]^{-1}b \\ x_{B^{(k)}} \geq 0_{\mathbb{R}^m};\ x_{N^{(k)}} \geq 0_{\mathbb{R}^{n-m}} \end{array}\right.$$

where $Y^{(k)} = \left[B^{(k)}\right]^{-1}.N$

Let $\nu_1^{(k)}, \nu_2^{(k)}, ..., \nu_m^{(k)}$ be the indices of the base variables of this iteration; the base matrix $B^{(k)}$ is: $B^{(k)} = \left\{A^{\nu_1^{(k)}}, A^{\nu_2^{(k)}}, ..., A^{\nu_m^{(k)}}\right\}$, where A^j is the jth column of A.

Let $\bar{x}^{(k)} = \begin{pmatrix} \bar{x}_{B^{(k)}}^{(k)} \\ \bar{x}_{N^{(k)}}^{(k)} \end{pmatrix} = \begin{pmatrix} \left[B^{(k)}\right]^{-1}.b \\ 0_{\mathbb{R}^{n-m}} \end{pmatrix}$.

Dantzig's first criterion:

– if $\Delta_{N^{(k)}}^{(k)} \geq 0_{\mathbb{R}^{n-m}}$, $\bar{x}^{(k)}$ is an optimal solution of (*LP*);

– otherwise, let $\Delta_{e^{(k)}}^{(k)} = \underset{j \in J_{N^{(k)}}}{Min} \left\{\Delta_j^{(k)}\right\}$, where $J_{N^{(k)}}$ is the set of indices of the non-base variables; the non-base variable $x_{e^{(k)}}$ becomes a base variable.

Dantzig's second criterion (pivoting): $x_{B^{(k)}} + Y^{(k)} x_{N^{(k)}} = \bar{x}_{B^{(k)}} \Rightarrow Y^{(k)}.x_{N^{(k)}} \leq \bar{x}_{B^{(k)}}$. Leaving from the point $\bar{x}^{(k)}$ and increasing only the variable $x_{e^{(k)}}$, we obtain the following two situations:

– if $Y^{(k)} \leq 0_{\mathbb{R}^m}$, the variable $x_{e^{(k)}}$ can increase indefinitely: the optimal solution is not bounded;

– otherwise, $Y^{e^{(k)}}.x_{N^{(k)}} \leq \bar{x}_{B^{(k)}}$, which implies:

$$x_{e^{(k)}} \leq \underset{i=1,...,m}{Min} \left\{ \frac{\bar{x}_{\nu_i^{(k)}}^{(k)}}{Y_i^{e^{(k)}}} / Y_i^{e^{(k)}} > 0_{\mathbb{R}^m} \right\} = \frac{\bar{x}_{\nu_s^{(k)}}^{(k)}}{Y_s^{e^{(k)}}}$$

$x_{e^{(k)}}$ takes the value $\frac{\bar{x}_{\nu_s^{(k)}}^{(k)}}{Y_s^{e^{(k)}}}$ and goes into the base, while the base variable, $\bar{x}_{\nu_s^{(k)}}^{(k)}$ becomes zero and leaves the base.

This pivoting operation provides a new feasible base solution (a new vertex), strictly improving the objective function if $x_{e^{(k)}} > 0$. The movement has taken place on an edge of the polyhedron P and has met a second edge, from which we obtain a new feasible vertex. In this way we obtain a new feasible base matrix:

$$B^{(k+1)} = \left\{A^{\nu_1^{(k)}}, ..., A^{\nu_{s-1}^{(k)}}, A^{e^{(k)}}, A^{\nu_{s+1}^{(k)}} ..., A^{\nu_m^{(k)}}\right\}$$

which differs from the old feasible matrix by only one column:

$$B^{(k)} = \left\{A^{\nu_1^{(k)}}, ..., A^{\nu_{s-1}^{(k)}}, A^{\nu_s^{(k)}}, A^{\nu_{s+1}^{(k)}} ..., A^{\nu_m^{(k)}}\right\}$$

7.3. Duality

Let us express the "primal" program:

$$(LP)\left\{\begin{array}{ll} Minimize\ z = c^t.x & \\ S.c. \quad \begin{array}{c} A.x = b \\ x \geq 0_{\mathbb{R}^n} \end{array} \end{array}\right\} (P)$$

The dual program of (*LP*) is the program:

$$(LP^*)\left\{\begin{array}{ll} \text{Maximize } z' = b^t.y & \\ S.c. \quad \begin{array}{c} A^t.y + s = c \\ s \geq 0_{\mathbb{R}^n} \end{array} \end{array}\right\} (P^*)$$

where (P^*) is the domain of the feasible solutions of $(LP^*), y \in \mathbb{R}^n,\ \ s \in \mathbb{R}^n$. Let $x \in P$ and $(y, s) \in P^*$. We call the real $DG(x, y, s)$, defined by $DG\,(x, y, s) = c^t.x - b^t.y$, the reality gap.

THEOREM 7.1.– *If (LP) has a finite optimal solution, (LP*) has a finite optimal solution and the values of the objective functions of (LP) and (LP*) at the optimum case are equal. If (LP) has a non-bounded optimal solution, (LP*) does not have a solution. If (LP*) has a non-bounded optimal solution, (LP) does not have a solution.*

PROPERTY 7.3.– Let $x \in P$ and $(y, s) \in P^*$. Therefore: $DG\,(x, y, s) = c^t.x - b^t.y = x^t.s \geq 0$; $c^t.x$ is an upper bound of z' and $b^t.y$ is a lower bound of z if (*LP*) and (LP^*) have finite optimal solutions; if x and (y, s), respectively, are finite optimal solutions of (*LP*) and (LP^*), the duality gap $DG(x, y, s)$ is zero.

COMMENT 7.1.– $(LP^*)^* = (LP)$. At the optimum, the vector π of the simplex multipliers is an optimal solution of the dual program (LP^*).

THEOREM 7.2.– *A necessary and sufficient condition for $\bar{x}$ and $\bar{y}$ to be optimal solutions of (LP) and (LP*) is that they satisfy the orthogonality conditions:*

$$\left\{\begin{array}{l} \bar{y}^t.\,(A.\bar{x} - b) = 0 \\ \left(c - A^t.\bar{y} - \bar{s}\right)^t.\bar{x} = 0 \end{array}\right.$$

Proof. Since $\bar{x}$ and $\bar{y}$ are feasible solutions of (*LP*) and (LP^*), $\bar{x}$ satisfies $\bar{x} \geq 0_{\mathbb{R}^n}$, $A.\bar{x} = b$ and $\bar{y}$, $\bar{s} \geq 0_{\mathbb{R}^n}$, $c - A^t.\bar{y} - \bar{s} = 0_{\mathbb{R}^n}$. The conditions of orthogonality are therefore satisfied. ∎

7.4. Khachiyan's algorithm

PROPERTY 7.4.– Let:

$$(LP)\left\{\begin{array}{ll} \text{Maximize } z = c^t.x & \\ S.c. \quad \left\{\begin{array}{l} A.x \leq b \\ x \geq 0_{\mathbb{R}^n} \end{array}\right. \end{array}\right.$$

and its dual:

$$(LP^*)\left\{\begin{array}{l}\text{Maximize } z' = b^t.y\\ S.c.\ \left\{\begin{array}{l}A^t.y \geq c\\ y \geq 0_{\mathbb{R}^m}\end{array}\right.\end{array}\right.$$

The solution of (*LP*) is equivalent to that of the linear system (K):

$$\left\{\begin{array}{l}A.x \leq b\\ x \geq 0_{\mathbb{R}^n}\end{array}\right., \left\{\begin{array}{l}A^t.y \geq c\\ y \geq 0_{\mathbb{R}^m}\end{array}\right., \left\{c^t.x \geq b^t.y\right.$$

Khachiyan's algorithm solves, in polynomial time, this linear system with integer coefficients transformed into a strict system, that is $O\left(L^2 n^6\right)$ elementary operations.

$L = \sum\limits_{i=1}^{m}\sum\limits_{j=1}^{n}\log_2\left(|a_{ij}|+1\right) + \sum\limits_{i=1}^{m}\log_2\left(|b_i|+1\right) + \log_2 mn + 1$ is the length of the binary record of all the data.

DEFINITION 7.4.– *Let $x' \in \mathbb{R}^n$ and B be a matrix defined as positive:*

$$E = \left\{x/\left(x - x'^t\right).B^{-1}.\left(x - x'\right) \leq 1\right\}$$

is an ellipsoid with center x'. $\frac{1}{2}E_a = E \cap \left\{x/a^t.\left(x - x'\right) \leq 0\right\}$ is a half-ellipsoid. $E' = \left\{y/\left(y - y'\right)^t.\left[B'\right]^{-1}.\left(y - y'\right) \leq 1\right\}$ where $y' = x' - \frac{1}{n+1}\frac{B.a}{a^t.B.a}$ and:

$$B' = \frac{n^2}{n^2-1}\left[B - \frac{2}{n+1}\frac{\left(B.a\right)\left(B.a\right)^t}{a^t.B.a}\right]$$

is an ellipsoid with center y'.

The algorithm is expressed as follows:

– **Step 1:** Initialization: $x^{(0)} = 0_{\mathbb{R}^{(n)}};\ B = 2^{2L}.Id\ k := 0$ where *Id* is the identity matrix;

– **Step 2:** Stopping criterion:
- if $x^{(k)}$ satisfies the system of equations (K), $x^{(k)}$ is an optimal solution;
- if $k < 4\left(n+1\right)^2 L$, go to step 3;
- if $k \geq 4\left(n+1\right)^2 L$, there is no feasible solution;

– **Step 3:** Iteration: choose a constraint violated in $x^{(k)}$: $A_i^t.x^{(k)} \geq b_i$, where A_i^t is the ith line of A and state: $x^{(k+1)} = x^{(k)} - \frac{1}{n+1}\frac{B^{(k)}.A_i}{\sqrt{A_i^t.B^{(k)}.A_i}}$:

$$B^{(k+1)} = \frac{n^2}{n^2-1}\left[B^{(k)} - \frac{2}{n+1}\frac{\left(B^{(k)}.A_i\right).\left(B^{(k)}.A_i\right)^t}{A_i^t.B^{(k)}.A_i}\right]$$

and $k := k+1$. Go to step 2.

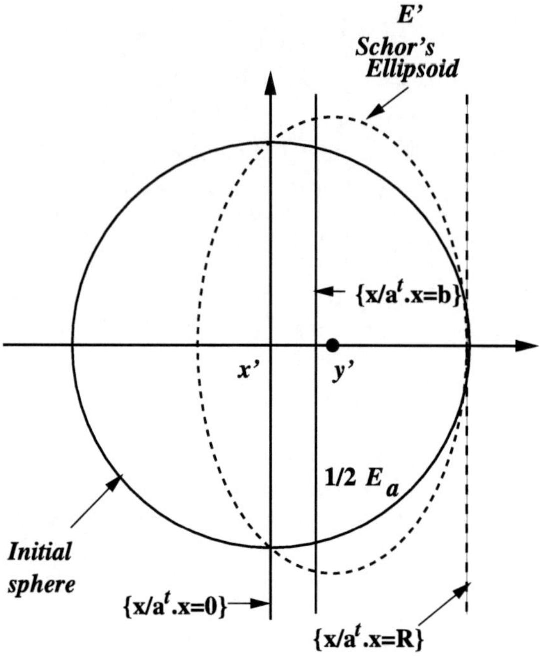

Figure 7.2. *Khachiyan's algorithm*

COMMENT 7.2.– The matrix A and the vector b have integer coefficients; the points inspected in the polyhedron P have rational components with bounded denominators.

Let us assume that $P \neq \emptyset$. There exists $S \subset P$, of a given volume and contained in a sphere with its center at the origin and of radius $R = 2^{2L}$. If the center of the sphere violates a constraint defined by $a^t.x > b$, we cut the ball with the help of a hyperplane parallel to this constraint and going through its center; only one of the two half-spheres defined in this way will contain feasible points.

Khachiyan suggests covering this with an ellipsoid of minimal size proposed by Shor; this procedure is repeated until an ellipsoid, the center of which satisfies the inequalities (K), is obtained. This method, supposed to solve any linear program in polynomial time, did not, for technical reasons, give the expected numerical results. However, its theoretical aspects have greatly contributed to the development of polynomial algorithms which are efficient in the solution of linear programs.

7.5. Interior methods

7.5.1. *Karmarkar's projective algorithm*

This presentation is drawn from N. Karmakar's research report (AT&T Bell Laboratories, Murray Hill, USA) and from a working document by J. Desrosiers. N. Karmarkar's method is particularly based on the works of Khachiyan and also on P. Huard's method of centers [HUA 63, HUA 67].

The linear program solved with the help of the initial version of Karmarkar's method is the following:

$$\begin{cases} \text{Minimize } z = c^t.x \\ S.C. \quad \left.\begin{array}{l} A.x = 0_{\mathbb{R}^n} \\ e^t.x = 1 \\ x \geq 0_{\mathbb{R}^n} \end{array}\right\} \text{where e} = \begin{pmatrix} 1 \\ \vdots \\ 1 \end{pmatrix} \end{cases}$$

Let $S = \left\{ x \in \mathbb{R}^n / x \geq 0_{\mathbb{R}^n}, \sum_{i=1}^{n} x_i = 1 \right\}$ and $\Omega = \{x \in \mathbb{R}^n / A.x = 0_{\mathbb{R}^n}\}$. The set of feasible points is defined by the constraints if $\Omega \cap S$.

By hypothesis, the optimal value of the objective function is zero and the optimal solution is non-trivial. Moreover, the center of the simplex $a_0 = \frac{1}{n}e$ must be feasible.

7.5.1.1. *The algorithm*

Initialization: $k := 0$; $x^{(0)} := a^0$; *ε the stopping test threshold is set.*
Iteration: *while* $c^t.x^{(k)} > \varepsilon$ *do*: $x^{(k+1)} = \varphi\left(x^{(k)}\right)$; $k := k+1$ where φ *is defined by the following series of operations:*
Let $D^{(k)} = diag\left(a_1^{(k)}, a_2^{(k)}, ..., a_n^{(k)}\right)$, *diagonal matrix.*

1) *Let* $B^{(k)} = \begin{bmatrix} A.D^{(k)} \\ e^t \end{bmatrix}$

2) *Calculate* $c_p^{(k)} = \left[Id - \left(B^{(k)}\right)^t . \left(B^{(k)} . \left(B^{(k)}\right)^t\right)^{-1} .B^{(k)}\right] .D^{(k)}.c$

3) *Calculate* $\hat{c}^{(k)} = \frac{c_p^{(k)}}{\left\| c_p^{(k)} \right\|}$ *unitary direction vector* $c_p^{(k)}$

4) *Calculate* $b' = x^{(k)} - \alpha r \hat{c}^{(k)}$*where* $r = \frac{1}{\sqrt{n(n-1)}}$ *is the radius of the biggest sphere included in the simplex and* $\alpha \in [0, 1]$

5) $x^{(k+1)} = \frac{D^{(k)}.b'}{e^t.D^{(k)}.b'}$

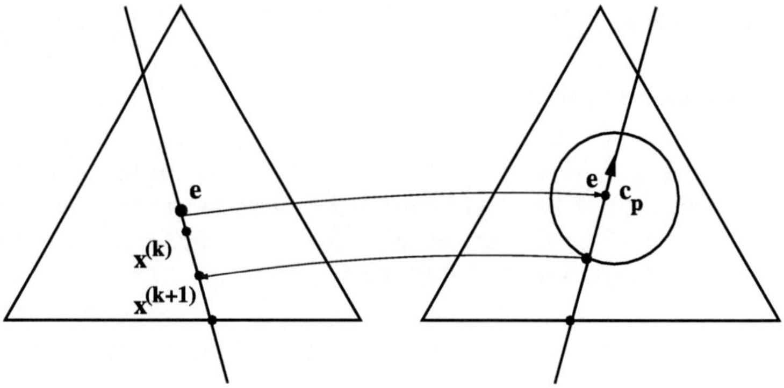

Figure 7.3. *Karmarkar's algorithm*

7.5.1.2. *Principle of the method*

The presentation of the method relies on the description of the above algorithm and Figure 7.3. In fact, to develop a clear description of Karmarkar's algorithm, we introduce, in this section, various underlying concepts and theoretical results.

7.5.1.2.1. Theoretical results

PROPOSITION 7.1.– *Let us consider the standard linear program:*

$$\begin{cases} Minimize\ z = c^t.x \\ S.C. \quad \begin{cases} A.x = b \\ x \geq 0_{\mathbb{R}^n} \end{cases} \end{cases} (P)$$

Let the polytope $P = \{x \in \mathbb{R}^n / A.x = b,\ x \geq 0_{\mathbb{R}^n}\}$, a_0 *be a point strictly interior to* P, E *be an ellipsoid with center* a_0 *contained in* P, E' *be an ellipsoid with center* a_0 *constructed from* E *with the help of a homothetic transformation with center* a_0, *and* ν *be a parameter sufficiently large that* $E' \supset P$. $E' = \nu E$. *We can therefore calculate, as a function of* ν, *a lower bound of the multiplicative factor of reduction of the objective function.*

Proof. Let $f_E, f_P, f_{E'}$ be the respective minimal values of f on E, P and E'; we have: $f(a_0) - f_E \leq f(a_0) - f_P \leq f(a_0) - f_{E'} = \nu [f(a_0) - f_E]$. We deduce from this: $\frac{f(a_0)-f_E}{f(a_0)-f_P} \geq \frac{1}{\nu}$, then $\frac{f_E-f_P}{f(a_0)-f_P} \leq 1 - \frac{1}{\nu}$. ∎

Note that the weaker the value of ν, the quicker the convergence of this algorithm.

We thus have the following properties:

1) The projective transformation T defined by $x' = T(x) = \frac{D^{-1}.x}{e^t.D^{-1}.x}$, where $a = (a_1, a_2, ..., a_n)$ is a point strictly interior to P, having positive components; $D = diag(x_1, ..., x_n)$; and T^{-1} is defined by $x = T^{-1}(x') = \frac{D.x'}{e^t.D.x}$. This can be written: $x' = \frac{x_i/a_i}{\sum\limits_{i=1}^{n} x_i/a_i}$, $i = 1, ..., n$ and $x'_{n+1} = 1 - \sum\limits_{i=1}^{n} x'_i = \frac{1}{1+\sum\limits_{i=1}^{n}(x_i/a_i)}$.

2) T^{-1}, the inverse of T, is defined by $T^{-1}(x') = x\ x_i = \frac{a_i x'_i}{1-\sum\limits_{i=1}^{n} x'_i}, \forall i \in \{1, ..., n\}$.

3) The strictly interior point a is projected by T to the center a_0 of the simplex.

4) The projective transformation T projects the affine space Ω onto another affine space Ω', as $a \in \Omega$, $a_0 \in \Omega'$.

5) Each face $F_i = \{x \in \mathbb{R}^n / x_i = 0\}$, i $\in \{1, ..., n\}$ of the positive orthant $P = \{x \in \mathbb{R}^n / x \geq 0_{\mathbb{R}^n}\}$ is projected onto a face of the simplex S of dimension n and the infinite points are projected onto the $(n+1)^{th}$ face of the simplex.

COROLLARY 7.1.– If $B(a_0, r)$ is the largest sphere with center a_0 included in S and $B(a_0, R)$ is the smallest sphere with center a_0 within the limits of S, then $\frac{R}{r} = n$ and $\nu = n$.

7.5.1.2.2. The working of the algorithm

Let us go back over Karmarkar's algorithm in a more detailed way by drawing from the discussion in the previous section.

First of all, T projects the current point a to the center of the simplex because $T(a) = \frac{1}{n+1}e = a_0$. We note that T transforms the linear objective function into a non-linear function. So, Karmarkar replaces the initial objective function by a "potential function" f that stays invariant through the projection T and further shows that if f strictly decreases, so does the linear objective function. f is defined by $f(x) = \sum\limits_{j=1}^{n} \log\left(\frac{|c^t.x|}{x_j}\right)$ and the transformed potential function, such that $f(x) = f'(T(x))$, is defined by $f'(y) = \sum\limits_{j=1}^{n} \log\left(\frac{|c'^t.y|}{y_j}\right) - \sum\limits_{j=1}^{n} \log(a_j)$, where $c' = D.c$. Furthermore, f, as a sort of barrier function, allows us to obtain well-centered points. The initial problem is transformed into a non-linear program:

$$(NLP) \left\{ \begin{array}{ll} \text{Minimize } f(x) = \sum\limits_{j=1}^{n} \log\left(\frac{c^t.x}{x_j}\right) & \\ S.C. & \left. \begin{array}{l} A.x = 0_{\mathbb{R}^n} \\ e^t.x = 1 \\ x \geq 0_{\mathbb{R}^n} \end{array} \right\} \end{array} \right.$$

in the same way as for the projected program:

$$(PNLP)\left\{\begin{array}{ll} \text{Minimize } f'(x) = \sum\limits_{j=1}^{n} \log\left(\frac{c^t.D_a.y}{y_j}\right) - \sum\limits_{j=1}^{n} \log(a_j) & \\ S.C. \quad \left.\begin{array}{l} A.D_a.y = 0_{\mathbb{R}^n} \\ e^t.y = 1 \\ y \geq 0_{\mathbb{R}^n} \end{array}\right\} & \end{array}\right.$$

By solving the program (*PNLP*), we advance the solution of the linear problem, because, as before, when f' strictly decreases, f does the same and it is possible to establish an upper bound for this difference. However, solution of (*PNLP*) is not straightforward; for this reason, we can establish only the improvement of the initial objective function in a measurable way.

To simplify the solving and to obtain a feasible solution, the set of feasible points is reduced by taking the intersection of the simplex with a sphere included in the simplex, smaller than the largest sphere included, and with the center of the simplex as its center. The problem obtained will be:

$$(BPNLP)\left\{\begin{array}{ll} Minimize\ f'(x) = \sum\limits_{j=1}^{n} \log\left(\frac{c^t.D_a.y}{y_j}\right) - \sum\limits_{j=1}^{n} \log(a_j) & \\ S.C. \quad \left.\begin{array}{l} A.D_a.y = 0_{\mathbb{R}^n} \\ e^t.y = 1 \\ y \in B(a', \alpha r),\ 0 < \alpha < 1 \end{array}\right\} & \end{array}\right.$$

where r is the radius of the largest sphere included in the simplex: $B(a', r)$.

With the aim of obtaining a strictly improved solution, we must have an efficient search direction; but the gradient of the objective function is exterior to the set of feasible points. We therefore use a classical orthogonal projection matrix of the gradient of the objective function of (*BPNLP*) linearized:

$$c_p^{(k)} = \left[Id - \left(B^{(k)}\right)^t.\left(B^{(k)}.\left(B^{(k)}\right)^t\right)^{-1}.B^{(k)}\right].D^{(k)}.c$$

We can then carry out a monodimensional search from $x^{(k)}$ in the direction of $c_p^{(k)}$ where, from its formulation in terms of normed vectors, $\hat{c}^{(k)}$: $b' = x^{(k)} - \alpha r \hat{c}^{(k)}$ with $0 < \alpha < 1$.

The obtained point b' is projected onto the initial domain Ω using the inverse projection T^{-1}.

The algorithm stops when $\frac{c^t.x^{(k)}}{c^t.a_0} \leq 2^{-O(L)}$, where L is the length of the binary record of the data.

Karmarkar showed that the number of iterations of the algorithm for a fixed step α of a monodimensional search in the included projected sphere is $O\,(nL)$, where L is the length of the binary record of the data. This has been proved in particular for $\alpha = 0.25$. As the cost of each iteration depends essentially on the calculation of the orthogonal projection matrix on the included sphere, polynomial if we use a Cholesky method, we conclude from this that Karmarkar's algorithm solves a linear program in polynomial time.

7.5.2. *Primal–dual methods and corrective predictive methods*

7.5.2.1. *Theoretical tools*

Let a point $x^{(0)} \in (P)$, the set of feasible points of (P). We say that d is a feasible direction if $\exists \lambda \in \mathbb{R}^* : x^{\lambda} = x^{(0)} + \lambda d \in (P)$. Let A be the matrix of the constraints of (P). The set of feasible directions of (P) is the kernel $N(A)$ of A where $N(A) = \{d \in \mathbb{R}^n / A.d = 0_{\mathbb{R}^m}\}$. Indeed, let $x^{\lambda} = x^{(0)} + \lambda d \in (P)$, where $x^{(0)} \in (P)$ and $\lambda \neq 0$. Then, $A.x^{\lambda} = A.x^{(0)} + \lambda A.d = b + \lambda A.d = b$. Therefore $A.d = 0_{\mathbb{R}^n}$.

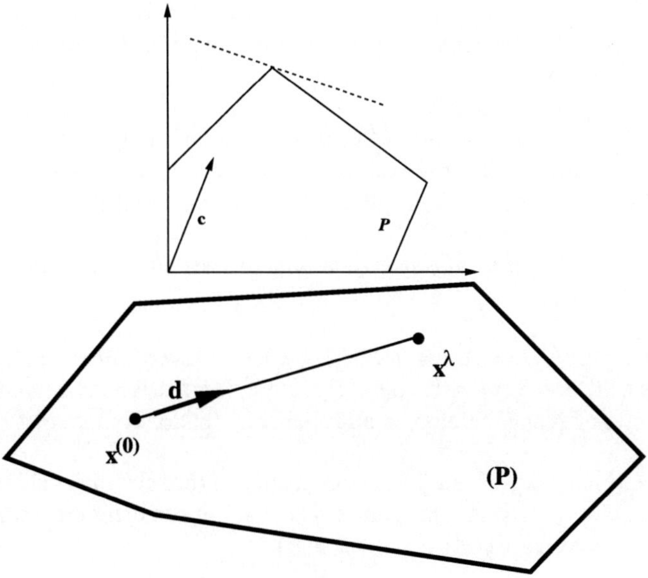

Figure 7.4. *Feasible direction*

The function $\tilde{g}_{\mu}$ is defined by: $\tilde{g}_{\mu}(x) = c^t.x - \mu \sum_{j=1}^{n} Log\,(x_j)$, $\mu \in \mathbb{R}_*^+$; this is called a barrier function.

Does the barrier function have an optimum? If this optimum exists, is it unique?

We first notice that $\tilde{g}_\mu$ is differentiable on $\mathbb{R}^n$ and:

$$\nabla\tilde{g}_\mu(x) = \begin{pmatrix} c_1 - \frac{\mu}{x_1} \\ \vdots \\ c_n - \frac{\mu}{x_n} \end{pmatrix} = c - \mu X^{-1}.e$$

where $X^{-1} = \begin{pmatrix} 1/x_1 & & \\ & \ddots & \\ 0 & & 1/x_n \end{pmatrix}$ and $e = \begin{pmatrix} 1 \\ \vdots \\ 1 \end{pmatrix}$.

The Hessian matrix of g_μ will be:

$$\nabla^2\tilde{g}_\mu(x) = \mu \begin{pmatrix} 1/x_1^2 & & 0 \\ & \ddots & \\ 0 & & 1/x_n^2 \end{pmatrix} = \mu X^{-2}$$

As x has positive components, $\nabla^2\tilde{g}_\mu(x)$ is a positively defined matrix (all its characteristic values are positive); therefore $\tilde{g}_\mu$ is strictly convex on $\mathbb{R}^n_{+*}$. We deduce from this that $\tilde{g}_\mu$ has a minimum if $x \in P^+ = \{x/\ A.x = b,\ x > 0_{\mathbb{R}^n}\}$ exists, such that $c - \mu X^{-1}.e \perp$ N(A), where $N(A)$ is the kernel of A. Indeed, this signifies that $\nabla\tilde{g}_\mu(x) = c - \mu X^{-1}.e$ is orthogonal in all the feasible directions of P; we cannot therefore further improve the value of $\tilde{g}_\mu$, therefore x is optimal.

LEMMA 7.1.– *The space defined by the linear combinations of the rows of A is orthogonal to $N(A)$.*

Indeed, $d \in N(A) \Leftrightarrow A.d = 0$. Let A_i be the ith row of A for $i = 1, \ldots, m$. Then $\forall d \in N(A)$, $A_i.d = 0$, $\forall i = 1, ..., m$. Therefore, any feasible direction is orthogonal to all the rows of A and therefore to all linear combinations of the rows of A.

If x is optimal, $\nabla\tilde{g}_\mu(x)$ is a linear combination of the rows of A. This follows from the fact that $\nabla\tilde{g}_\mu(x) \perp N(A)$ and from lemma 7.1. We can therefore write: $\nabla\tilde{g}_\mu(x) = c - \mu X^{-1}.e = A^t.y\ (= y_1A_1^t + ... + y_mA_m^t)$.

Let us write $s = \mu X^{-1}.e$. The following property follows on from this.

PROPERTY 7.5.– $\tilde{g}_\mu$ has a minimum if and only if x,y *and* s exist such that:

$$\left\{\begin{array}{l} A.x = b,\ \mathrm{x} > 0 \\ \mathrm{A}^t.y + s = c,\ \mathrm{s} > 0 \\ \mathrm{X.s} = \mu\mathrm{e} \end{array}\right\} (KKT)$$

The (KKT) system is none other than the optimality conditions for Karush, Kuhn and Tucker's system for the logarithmic barrier function minimization problem under the hypothesis of satisfying the constraints of the primal and dual programs.

We can replace $x > 0$ and $s > 0$ by $x \geq 0$ and $s \geq 0$, because the expression of the logarithmic barrier function must be strictly positive. The first two equations are therefore constraints of both the primal and the dual.

7.5.2.1.1. Condition of existence of an interior point

If the (KKT) system has a solution for $\mu > 0$, then P contains $x > 0$ and D contains (y, s), where the gap variable s is positive. The reverse is true from theorem 7.3.

THEOREM 7.3.– *Let $\mu > 0$; the following assertions are equivalent:*

1) P and D contain a vector with strictly positive components;

2) there exists a unique x minimizing $\tilde{g}_\mu$ on P^+;

3) the (KKT) system has one and only one solution.

7.5.2.1.2. Condition of existence of an interior point

This concerns condition (1), which was discussed above: P and D contain a vector with strictly positive components.

COMMENT 7.3.– The unicity of the optimal point comes from the strict convexity of $\tilde{g}_\mu$.

7.5.2.1.3. Lagrangian and (*KKT*) conditions

Let us consider the non-linear program (LP_μ):

$$(LP_\mu) \begin{cases} \text{Minimize } \tilde{g}_\mu(x) = c^t.x - \mu \sum\limits_{j=1}^{n} Log\,(x_j) \\ \text{A.x} = \text{b} \end{cases}$$

(P) is the "limit" when $\mu \to 0$ of (LP_μ) and its solution is the limit when $\mu \to 0$ of the solution $x(\mu)$ of (LP_μ). However, when μ tends towards 0, it becomes technically difficult to generate points close to the frontiers of P.

According to comments made in the preceding sections, the concept of (LP_μ) is interesting because it contributes to the construction of different solution methods of (P) using the idea of an interior point on the model of a generic algorithm.

7.5.2.1.4. The Lagrangian of $\tilde{g}_\mu$ and conditions of optimality of the first order

The Lagrangian $L_\mu(x, \lambda)$ associated with (LP_μ) is expressed as:

$$L_\mu(x, \lambda) = \tilde{g}_\mu(x) - \lambda\,(A.x - b) = c^t.x - \mu \sum_{j=1}^{n} Log\,(x_j) - \lambda\,(A.x - b)$$

The positivity constraints of the variables are absent because they are implicit in the definition of $\tilde{g}_\mu$.

Necessary and sufficient conditions of optimality are expressed as:

$$\begin{cases} \nabla_x L_\mu(x,\lambda) = 0 \Leftrightarrow c - \mu X^{-1}.e - A^t.\lambda = 0 \\ \nabla_\lambda L_\mu(x,\lambda) = 0 \Leftrightarrow \text{A.x-b} = 0 \end{cases}$$

because $\tilde{g}_\mu$ is strictly convex.

By analogy with the dual program (D), we state $y = \lambda$ and $s = \mu X^{-1}.e$. In this way we obtain Karush, Kuhn and Tucker's conditions:

$$\begin{cases} (1) & s - \mu X^{-1}.e = 0 \\ (2) & c - A^t.y - s = 0 \\ (3) & A.x - b = 0 \end{cases}$$

where $Xs = \mu e$, $c - A^t.y - s = 0$, and $A.x - b = 0$. The (*KKT*) conditions are necessary conditions of optimality for (*LP*$_\mu$) depending on the parameter μ; their solution will allow us to find a point $(x(\mu), y(\mu), s(\mu))$ that satisfies them, and, by controlled decreasing of μ towards 0, to find an optimal solution of (P).

7.5.2.1.5. Analytical centers, μ-centers and central trajectories

DEFINITION 7.5.– *Let T be a non-empty bounded set, the intersection of an affine space of $\mathbb{R}^n$ with the non-negative orthant of $\mathbb{R}^n$. We express as $\sigma(T)$, and call the support of T, the subset of indices $\{1,\ldots,n\}$ defined by $\sigma(T) = \{i : \exists x \in T, x_i > 0\}$. The analytical center of T is 0 if $\sigma(T) = \emptyset$, otherwise the analytical center of T is the vector that maximizes $\prod_{i\in\sigma(T)} x_i$ for $x \in T$.*

The condition of existence of an interior point is independent of the parameter μ: P and D each contain a point with positive components. According to the preceding theory, this condition is equivalent to the existence of a unique minimum of $\tilde{g}_\mu$ for all $\mu > 0$. This therefore guarantees that the (KKT) system has only one solution for each value of μ.

DEFINITION 7.6.– *We call the μ-center of P, $x(\mu)$, and the μ-center of D, $(y(\mu),s(\mu))$, the triplet $(x(\mu),y(\mu),s(\mu))$ satisfying the (KKT) conditions with fixed μ. The set of all the μ-centers $x(\mu)$ of P is called the central trajectory of P. The set of all the μ-centers $(y(\mu),s(\mu))$ of D is called the central trajectory of D.*

Minimization of $\tilde{g}_\mu$ provides a vector $x(\mu)$ of which all the components are strictly positive. When we make μ tend towards 0, the dependent term of μ gradually loses its importance as μ decreases; this allows us to get closer to the edge of P in proximity to the optimum.

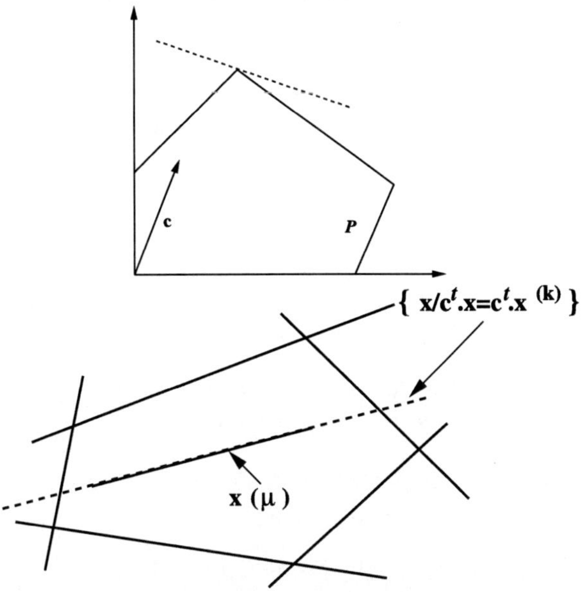

Figure 7.5. *Analytical center, solution of the (KKT) conditions*

We also note that $\sum_{j=1}^{n} Log\,(x_j) = Log\left(\prod_{j=1}^{n} x_j\right)$. Minimization of the term $-\sum_{j=1}^{n} Log\,(x_j)$ is thus "equivalent" to maximization of $\prod_{j=1}^{n} x_j$ with $x_j > 0$, $\forall j \in \{1, 2, ..., n\}$. We deduce from this that the μ-center of P is none other than the analytical center of the polyhedron obtained by cutting P with the hyperplane defined by $\{x/c^t.x = c^t.x^{(k)}\}$.

We can also see that what is valid for the positivity constraints of the variables is also valid for the inequality constraints transformed into equality constraints by the addition or removal of positive gap variables; there is therefore also recentering in relation to these types of constraints.

The preceding remarks are valid when pertaining to D and its μ-center $(y(\mu), s(\mu))$.

PROPERTY 7.6.– The duality gap is equivalent to: $x^t.s = n\mu$.

Proof: $s = \mu X^{-1}.e = \mu \begin{pmatrix} 1/x_1 & \cdots & 0 \\ \vdots & \ddots & \vdots \\ 0 & \cdots & 1/x_n \end{pmatrix} \begin{pmatrix} 1 \\ \vdots \\ 1 \end{pmatrix} = \begin{pmatrix} \mu/x_1 \\ \vdots \\ \mu/x_n \end{pmatrix}$. As a consequence, $x^t.s = (x_1, ..., x_n) . \begin{pmatrix} \mu/x_1 \\ \vdots \\ \mu/x_n \end{pmatrix} = n\mu$.

We note that the duality gap tends towards 0 with μ.

Along the primal central trajectory, $c^t.x(\mu)$ is strictly decreasing, and along the dual central trajectory, $b^t.y(\mu)$ is strictly increasing.

7.5.2.2. *Construction of a generic solution method*

It is clear that the solution of the system (S):

$$\left\{ \begin{array}{ll} (1) & X.s = \mu e \\ (2) & Ax - b = 0 \\ (3) & A^t.y + s = c \end{array} \right\}$$

is of prime importance in the execution of an algorithm of a linear program solution using a barrier function procedure. Since the equations (1) are not linear, it is relatively awkward to find a solution to them. The Newton–Raphson method is commonly used to this end.

7.5.2.3. *Principle of the Newton–Raphson method*

Let $f = \begin{pmatrix} f_1 \\ \vdots \\ f_n \end{pmatrix}$ where f_i of $\mathbb{R}^n$ in $\mathbb{R}$ of the class C_2.

We initially wish to solve the system $f(x) = 0$. For this, we develop each first-order function to the proximity of an initial point $x^{(0)}$: for each $i = 1, \ldots, n$:

$$f_i(x) = f_i\left(x^{(0)}\right) + \left(x - x^{(0)}\right)^t .\nabla f\left(x^{(0)}\right) + o\left(x - x^{(0)}\right)$$

By removing the corrective term, we obtain a linearization of $f_i(x)$ and we can thus use an approximate solution method.

Let us state $\Delta_1 = \left(x_1 - x_1^{(0)}\right), ..., \Delta_n = \left(x_n - x_n^{(0)}\right)$. We will thus have for any $i = 1, \ldots, n$:

$$f_i(x) = \frac{\partial f_i\left(x^{(0)}\right)}{\partial x_1}\Delta_1 + \cdots + \frac{\partial f_i\left(x^{(0)}\right)}{\partial x_n}\Delta_n + f_i\left(x^{(0)}\right)$$

The matrix:

$$J\left(x^{(0)}\right) = \left(\frac{\partial f_i\left(x^{(0)}\right)}{\partial x_j}\right)_{\substack{i=1,...,n \\ j=1,...,n}}$$

is called a Jacobian matrix in $x^{(0)}$ and its determinant is called the Jacobian of f in $x^{(0)}$.

We next solve the linear system:

$$\begin{cases} \frac{\partial f_1\left(x^{(0)}\right)}{\partial x_1}\Delta_1 + \cdots + \frac{\partial f_1\left(x^{(0)}\right)}{\partial x_n}\Delta_n + f_1\left(x^{(0)}\right) = 0 \\ \ldots \\ \frac{\partial f_n\left(x^{(0)}\right)}{\partial x_1}\Delta_1 + \cdots + \frac{\partial f_n\left(x^{(0)}\right)}{\partial x_n}\Delta_n + f_n\left(x^{(0)}\right) = 0 \end{cases}$$

The iterate $x^{(1)}$ will be $x^{(1)} = x^{(0)} + \Delta^{(0)}$ where $\Delta^{(0)} = \begin{pmatrix} \Delta_1 \\ \vdots \\ \Delta_n \end{pmatrix}$.

The following iterates are calculated using an analog process that is stopped when the residual vector $\Delta^{(k)}$ is normally less than a threshold set in advance.

7.5.2.4. *Solution*

Let the system to solve be:

$$(S) \quad \left\{ \begin{array}{ll} (1) & X.s = \mu e \\ (2) & Ax - b = 0 \\ (3) & A^t.y + s = c \end{array} \right\}$$

The Jacobian matrix associated with (S) is: $\begin{pmatrix} S & 0 & X \\ A & 0 & 0 \\ 0 & A^t & Id \end{pmatrix}$ where:

$$S = \begin{pmatrix} s_1 & & 0 \\ & \ddots & \\ 0 & & s_n \end{pmatrix}, \quad X = \begin{pmatrix} x_1 & & 0 \\ & \ddots & \\ 0 & & x_n \end{pmatrix}$$

and *Id* is the identity matrix.

Let $(\bar{x}, \bar{y}, \bar{s})$ be the starting point of the procedure. The constant part of the system is:

$$\begin{pmatrix} \bar{X}.\bar{s} - \bar{\mu}e \\ A\bar{x} - b \\ A^t\bar{y} + \bar{s} - c \end{pmatrix}$$

The system to solve will thus be:

$$\begin{pmatrix} \bar{S} & 0 & \bar{X} \\ A & 0 & 0 \\ 0 & A^t & Id \end{pmatrix} \begin{pmatrix} \Delta x \\ \Delta y \\ \Delta s \end{pmatrix} + \begin{pmatrix} \bar{X}\bar{s} - \bar{\mu}e \\ A.\bar{x} - b \\ A^t.\bar{y} + \bar{s} - c \end{pmatrix} = 0$$

This is equivalent to the system:

$$\begin{cases} (1) & \bar{S}.\Delta x + \bar{X}.\Delta s = -\left(\bar{X}.\bar{s} - \mu e\right) \\ (2) & A.\Delta \mathrm{x} \ = -\left(A.\bar{x} - b\right) \\ (3) & A^t.\Delta y + \Delta s = -\left(A^t\bar{y} + \bar{s} - c\right) \end{cases}$$

Let $r_b = A\bar{x} - b$ and $r_c = A^t.\bar{y} + \bar{s} - c$ be the residuals corresponding to the constraints of (P) and (D). The system becomes:

$$\begin{cases} (1) & \bar{S}.\Delta x + \bar{X}.\Delta s = -\left(\bar{X}.\bar{s} - \mu e\right) \\ (2) & A.\Delta x = -r_b \\ (3) & A^t.\Delta y + \Delta s = -r_c \end{cases}$$

By multiplying (1) by $\bar{S}^{-1}$, we obtain: $\Delta x = -\bar{S}^{-1}.\bar{X}.\bar{s} + \mu\bar{S}^{-1}.e - \bar{S}^{-1}\bar{X}.\Delta s$. $\bar{X}$ and $\bar{S}^{-1}$ being diagonal matrices, we can permutate them and, knowing that $\bar{S}^{-1}.s = e$, we obtain: $(1') : \Delta x = -\bar{X}.e + \mu\bar{S}^{-1}.e - \bar{X}.\bar{S}^{-1}.\Delta s = -\bar{x} + \mu\bar{S}^{-1}.e - \bar{X}.\bar{S}^{-1}.\Delta s$. By multiplying $(1')$ by A and replacing $A.\Delta x$ with $-r_b$, we obtain: $-A.\bar{x} + \mu A.\bar{S}^{-1}.e - A.\bar{X}.\bar{S}^{-1}.\Delta s = -r_b$, where $\mu A.\bar{S}^{-1}.e - A.\bar{X}.\bar{S}^{-1}.\Delta s = A.\bar{x} - r_b$. But *(3)* $\Rightarrow \Delta.s = -r_c - A^t.\Delta y$. Therefore $\mu A.\bar{S}^{-1}.e - A.\bar{X}.\bar{S}^{-1}.\left(-r_c - A^t.\Delta y\right) = A.\bar{x} - r_b$, where $\mu A.\bar{S}^{-1}.e + A.\bar{X}.\bar{S}^{-1}.r_c + A.\bar{X}.\bar{S}^{-1}.A^t.\Delta y = A.\bar{x} - r_b$. Thus:

$$\Delta y = \left(A.\bar{X}.\bar{S}^{-1}.A^t\right)^{-1}.\left[A\bar{x} - \mu A.\bar{S}^{-1}.e - A.\bar{X}.\bar{S}^{-1}r_c - r_b\right]$$

Let us replace r_b and r_c by their expressions in these three formulas:

$$\begin{aligned} \Delta y &= \left(A.\bar{X}.\bar{S}^{-1}.A^t\right)^{-1}.\left[A\bar{x} - \mu A.\bar{S}^{-1}.e - A.\bar{X}.\bar{S}^{-1}\left(A^t.\bar{y} + \bar{s} - c\right)\right. \\ &\quad \left. - \left(A.\bar{x} - b\right)\right] \\ &= \left(A.\bar{X}.\bar{S}^{-1}.A^t\right)^{-1}.\left[b - \mu A.\bar{S}^{-1}.e + A\bar{X}.\bar{S}^{-1}.c - \right. \\ &\quad \left. A.\bar{X}.\bar{S}^{-1}.A^t.\bar{y} - A.\bar{X}.\bar{S}^{-1}.\bar{s}\right] \end{aligned}$$

Since $\bar{S}^{-1}.\bar{s} = e$, and $\bar{X}.e = \bar{x}$, Δy is expressed as:

$$\Delta y = -\bar{y} + \left[A.\bar{X}.\bar{S}^{-1}.A^t\right]^{-1}.\left[b - A\bar{x} - \mu A.\bar{S}^{-1}.e + A.\bar{X}.\bar{S}^{-1}.c\right]$$

In the same way, $\Delta.s = -\bar{s}+c-A^t.\left[\bar{y}+\Delta y\right]$ and $\Delta x = -\bar{x}+\mu\bar{S}^{-1}.e-\bar{X}.\bar{S}^{-1}.\Delta s$. Thus we can establish $\bar{\bar{x}}, \bar{\bar{y}}, \bar{\bar{s}}$ by:

$$\begin{cases} \bar{\bar{x}} = \bar{x} + \Delta x \\ \bar{\bar{y}} = \bar{y} + \Delta y \\ \bar{\bar{s}} = \bar{s} + \Delta z \end{cases}$$

In reality, one iteration of this method is not sufficient to find a good approximation of the solution, which is why we use a versatile and low-cost Newton–Raphson method that calculates a good estimate of the solution at less cost.

7.5.2.5. *Versatile Newton–Raphson method*

We consider that the directions Δx, Δy and Δs are directions for improving the solution of the system (S); obviously we must satisfy the positivity constraint of the variables.

We will calculate a maximal movement step in the primal direction Δx, and dual directions Δy and Δs, which maintains the positivity property of the variables:

$$\begin{cases} \alpha_P = Max\left\{\alpha/\left(\bar{x}+\alpha\Delta x\right)_j \geq 0, \forall j=1,...,n\right\} \\ \alpha_D = Max\left\{\alpha/\left(\bar{s}+\alpha\Delta s\right)_j \geq 0, \forall j=1,...,n\right\} \end{cases}$$

7.5.2.6. *Generic interior point algorithm model*

Step 0: *Initialization*: let $x^{(0)}, y^{(0)}, s^{(0)}$ be the starting points, $\mu^{(0)}$ be reasonably large, γ be the coefficient of reduction of μ (≈ 0.95 or 0.99), and $k=0$ be the number of iterations. Let $\varepsilon_s, \varepsilon_P, \varepsilon_D$ be the thresholds of primal and dual feasibility ($\approx$ 10-8).

Step 1: *Calculate directions for improving the solution of (S)*: dual directions:

$$\begin{aligned} \Delta_y^{(k)} &= -y^{(k)} + \left(AX_k.S_k^{-1}.A^t\right)^{-1}\left[b - A.x^{(k)} - \mu^{(k)}A.S_k^{-1}.e \right. \\ &\quad \left. + A.X_k.S_k^{-1}.c\right] \\ \Delta_s^{(k)} &= -s^{(k)} + c - A^t\left(y^{(k)} + \Delta_y^{(k)}\right) \end{aligned}$$

primal direction: $\Delta_x^{(k)} = -x^{(k)} + \mu^{(k)}S_k^{-1}e - X_kS_k^{-1}\Delta_s^{(k)}$.

Step 2: *Calculate maximal movement steps*:

$$\begin{aligned} \alpha_P &= Max\left\{\alpha/\left(\bar{x}+\alpha\Delta x\right)_j \geq 0, \forall j=1,...,n\right\} \\ \alpha_D &= Max\left\{\alpha/\left(\bar{s}+\alpha\Delta s\right)_j \geq 0, \forall j=1,...,n\right\}. \end{aligned}$$

Step 3: *Calculate new iterates*:

$$\begin{aligned} x^{(k+1)} &= x\left(^{k}\right) + \gamma\alpha_P\Delta_x^{(k)} \\ y^{(k+)1} &= y^{(k)} + \gamma\alpha_D\Delta_y^{(k)} \\ s^{(k+1)} &= s^{(k)} + \gamma\alpha_D\Delta_s^{(k)}. \end{aligned}$$

Step 4: *Stopping tests*: if

$$\begin{aligned} \frac{\left|c^t.x^{(k+1)} - b^t y^{(k+1)}\right|}{Max\left\{1, c^t x^{(k+1)}\right\}} &\leq \varepsilon_s \\ \frac{\left\|A.x^{(k+1)} - b\right\|_2}{Max\left\{1, \|b\|_2\right\}} &\leq \varepsilon_P \\ \frac{\left\|c - A^t y^{k+1}\right\|_2}{Max\left\{1, \|c\|_2\right\}} &\leq \varepsilon_D \end{aligned}$$

then $x^{(k+1)}$ and $y^{(k+1)}$ can be considered as ε-optimal solutions. We can stop the process and go to Step 6. Otherwise go to Step 5.

Step 5: Establish $\mu^{(k+1)} < \mu^{(k)}$; state $k := k + 1$ and go to Step 1.

Step 6: END

7.5.2.7. *Performance of the generic algorithm*

7.5.2.7.1. Property of the versatile Newton–Raphson method

Because of the characteristics of the system to be solved, the Newton–Raphson method is of super-linear convergence. If we use a procedure with large steps α_P and α_D, then as long as we stay in the convergence domain of the iterative formula, this method gives a very good approximate solution of the system (S) within a very small number of iterations; for example, with steps in the order of 0.5, this method generally converges in less than six iterations with a precision in the order of 10^{-16}.

7.5.2.7.2. Property of the directions Δx, Δy and Δs

$$\begin{aligned}\Delta y &= \left(A.\bar{X}.\bar{S}^{-1}.A^t\right)^{-1}.A\bar{x} - \mu\left(A.\bar{X}.\bar{S}^{-1}.A^t\right)^{-1}A.\bar{S}^{-1}.e \\ &\quad + \left(A.\bar{X}.\bar{S}^{-1}.A^t\right)^{-1}\left[A.\bar{X}.\bar{S}^{-1}r_c - A.\bar{x} + r_b\right]\end{aligned}$$

where $\Delta y = \left(A.\bar{X}.\bar{S}^{-1}.A^t\right)^{-1}\left[A.\bar{X}.\bar{S}^{-1}r_c + r_b\right] - \mu\left(A.\bar{X}.\bar{S}^{-1}.A^t\right)^{-1}A.\bar{S}^{-1}.e.$ The first term helps to reduce the residuals r_b and r_c and therefore helps us to obtain a solution satisfying both the primal and the dual constraints. The second term weighted by the parameter μ makes the current point advance towards the μ-center of the μ-optimal cut of the polyhedron of the dual feasible points D.

7.5.2.7.3. Complexity of each iteration

The calculation of the matrix $\left(A.\bar{X}.\bar{S}^{-1}.A^t\right)^{-1}$ is obviously the most costly part of each iteration. It needs $m^2n + 2m^2$ multiplications to calculate the matrix to be inverted and $O(m^3)$ multiplications for its inversion or calculation of a triangular decomposition. It is therefore in $O(m^2n + m^3)$. The use of a triangular decomposition of the Cholesky type (since this matrix is symmetrical) is particularly suitable because it allows us to to take into account the scattered nature of the constraint matrices if most of the linear programs are large and therefore to make some substantial savings in memory space and calculations.

7.5.2.7.4. Convergence of the generic algorithm

We can establish the convergence properties of the algorithm and the number of iterations by studying the convergence towards 0 of $\left(x^{(k)}\right)^t.s^{(k)}$. Because the algorithm is of an iterative type, we can study its convergence type in the same way as for non-linear optimization algorithms:

– super-linearity: $\lim\limits_{k\to\infty} \dfrac{\left(x^{(k+1)}\right)^t.s^{(k+1)}}{\left(x^k\right)^t.s^k} = 0;$

– quadraticity: $\limsup\limits_{k\to\infty} \dfrac{\left(x^{(k+1)}\right)^t.s^{(k+1)}}{\left[\left(x^{(k+1)}\right)^t.s^{k(+1)}\right]^2} = M$, where M is a real finite number.

In reality, we limit the number of iterations by setting the precision with the help of thresholds. For example, in a quadratic convergence method, the number of significant figures of the solution doubles with each iteration, on the condition of being near enough to the solution that is sought. If we want eight exact significant figures, it is necessary that $\left(x^{(k)}\right)^t.s^{(k)} < 10^{-8}$.

For a quadratic convergence method, four to five iterations may be sufficient to obtain this result as long as we have a point that is "sufficiently" close to the optimal solution. The proofs of convergence cited below are based on the case of a primal–dual method such as the one introduced here.

7.5.2.7.5. Kojima, Mizuno and Yoshise's proof of convergence [KOJ 89]

We assume that $x^{(k)}, y^{(k)}$ and $s^{(k)}$ are feasible $\forall k > 0$. By making a step α in the directions $\Delta^{(k)}x$, $\Delta^{(k)}y$ and $\Delta^{(k)}s$ of the model, the duality gap of the $(k+1)^{th}$ iteration is:

$$\left(x^{(k+1)}\right)^t.s^{(k+1)} = \left(x^{(k)}\right)^t.s^{(k)} - \left[\left(x^{(k)}\right)^t.s^{(k)} - \eta\mu^{(k+1)}\right]\alpha$$

If the value of α is sufficiently small and the chosen $\mu^{(k+1)}$ is strictly less than $\mu^{(k)}$, we have: $\left(x^{(k+1)}\right)^t.s^{(k+1)} < \left(x^{(k)}\right)^t.s^{(k)}$. The series of duality gaps converges towards 0 in a finite number of iterations. By taking $\mu^{(k+1)} = \frac{\left(x^{(k)}\right)^t.s^{(k)}}{n}$ and α sufficiently small, Kojima *et al.* show that the algorithm is in $O\left(\sqrt{n}L\right)$ iterations, therefore polynomial in time.

A proof of convergence that does not take into account the feasibility hypothesis was proposed by Kojima, Meggido and Mizuno in 1991 [KOJ 93].

COMMENT 7.4.– The Newtonian directions $\Delta^{(k)}x$, $\Delta^{(k)}y$ and $\Delta^{(k)}s$ can be improved by approximating the functions that define the system (S) to the second order. This can lead to a better approximation of the μ-centers. This idea has been successfully used, particularly by Mehrotra in his predictive–corrective algorithm [MEH 89, MEH 90].

7.5.2.8. *Search for a feasible solution*

7.5.2.8.1. Use of an artificial variable

We can solve the following program (P_λ), where λ is an artificial variable:

$$(P_\lambda)\begin{cases} Min\lambda \\ A.x - \lambda\left(A.x^{(0)} - b\right) = b \\ x \geq 0, \lambda \in \mathbb{R}^+, \text{ any} x^{(0)} \in \mathbb{R}^n \end{cases}$$

The stopping criteria will be: $\frac{\lambda^{(k+1)} - \lambda^{(k)}}{Max(1, \lambda^{(k)})} < \varepsilon(\varepsilon \approx 10^{-8})$. For $\lambda^{(0)} = 1$, the point $x = x^{(0)}$ is feasible.

7.5.2.8.2. M method

We can solve the linear program:

$$(P_M)\begin{cases} Minc^t.x - M\lambda \\ A.x - \lambda\left(A.x^{(0)} - b\right) = b \\ x \geq 0, \lambda \in \mathbb{R}^+ \end{cases}$$

M must be chosen to be much bigger than $\underset{j}{Max}\{c_j\}$ to quickly eliminate the artificial variable.

7.5.2.9. *Reducing the number of iterations of a primal–dual method*

The directions of two successive centers in the primal and the dual which improve their objective functions are strict optimization directions in both cases. Let $(x(\mu_1), y(\mu_1), s(\mu_1))$ and $(x(\mu_2), y(\mu_2), s(\mu_2))$ be two successive centers. Thus:

– $x(\mu_2) - x(\mu_1)$ is a strict optimization direction in P;

– $(y(\mu_{k+1}) - y(\mu_k), s(\mu_{k+1}) - s(\mu_k))$ is a strict optimization direction in D.

They make up the direction of a chord of the primal central path and of the dual central path, respectively.

As we have seen previously, the number of iterations of a primal–dual algorithm depends essentially on the chosen number μ. The following algorithm allows us to calculate a value of μ corresponding to a good improvement of the objective functions of the primal and of the dual with each iteration. It is assumed that we have two successive centers of the primal and of the dual.

Let: $\left\{ \begin{array}{l} m_1 = x(\mu_1) \\ m_1^* = (y(\mu_1), s(\mu_1)) \end{array} \right.$ and $\left\{ \begin{array}{l} m_2 = x(\mu_2) \\ m_2^* = (y(\mu_2), s(\mu_2)) \end{array} \right.$

Step 1: Do a monodimensional search in the primal from m_2 in the direction $d = x(\mu_2) - x(\mu_1)$ in such a way as to find the intersection of the ray of origin m_2 and of direction d with P: $x(\mu_3) = m_2 + \lambda_{\mu_3} d, \lambda_{\mu_3}\mu_3 > 0$.

Step 2: Do a monodimensional search in the dual from m_2^* in the direction $d^* = (y(\mu_2) - y(\mu_1), s(\mu_2) - s(\mu_1))$ in such a way as to find the intersection of the ray of origin m_2^* and of direction d^* with D: $(y(\mu_3), s(\mu_3)) = m_2^* + \lambda_{\mu_3}^* d^*, \lambda_{\mu_3} > 0$.

Step 3: *Calculation of* μ_3. The duality gap is $SD(\mu_3) = c^t.x(\mu_3) - b^t.y(\mu_3) = n\mu_3$. Thus, $\mu_3 = \frac{SD(\mu_3)}{n}$. In this way we obtain a realistic prediction of μ_3 that improves the two objective functions in the global direction of the central trajectory: in effect, it appreciably reduces the duality gap.

We can apply this method to any primal–dual interior method by solving the Karush, Kuhn and Tucker conditions for $\mu = \mu_3$.

7.5.3. *Mehrotra predictor–corrector method*

The aim is to advance towards an optimal solution with more refined search directions than those derived from a first-order development of the functions associated with Karush, Kuhn and Tucker's system of optimality conditions.

7.5.3.1. *Problem examined*

This concerns a linear program with bounded variables:

$$(LP)\begin{cases} Minimize f(x) = c^t.x \\ A.x = b \\ 0 \leq x \leq u \end{cases}$$

7.5.3.2. *Application of the logarithmic barrier function principle*

(*LP*) can be expressed either in the previous form, or as:

$$\begin{cases} Minimize f(x) = c^t.x \\ A.x = b \\ -x - s = -u \\ x \geq 0; s \geq 0 \end{cases}$$

In this case, the logarithmic barrier function is defined by:

$$\tilde{g}_\mu(x) = c^t.x - \mu \sum_{j=1}^{n} Log(x_j) - \mu \sum_{j=1}^{n} Log(s_j)$$

If $L(x,s,y,w,\mu)$ is the Lagrangian associated with the problem:

$$\begin{cases} Min \tilde{g}_\mu(x) \\ A.x = b \\ x + s = u \end{cases}$$

then:

$$\begin{aligned} L(x, s, y, w, \mu) &= c^t.x - \mu \sum_{j=1}^{n} Log(x_j) - \mu \sum_{j=1}^{n} Log(s_j) \\ &\quad - y^t.(A.x - b) - w^t.(u - x - s) \end{aligned}$$

The conditions of optimality are expressed as:

$$(S) \begin{cases} X.z = \mu e \\ S.w = \mu e \\ A.x - b = 0 \\ x + s - u = 0 \\ A^t.y + z - w - c = 0 \end{cases} \quad \text{because :} \quad \begin{aligned} \frac{\partial L}{\partial x} &= c - \mu X^{-1}e - A^t.y + w = 0 \\ \frac{\partial L}{\partial y} &= A.x - b = 0 \\ \frac{\partial L}{\partial w} &= u - x - s = 0 \\ \frac{\partial L}{\partial s} &= -\mu S^{-1}e + w = 0 \end{aligned}$$

By stating $z = \mu X^{-1}.e$, we obtain $X.z{=}\mu e$ and the system (S).

Let Δx, Δy, Δz, Δs, Δw be the increments of x,y,z,s,w. They must satisfy:

$$\begin{cases} (X+\Delta X).(Z+\Delta Z).e = \mu e \\ (S+\Delta S).(W+\Delta W).e = \mu e \\ A.(x+\Delta x) - b = 0 \\ (x+\Delta x) + (s+\Delta s) - u = 0 \\ A^t.(y+\Delta y) + (z+\Delta z) - (w+\Delta w) - c = 0 \end{cases}$$

or

$$(\tilde{S}) \begin{cases} X\Delta z + Z\Delta x = \mu e - X.z{-}\Delta X.\Delta Z.e \\ S\Delta w + W\Delta s = \mu e - Sw{-}\Delta S.\Delta W.e \\ A.\Delta x = b - A.x \\ \Delta x + \Delta s = u - x - s \\ A^t.\Delta y + \Delta z - \Delta w = c - A^t y + w - z \end{cases}$$

Note that the system $(\tilde{S})$ contains second-degree terms: $\Delta X.\Delta Z.e$ and $\Delta S.\Delta W.e$. To denote this system, we have used the notation:

$$\Delta T = \begin{pmatrix} \Delta t_1 & & 0 \\ & \ddots & \\ 0 & & \Delta t_n \end{pmatrix} \quad \text{for } \Delta t = \begin{pmatrix} \Delta t_1 \\ \vdots \\ \Delta t_n \end{pmatrix}$$

7.5.3.3. *Calculation of the directions of acceleration in the Newton–Raphson method*

7.5.3.3.1. Calculation of Δx, Δy, Δz, Δw, Δs

Iteration of the Mehrotra method is carried out as follows:

– We state that $\mu = 0$ in $(\tilde{S})$ and we remove second-order terms.

– We calculate, from the current points $x^{(k)}, y^{(k)}, s^{(k)}, w^{(k)}, z^{(k)}$, the following directions: $\Delta^{(k)}x, \Delta^{(k)}y, \Delta^{(k)}s, \Delta^{(k)}w, \Delta^{(k)}z$, using the Newton–Raphson method applied to $(\tilde{S})$ with $\mu = 0$:

$$(\tilde{S}_{\mu=0}) \begin{cases} X^{(k)}\hat{\Delta} z + Z^{(k)}\hat{\Delta} x = X^{(k)}.z^{(k)} \\ S^{(k)}\hat{\Delta} w + W^{(k)}\hat{\Delta} s = -S^{(k)}w^{(k)} \\ A.\hat{\Delta} x = b - A.x^{(k)} \\ \hat{\Delta} x + \hat{\Delta} s = u - x^{(k)}{-}s^{(k)} \\ A^t.\hat{\Delta} y + \hat{\Delta} z - \hat{\Delta} w = c - A^t y^{(k)} + w^{(k)} - z^{(k)} \end{cases}$$

– We state that $D_k^2 = \left(\left(X^{(k)}\right)^{-1} Z^{(k)} + \left(S^{(k)}\right)^{-1} W^{(k)}\right)^{-1}$. Let $r_b^{(k)} = b - Ax^{(k)}; r_u^{(k)} = u - x^{(k)} - s^k; r_c^{(k)} = c - A^t y^{(k)} + w^{(k)} - s^{(k)}$ be the primal and dual feasibility gaps. The search directions are expressed by:

$$\begin{aligned}
\hat{\Delta}_y^{(k)} &= -\left(A.D_2^{(k)}A^t\right)\left[A.D_2^{(k)}\left(W^{(k)}-Z^{(k)}\right)e\right. \\
&\left.\quad +A.D_2^{(k)}r_c^{(k)}-r_b^{(k)}-A.D_2^{(k)}\left(S^{(k)}\right)^{-1}W^{(k)}r_u^{(k)}\right] \\
\hat{\Delta}_x^{(k)} &= D\left(\begin{smallmatrix}k\\2\end{smallmatrix}\right)\left[A^t\hat{\Delta}_y^{(k)}-z^{(k)}+w^{(k)}-r_c^{(k)}-\left(S^{(k)}\right)^{-1}W^{(k)}r_u^{(k)}\right] \\
\hat{\Delta}_s^{(k)} &= r_u^{(k)}-\hat{\Delta}_x^{(k)} \\
\hat{\Delta}_w^{(k)} &= -w^{(k)}-\left(S^{(k)}\right)^{-1}W^{(k)}\left(r_u^{(k)}-\hat{\Delta}_x^{(k)}\right) \\
\hat{\Delta}_z^{(k)} &= -z^{(k)}-\left(X^{(k)}\right)^{-1}Z^{(k)}\hat{\Delta}_x^{(k)}
\end{aligned}$$

– We note that $\hat{\Delta}^{(k)}x, \hat{\Delta}^{(k)}s, \hat{\Delta}^{(k)}w, \hat{\Delta}^{(k)}z$ are wholly determined by the calculation of $\hat{\Delta}^{(k)}y$.

An optimization in the above directions allows us to establish a new duality gap.

7.5.3.4. *Calculation of the duality gap*

Let (*LP*) $\left\{\begin{array}{l} Minimize f(x) = c^t.x \\ A.x = b \\ x + s = u \\ x \geq 0; s \geq 0 \end{array}\right.$ and its dual (*LPD*) $\left\{\begin{array}{l} Max b^t.y - u^t w \\ A^t.y - w + z = c \end{array}\right.$

In this case, the expression of the duality gap is $DG = c^t.x - b^t.y + u^t.w$. Since $z = c - A^t y + w, x^t z = c^t x - x^t A^t y + x^t w = c^t x - b^t y + x^t w$. But $x = u - s$; therefore $x^t z = c^t x - b^t y + u^t w - s^t w$. The duality gap is therefore: $DG = c^t.x - b^t.y + u^t.w = x^t z + s^t w$. Following a step of the Newton–Raphson method in the direction $\hat{\Delta}^{(k)} = \left(\hat{\Delta}^{(k)}x, ..., \hat{\Delta}^{(k)}z\right)^t$, the duality gap obtained is:

$$\begin{aligned}
DG^{(k)} &= \left(x^{(k)}+\gamma\tilde{\alpha}_P\hat{\Delta}^{(k)}x\right)^t\left(z^{(k)}+\gamma\tilde{\alpha}_D\hat{\Delta}^{(k)}z\right) \\
&\quad +\left(s^{(k)}+\gamma\tilde{\alpha}_P\hat{\Delta}^{(k)}z\right)^t\left(w^{(k)}+\gamma\tilde{\alpha}_D\hat{\Delta}^{(k)}w\right)
\end{aligned}$$

This first procedure is called *predictor step*.

7.5.3.5. *Finding the weight μ of the logarithmic barrier function*

The primal–dual trajectory must not diverge excessively from the central path so that the algorithm may converge towards the optimum. The point obtained at the end of the predictor procedure does not exactly satisfy the system $(\tilde{S})$; it is therefore not exactly on the ideal trajectory.

Mehrotra therefore calculates μ according to the prediction of the corrective terms of $\hat{\Delta}$. If the point obtained is off-center, we must give μ a greater value than that which is predetermined in such a way as to move the current point closer to the center of the polyhedron. In the opposite case, we can reduce μ to take into account the satisfactory progress of the algorithm. For this, we can predict the position of the current point in the polyhedron with the help of the relation $\frac{DG^{(k)}}{DG}$, where $DG^{(k)}$ is the duality gap at the current point. In effect, if at the iteration $k-1$, $DG^{(k-1)}$ is reduced by a polynomial factor proportional to $\frac{1}{n}$, $\frac{DG^{(k)}}{DG^{(k-1)}}$ is in the order of $\frac{1}{n}$. If we obtain this with each iteration, the algorithm converges in a number of iterations in $O\left(\sqrt{n}L\right)$. Mehrotra proposes as a "good value of $\mu^{(k)}$", $\hat{\mu}^{(k)} = \left(\frac{SD^{(k)}}{SD^{(k-1)}}\right)^2 \frac{SD^{(k)}}{n}$. This choice constitutes a predictor–corrector heuristic, because it adapts itself to the progression of the algorithm.

7.5.3.6. *Calculation of the corrector terms*

We remove from $(\tilde{S})$ the affine terms of the second members and we set $\Delta X.\Delta Z.e$ and $\Delta S.\Delta W.e$. The corrector terms c_x, c_y, c_z, c_z, c_w of the corresponding directions are solutions of the system:

$$(\tilde{S}_{\mu=\mu*}) \begin{cases} X^{(k)}c_z + Z^{(k)}c_x = \mu e - \hat{\Delta}X^{(k)}\hat{\Delta}Z^{(k)}e \\ S^{(k)}c_w + W^{(k)}c_s = \mu e - \hat{\Delta}S^{(k)}\hat{\Delta}W^{(k)}e \\ A.c_x = 0 \\ c_s + c_x = 0 \\ A^t c_s + c_z - c_w = 0 \end{cases}$$

From this we get:

$$
\begin{aligned}
c_y^{(k)} &= \left(AD_k^2A^t\right)^{-1} AD_k^2 \Big[\left(X^{(k)}\right)^{-1} \hat{\Delta} X^{(k)} \hat{\Delta} Z^{(k)} \\
&\quad - \left(S^{(k)}\right)^{-1} \hat{\Delta} S^{(k)} \hat{\Delta} W^{(k)} + \hat{\mu}^{(k)} \left(\left(S^{(k)}\right)^{-1} - \left(X^{(k)}\right)^{-1} \right) \Big] e \\
c_x^{(k)} &= D_k^2 \Big[A^t c_y^{(k)} - \hat{\mu}^{(k)} \left(\left(S^{(k)}\right)^{-1} \right) - \left(X^{(k)}\right)^{-1} \\
&\quad - \left(X^{(k)}\right)^{-1} \hat{\Delta} X^{(k)} \hat{\Delta} Z^{(k)} + \left(S^{(k)}\right)^{-1} \hat{\Delta} S^{(k)} \hat{\Delta} W^{(k)} \Big] \mu e \\
c_s^{(k)} &= -c \binom{k}{x} \\
c_w^{(k)} &= \left(S^{(k)}\right)^{-1} \left[\hat{\mu} e + W^{(k)} c_x^{(k)} e - \hat{\Delta} S^{(k)} \hat{\Delta} W^{(k)} e \right] \\
c_z^{(k)} &= c_w^{(k)} - A^t c_y^{(k)}
\end{aligned}
$$

The new Mehrotra directions and the new iterates are, respectively:

– $R_x^{(k)} = \hat{\Delta}^{(k)} x + c_x^{(k)}$, $x^{(k+1)} = x^{(k)} + \gamma \alpha_P R_x^{(k)}$;

– $R_s^{(k)} = \hat{\Delta}^{(k)} s + c_s^{(k)}$, $s^{(k+1)} = s^{(k)} + \gamma \alpha_D R_s^{(k)}$;

– $R_y^{(k)} = \hat{\Delta}_y^{(k)} + c_y^{(k)}$,$y^{(k+1)} = y^{(k)} + \gamma \alpha_D R_y^{(k)}$;

– $R_w^{(k)} = \hat{\Delta}_w^{(k)} + c_w^{(k)}$,$w^{(k+1)} = w^{(k)} + \gamma \alpha_D R_w^{(k)}$

– $R_z^{(k)} = \hat{\Delta}^{(k)} z + c_z^{(k)}$,$z^{(k+1)} = z^{(k)} + \gamma \alpha_D R_z^{(k)}$.

We can find an implementation of the Mehrotra predictor–corrector method in an article by I. J. Lustig, R. E. Marsten and F. Shanno [LUS 90].

7.6. Conclusion

Since the creation of the first interior algorithm, many programs have been created; they include procedures for taking into account bound constraints [TOD 94], procedures for pre-optimization, for sensitivity analysis and for parametrization, already present in simplistic programs. They often use the principles of primal, dual and primal–dual algorithms. H. Leterrier [LET 97] has shown practically that Gonzaga's dual algorithm [GON 89] could, subject to a few modifications, achieve the performances of primal–dual algorithms.

Interior methods are now used to solve combinatorial optimization problems of great size [PLA 01]. Furthermore, they have been generalized to the solution of non-linear programs and to semi-definite optimization. The site http://www.optimization-online.org lists many research articles in these domains. Finally, the fundamental concepts of interior methods are set out in the works of Roos *et al.* [ROO 97], Terlaky [TER 96], and Ye [YE 97].

7.7. Bibliography

[DAN 63] DANTZIG G.B., *Linear Programming and Extensions*, Princeton University Press, 1963.

[GON 89] GONZAGA C.C., "An algorithm for solving linear programming problems in $O(n^3N)$ operations", MEGGIDO N., Ed., *Progress in Mathematical Programming: Interior Points and Related Methods*, Springer-Verlag, New York, 1989.

[HUA 63] HUARD P., "Programmes mathématiques à contraintes non linéaires", Rapport HX 0/868/402 PH/JT, DER/EDF, 1963.

[HUA 67] HUARD P., "Programmation mathématique convexe", *RIRO*, vol. 7, 1967.

[KAR 84] KARMARKAR N.K., A new polynomial algorithm for linear programming, *Combinatorica*, vol. 4, p. 373–395, 1984.

[KHA 79] KHACHIYAN L.G., "A polynomial algorithm in linear programming", *Doklady Akademiia Nauk*, vol. 244, 1979, Translated into English in *Soviet Mathematics Doklady* vol. 20, p. 191–194.

[KLE 72] KLEE V., MINTY G.J., "How good is the simplex method?", SHISHA O., Ed., *Inequalities III*, Academic Press, New York, 1972.

[KOJ 89] KOJIMA M., MIZUNO S., YOSHISE A., "A primal-dual interior point algorithm for linear programming", MEGIDDO N., Ed., *Progress in Mathematical Programming: Interior Point and Related Methods*, Springer-Verlag, New York, 1989.

[KOJ 93] KOJIMA M., MEGIDDO N., MIZUNO S., "A primal-dual infeasible-interior-point algorithm for linear programming", *Mathematical Programming*, vol. 61, p. 263–280, 1993.

[LET 97] LETERRIER H., Méthodes intérieures en programmation linéaire: élaboration et mise en œuvre d'algorithmes rapides et robustes, PhD thesis, Paris Dauphine University, 1997.

[LUS 90] LUSTIG I.J., MARSTEN R.E., SHANNO D.F., On implementing Mehrotra's predictor-corrector interior point method for linear programming, IBM technical report, SQR 90-03, 1990.

[MEH 89] MEHROTRA S., On finding a vertex solution using interior point methods, Technical report 89-22, Dept. of Industrial Engineering and Management Sciences, Northwestern University, Evanston, 1989.

[MEH 90] MEHROTRA S., On the implementation of a primal dual interior point method, Technical report 90-03, Dept. of Industrial Engineering and Management Sciences, Northwestern University, Evanston, 1990.

[PLA 01] PLATEAU A., TACHAT D., TOLLA P., "A hybrid search combining interior point methods and metaheuristics for 0-1 programming", *ITOR*, vol. 9, 2001.

[ROO 97] ROOS C., TERLAKY T., VIAL J.-PH., *Theory and Algorithms for Non Linear Optimization: An Interior Point Approach*, Wiley, 1997.

[SCH 85] SCHRIJVER A., "The linear programming method of Karmarkar", *CWI Newsletter*, num. 8, September 1985.

[TER 96] TERLAKY T., *Interior Point Methods Of Mathematical Programming*, Kluwer Academic Publishers, 1996.

[TOD 94] TODD M.J., "Analysis of interior-point methods for linear programming problems with variable upper bounds", GOMEZ S., HENNART J.-P., Eds., *Advances in Optimization and Numerical Analysis*, Kluwer Academic Publishers, p. 1–23, 1994.

[TOL 99] TOLLA P., "Improvements of primal-dual interior point methods in linear programming", *IFORS'99*, 1999.

[YE 97] YE Y., "Interior point algorithms: theory and analysis", *Wiley-Interscience Series in Discrete Mathematics and Optimization*, Wiley, 1997.

Chapter 8

Quadratic Optimization in 0–1 Variables

8.1. Introduction

A great number of combinatorial optimization problems can be formulated as the optimization of a quadratic function of bivalent variables:

$$f(x) = \sum_{i=1}^{n} c_i x_i + \sum_{i=1}^{n-1} \sum_{j=i+1}^{n} c_{ij} x_i x_j$$

subject, or not, to the linear constraints $\sum_{i=1}^{n} a_{ki} x_i \leq b_k$ $(k \in K_1)$ and $\sum_{i=1}^{n} a_{ki} x_i = b_k$ $(k \in K_2)$. For example, many graph theory problems can be formulated in this way. This is the case for the maximum weight stable set problem, for many partition problems, for certain coloring problems, and for certain grouping problems such as the one that consists of establishing a subgraph of k vertices having the largest possible number of edges, etc. Deterministic or probabilistic logical expression satisfaction problems, certain optimization problems in the finance domain, certain scheduling and localization problems and many other problems in operations research can also be formulated in this way. Because of this, knowing how to solve quadratic programs in bivalent variables is valuable knowledge and this is why a large number of studies have been made of the subject over the last 30 years. Furthermore, we can show that the problem of optimizing functions of a degree higher than 2 is easily reduced to a quadratic function optimization problem. This increases the problem's interest, at least from a theoretical point of view. In this chapter we introduce some theoretical and algorithmic aspects of quadratic optimization in bivalent variables, which is, in general, an **NP**-hard problem (see Chapter 1). This introduction is far from being

Chapter written by Alain BILLIONNET.

complete because, as we have said, there is an abundance of literature on this subject. The first part of the chapter is devoted to the, in our opinion, central problem of quadratic optimization in 0–1 variables without constraints. In the second part, we introduce some results in the constraint case.

8.2. Pseudo-Boolean functions and set functions

Let us consider n variables $x_1, ..., x_n$ which can only take the values 0 or 1 and the function $f(x_1, ..., x_n)$, of $\{0,1\}^n$ in the set of real numbers $\mathbb{R}$. The variables $x_1, ..., x_n$ are known as 0–1 variables, binary variables, bivalent variables or even Boolean variables. Given a finite set $V = \{1, ..., n\}$, a set function $f(S)$, $S \subseteq V$, is a function of the parts of V in $\mathbb{R}$. Let us note that a set function $f(S)$ can be considered as a function $f(x)$ of $\{0,1\}^n$ in $\mathbb{R}$ and *vice versa.* It is enough to define x as being the characteristic vector of the set S. These (set or $\{0,1\}^n$ in $\mathbb{R}$) functions can be defined in an implicit manner (oracle). This is the case in the following example: given a graph $G = (X, E)$, for each $S \subseteq X$, $f(S)$ gives the chromatic number of the subgraph of G generated by S. In many cases, these functions can be defined by an algebraic expression. Let us consider, for example, the set function f which associates the cardinal of the set S with each subset S of $V = \{1, ..., n\}$. We can express this function in the algebraic form $f(S) = x_1 + ... + x_n$, where $(x_1, ..., x_n)$ is the characteristic vector of the set S. Here is another example of such a function of $\{0,1\}^3$ in $\mathbb{R}$: $f(x_1, x_2, x_3) = 2x_1 + 3x_2 - x_3 + 4x_1x_2 - 3x_2x_3$. The operations used in these functions are the usual arithmetical operations and this is why we call them pseudo-Boolean functions (pBf). We can easily check that for the previous function $f(0,1,1) = -1$. Let us now consider the function in three variables $f(x_1, x_2, x_3)$, for which the values are given in Table 8.1. Let us associate with each vector $(\tilde{x}_1, \tilde{x}_2, \tilde{x}_3)$ of $\{0,1\}^3$ the monomial $y_1y_2y_3$, where y_i represents a literal, that is the variable x_i or the variable $\bar{x}_i = 1 - x_i$. Let us state $y_i = x_i$ if $\tilde{x}_i = 1$ and $y_i = \bar{x}_i$ if $\tilde{x}_i = 0$. We can easily see that the monomial $y_1y_2y_3$ is equal to 1 if and only if $(x_1, x_2, x_3) = (\tilde{x}_1, \tilde{x}_2, \tilde{x}_3)$. By applying this property to the previous function, we immediately obtain an algebraic expression of f as a function of the literals x_i and $\bar{x}_i$: $f(x_1, x_2, x_3) = -5\bar{x}_1\bar{x}_2\bar{x}_3 - 3x_1\bar{x}_2\bar{x}_3 + 8x_1\bar{x}_2x_3 + 6x_1x_2x_3$. By then replacing $\bar{x}_i$ with $1-x_i$ and by grouping identical terms, we eventually express f only as a function of the variables in their direct form: $f(x_1, x_2, x_3) = -5 + 2x_1 + 5x_2 + 5x_3 - 2x_1x_2 + 6x_1x_3 - 5x_2x_3$. This expression of f is called the polynomial form. An interesting property of this polynomial (multilinear) form is that it is unique [HAM 68].

THEOREM 8.1.– *Any pseudo-Boolean function $f(x_1, ..., x_n)$ allows a unique multilinear polynomial form $f(x_1, ..., x_n) = \sum_{S \subseteq V} c_S \Pi_{j \in S}\, x_j$, where the coefficients c_S are real numbers and $V = \{1, 2, ..., n\}$.*

The reference [HAN 79] introduces one of the first surveys of non-linear programming in 0–1 variables; it lists about 100 references.

x_1	x_2	x_3	$f(x_1, x_2, x_3)$
0	0	0	-5
0	0	1	0
0	1	0	0
0	1	1	0
1	0	0	-3
1	0	1	8
1	1	0	0
1	1	1	6

Table 8.1. *Enumeration of the values of the function* $f(x_1, x_2, x_3)$

8.3. Formalization using pseudo-Boolean functions

Let us consider a set of actions $\{a_1, a_2, ..., a_n\}$. The term *action* is deliberately general. It can represent very different things, such as the assignment of a person to a post, the choice of an arc to make up a path in a graph, or the opening of a client service center. With each action a_i we associate a bivalent variable x_i and we use the following coding: $x_i = 1$ if and only if the action a_i is carried out. In a very general way, we can say that an optimization problem consists of optimizing an objective function for which the variables are subject to a set of constraints. Let us give some examples of objective function and constraint formalization by pseudo-Boolean functions. Let us first of all look at the objective functions case. The quantity $c_S \Pi_{i|a_i \in S}\, x_i$ expresses the fact that a cost or a gain c_S is to be taken into account if and only if all the actions in the set $S \subseteq \{a_1, ..., a_n\}$ are carried out. The quantity $c_S(1 - \Pi_{i|a_i \in S}\, x_i)$ expresses the fact that a cost or a gain c_S is to be taken into account if and only if at least one of the actions belonging to $S \subseteq \{a_1, ..., a_n\}$ is not carried out. The quantity $c_{ij}x_i(1 - x_j)$ expresses the fact that a cost or a gain c_{ij} is to be taken into account if and only if the action a_i is carried out and the action a_j is not carried out.

Let us now look at the constraint case. The constraint $\sum_{i=1}^{n} x_i = 1$ expresses the fact that, among the actions $\{a_1, ..., a_n\}$, one and only one must be carried out. The constraint $\Pi_{i|a_i \in T}\, x_i \geq \Pi_{i|a_i \in S}\, x_i$ expresses the fact that if all the actions belonging to $S \subseteq \{a_1, ..., a_n\}$ are carried out, then all the actions belonging to $T \subseteq \{a_1, ..., a_n\}$ must be carried out. Let $S_1, ..., S_K$ be sets of actions included in $\{a_1, ..., a_n\}$. The constraint $\sum_{k=1}^{K} \Pi_{i|a_i \in S_k}\, x_i = 1$ expresses the fact that one and only one set of these actions must be carried out (in full) and the constraint $\sum_{k=1}^{K} \Pi_{i|a_i \in S_k}\, x_i \geq 1$ expresses the fact that at least one set of these actions must be carried out.

8.4. Quadratic pseudo-Boolean functions (qpBf)

We say that a pseudo-Boolean function is of degree d if, when it is written in its unique polynomial form, the maximum number of distinct variables used in the same monomial is equal to d. A quadratic pseudo-Boolean function is a function of degree 2, that is, written in its unique polynomial form, it contains only constant terms, linear terms, and quadratic terms. In general, we will express these functions by: $f(x) = \sum_{i=1}^{n} c_i x_i + \sum_{i=1}^{n-1} \sum_{j=i+1}^{n} c_{ij} x_i x_j$. Here is a qpBf in five variables:

$$f(x) = 2x_1 + x_3 + 3x_5 - 5x_1x_5 + x_2x_4 - 2x_2x_5$$

As we will see later, many combinatorial optimization problems can be formulated as the optimization of a qpBf with or without constraints. Kochenberger and Glover [GLO 02] even talk of "a royal route for combinatorial optimization". Furthermore, it is easy to show that the optimization of a pseudo-Boolean function of any degree can be reduced to the optimization of a qpBf [ROS 75].

THEOREM 8.2.– *The optimization of a pseudo-Boolean function of any degree can be polynomially reduced to the optimization of a quadratic pseudo-Boolean function.*

The proof of this theorem is based on the following property: given three bivalent variables x, y and z, $xy = z$ if and only if $xy - 2xz - 2yz + 3z = 0$ and $xy \neq z$ if and only if $xy - 2xz - 2yz + 3z > 0$. This property allows us to replace the product of two variables xy by a new variable z, provided that we add to the function to be optimized the "penalty" $M(xy - 2xz - 2yz + 3z)$, where M is a sufficiently large positive coefficient in the case of a minimization problem, and a sufficiently small negative coefficient in the case of a maximization problem. Let us consider as an example the problem that consists of minimizing the cubic function of four variables $f(x) = 2x_1x_3 + x_2x_3x_4 - 5x_1x_2x_4$. Let us replace the product x_2x_4 in f with a new bivalent variable x_5. The initial problem can now be stated as the minimization of the qpBf in five variables, $g(x) = 2x_1x_3 + x_3x_5 - 5x_1x_5 + 10\,(x_2x_4 - 2x_2x_5 - 2x_4x_5 + 3x_5)$. It is to be noted that theorem 8.2 is mostly of theoretical interest. In practice it will not often be worthwhile trying to minimize a pseudo-Boolean function of degree greater than 2 by reducing the problem to the minimization of a qpBf.

Let us now look at some examples of classical combinatorial optimization problems that can be formulated as the maximization or the minimization of pseudo-Boolean functions without constraints and, therefore, according to theorem 8.2, as the optimization of quadratic functions. Let us consider a graph $G = (V, E)$, where $V = \{v_1, ..., v_n\}$ refers to the set of vertices and E to the set of edges. A subset S of vertices of V makes a stable set if any two vertices of this subset are not adjacent. The number of vertices that make up the stable set of maximum cardinality, S^*, is called the stability number G. Establishing this stability number can be stated as the maximization of the qpBf $\sum_{i=1}^{n} x_i - \sum_{i=1}^{n-1} \sum_{j=i+1}^{n} \delta_{ij} x_i x_j$, where δ_{ij} is a coefficient that is worth 1 if the edge $[v_i, v_j] \in E$ and 0 if it is not. Let us consider one of the basic problems of non-linear optimization in bivalent variables under constraints, the maximization of a quadratic pseudo-Boolean function under a capacity constraint:

$$\min\left\{\sum_{i=1}^{n} c_i x_i + \sum_{i=1}^{n-1}\sum_{j=i+1}^{n} c_{ij} x_i x_j : \sum_{i=1}^{n} a_i x_i \leq d, x \in \{0,1\}^n\right\}$$

where c_i, c_{ij}, a_i and d are real coefficients. We can tackle this problem using the classical technique of Lagrangian duality. Let us associate the Lagrange multiplier $\lambda \geq 0$ with the constraint $\sum_{i=1}^{n} a_i x_i \leq d$. Calculating the dual function consists of, for a given value of the multiplier λ, minimizing a qpBf without constraints and the dual problem, to solve:

$$\max_{\lambda \geq 0}\left\{\min\left\{-\lambda\, d + \sum_{i=1}^{n} x_i(c_i + \lambda\, a_i) + \sum_{i=1}^{n-1}\sum_{j=i+1}^{n} c_{ij} x_i x_j : x \in \{0,1\}^n\right\}\right\}$$

Let us now consider the maximum cut problem. Given a graph $G = (V, E)$, establish $C^* \subseteq V$ such that the number of edges incident to C^* is maximal. By accepting that $x_i = 1$ if and only if the vertex v_i belongs to C^*, we can state the maximum cut problem as being that of maximizing the function $\sum_{i=1}^{n-1}\sum_{j=i+1}^{n} \delta_{ij}(x_i\bar{x}_j + \bar{x}_i x_j)$. Indeed, the edge $[v_i, v_j]$ belongs to the cut if and only if $x_i\bar{x}_j + \bar{x}_i x_j = 1$. In the MAX-SAT problem, we consider n bivalent variables $x_1, x_2, ..., x_n$ and the logical expression $E = C_1 \wedge C_2 \wedge ... \wedge C_m$, where each clause C_i is a disjunction of literals $u_{i_1} \vee u_{i_2} \vee ... \vee u_{i_{k(i)}}$. A literal u_{i_q} is a variable in its direct form or in its complemented form. Each variable can take the value "true" or "false" and we seek to satisfy a maximum number of clauses. We say that the clause C_i is satisfied if at least one of its literals is set to the value "true" and the expression E is satisfied if all its clauses are satisfied. Let us note first of all that the problem that consists of assigning the values "true" or "false" to each variable in such a way as to satisfy a maximum number of clauses is equivalent to the problem that consists of assigning the values "true" or "false" to the different variables in such a way as to minimize the number of unsatisfied clauses. By making the values "true" and 1 and the values "false" and 0 correspond to each other, we can state the problem of minimizing the number of unsatisfied clauses and thus the MAX-SAT problem as minimizing the pBf $\sum_{i=1}^{m}\prod_{q=1}^{k(i)} \bar{u}_{i_q}$. This function is of any degree. It is of degree 2 in the case of MAX-2-SAT, where each clause of E contains at most two literals.

EXAMPLE 8.1.– $E = (x_1 \vee \bar{x}_2) \wedge (x_1 \vee x_3) \wedge (\bar{x}_2 \vee \bar{x}_3) \wedge (x_2 \vee \bar{x}_4) \wedge (\bar{x}_3 \vee x_4)$. Finding the solution that maximizes the number of unsatisfied clauses comes down to minimizing the qpBf $\bar{x}_1 x_2 + \bar{x}_1\bar{x}_3 + x_2 x_3 + \bar{x}_2 x_4 + x_3\bar{x}_4$. We can easily see that the vector $x = (1, 0, 0, 0)$ gives the value 0 to this function. This means that the solution x_1= "true", x_2= "false", x_3= "false", and x_4= "false" satisfies all the clauses of E. Let us remember that the 2-SAT problem is polynomial (see section 8.7.2) and that the problem central to the theory of algorithmic complexity, 3-SAT, is **NP**-complete.

The reference [BIL 92e] introduces an efficient algorithm for this problem. Let us also mention the solution of network flow problems by the optimization of pseudo-Boolean functions [HAM 65] and an application for qpBf in the medical domain [IAS 01].

Let us now look at a different example illustrating the problem of optimizing a qpBf. Let us consider the fractional combinatorial optimization problem that consists of maximizing the ratio $r(x) = f(x)/g(x)$, where $f(x)$ and $g(x)$ are two pseudo-Boolean functions of any degree. We will assume that $g(x) > 0, \forall x \in \{0,1\}^n$. For example, $r(x) = (2 + x_1 + 4x_1x_2x_3 + 3x_2x_4)/(5 + 3x_1 + 2x_1x_3 + 4x_2x_4)$. This problem can be solved by Dinkelbach's algorithm [DIN 67], described in algorithm 8.1, where λ_0 is an feasible value in the ratio $r(x)$. The number of iterations of Dinkelbach's algorithm (or Newton's method) is polynomial [RAD 98].

Algorithm 8.1

```
1: λ ← λ0
2: calculate v(λ) = max { f(x) − λg(x) : x ∈ {0,1}^n }
3: let x_λ be such that v(λ) = f(x_λ) − λ g(x_λ)
4: if v(λ) > 0 then
5:    λ ← f(x_λ)/g(x_λ)
6:    Go to 2
7: else
8:    x_λ is an optimal solution
9: end if
```

The difficult part of this algorithm is the optimization problem in step 2, which consists of maximizing a pseudo-Boolean function of any degree. In the end, since the optimization of a pBf of any degree may be reduced to the optimization of a qpBf, we can say that the maximization of the ratio of two pBfs of any degree can be reduced to a series of maximizations of qpBfs (see also [LI 94]).

8.5. Integer optimum and continuous optimum of qpBfs

We will see below a property of pseudo-Boolean functions that is relatively rare in optimization [ROS 72].

THEOREM 8.3.– *Let the function* $f(x) = \sum_{i=1}^{n} c_i x_i + \sum_{i=1}^{n-1} \sum_{j=i+1}^{n} c_{ij} x_i x_j$. *For each* $x \in [0,1]^n$, *algorithm 8.2 allows us to construct, in polynomial time, a vector* $\tilde{x} \in \{0,1\}^n$ *such that* $f(\tilde{x}) \geq f(x)$.

Note that the validity of the algorithm comes from the fact that the functions under consideration do not contain any x_i^2 terms. A direct consequence of theorem 8.3 is that the search for a continuous optimum ($\in [0,1]^n$) is equivalent to the search for

Algorithm 8.2

let $x \in [0,1]^n$
$\tilde{x} \leftarrow x$
for i=1 to n **do**
  calculate $\varphi(i) = c_i + \sum_{j=1}^{i-1} c_{ji}\tilde{x}_j + \sum_{j=i+1}^{n} c_{ij}\tilde{x}_j$
  if $\varphi(i) > 0$ **then**
    $\tilde{x}_i \leftarrow 1$
  else
    $\tilde{x}_i \leftarrow 0$
  end if
end for

an integer optimum ($\in \{0,1\}^n$) in the following sense: $\max\{\, f(x) : x \in [0,1]^n\} = \max\{\, f(x) : x \in \{0,1\}^n\}$. It is easy to extend theorem 8.3, and consequently this last property, to pseudo-Boolean functions of any degree. Note that algorithm 8.2 is easily adapted to construct, in polynomial time, a vector $\tilde{x} \in \{0,1\}^n$ such that $f(\tilde{x}) \leq f(x)$.

8.6. Derandomization

Let us consider a qpBf $f(x)$ and let us assume that the variables x_i are random independent variables taking the value 1 with probability p_i and the value 0 with probability $1 - p_i$. The property below gives the mathematical expectation of the value of the function f.

THEOREM 8.4.– *Let us consider the qpBf* $f(x) = \sum_{i=1}^{n} c_i x_i + \sum_{i=1}^{n-1} \sum_{j=i+1}^{n} c_{ij} x_i x_j$, *where the variables* x_i *are random independent variables that take the value 1 with probability* p_i *and the value 0 with probability* $1 - p_i$. *The mathematical expectation of the value of the function* f *is equal to* $f(p_1, p_2, ..., p_n)$.

If the mathematical expectation $E(f)$ of the value of f, is $f(p_1, p_2, ..., p_n)$, this means that at least one vector $\tilde{x}$ of $\{0,1\}^n$ exists such that $f(\tilde{x}) \geq E(f)$. This reasoning is often used, for example, in combinatorics or to establish guarantees of performance for approximation algorithms. An important question that then arises is: "Can we find, such a vector $\tilde{x}$ in an efficient way?". Here, we find the vector by directly using algorithm 8.2, which gives, from a vector of $[0,1]^n$, a vector of $\{0,1\}^n$ with at least as good a value for the function $f(x)$. We will take as the vector of $[0,1]^n$ the vector $(p_1, p_2, ..., p_n)$.

8.7. Posiforms and quadratic posiforms

Given a set of bivalent variables $x_1, ..., x_n$, a posiform is a function $\varphi(x, \bar{x})$ of the literals $x_1, ..., x_n, \bar{x}_1, .., \bar{x}_n$ which is expressed in the following form: $\varphi(x, \bar{x}) = c + c_1T_1 + c_2T_2 + ... + c_mT_m$, where c is a real number constant, c_i, $i = 1, ..., m$, a positive real coefficient, and T_i, $i = 1, ..., m$, a literal or a product of literals. Given a pBf $f(x)$, we say that $\varphi(x, \bar{x})$ is a posiform associated with $f(x)$ if and only if $f(x) = \varphi(x, \bar{x})$ for each $x \in \{0, 1\}^n$. Note that several posiforms may be associated with one pseudo-Boolean function. A quadratic posiform is a posiform where T_i, $i = 1, ..., m$ is a literal or a product of two literals.

THEOREM 8.5.– *Any pseudo-Boolean function can be written as a posiform by using the relationship $x_i = 1 - \bar{x}_i$ and a pseudo-Boolean function can have several posiforms.*

EXAMPLE 8.2.– The function $f(x) = 2x_1 - 5x_2 + 4x_3 - 2x_1x_2 + 4x_1x_3 - 6x_2x_3$ can be expressed in the form $\varphi_1(x, \bar{x}) = -9 + 2x_1 + 7\bar{x}_2 + 2\bar{x}_3 + 2\bar{x}_1x_2 + 4x_1x_3 + 6\bar{x}_2x_3$, or even in the form $\varphi_2(x, \bar{x}) = -11 + 11\bar{x}_2 + 4x_3 + 2x_1\bar{x}_2 + 4x_1x_3 + 6x_2\bar{x}_3$.

8.7.1. *Posiform maximization and stability in a graph*

Let us associate with the posiform $\varphi(x, \bar{x}) = c + c_1T_1 + ... + c_mT_m$ the graph $G_\varphi = (T, E)$, where $T = \{T_1, ..., T_m\}$ is the set of vertices and where the edge $[T_i, T_j]$ belongs to E if and only if a literal $u \in T_i$ exists for which $\bar{u} \in T_j$. This means that the two monomials T_i and T_j cannot simultaneously take the value 1. For this reason, the graph $G_\varphi = (T, E)$ is called a *conflict graph.* Let us now associate the weight c_i with each vertex of the graph. Theorem 8.6 gives the relationship between the posiform maximization problem and that of the search for a stable set of maximum weight in a graph.

THEOREM 8.6.– *If G_φ is the conflict graph associated with the posiform $\varphi(x, \bar{x}) = c + c_1T_1 + ... + c_mT_m$, then the maximum of $\varphi(x, \bar{x})$ is equal to the constant c added to the weight of the maximum weight stable set of G_φ.*

EXAMPLE 8.3.– $\varphi(x, \bar{x}) = 8 + 11\bar{x}_2 + 4x_3 + 2x_1\bar{x}_2 + 4x_1x_3 + 6x_2\bar{x}_3 + 2\bar{x}_1x_2x_3$. The maximum weight stable set of the conflict graph associated with this posiform is made up of the vertices associated with the monomials $11\bar{x}_2$, $4x_3$, $4x_1x_3$, and $2x_1\bar{x}_2$. The weight of this stable set is equal to $11 + 4 + 4 + 2 = 21$, and we deduce that the maximum value of $\varphi(x, \bar{x})$ is equal to $21 + 8 = 29$. The optimal solution itself is established by setting the variables in such a way that the monomials associated with the vertices of the stable set are equal to c_i. In this way we obtain $x_1 = 1$, $x_2 = 0$ and $x_3 = 1$.

8.7.2. *Implication graph associated with a quadratic posiform*

Let us consider the quadratic posiform $\varphi(x, \bar{x}) = c_1 T_1 + c_2 T_2 + ... + c_m T_m$, and let us state that $T_k = y_1^k\, y_2^k$. The literals y_1^k and y_2^k are therefore the two literals that make up the monomial T_k. Let us associate the graph $G = (V, U)$ with this posiform, where the set of vertices V is equal to $\{x_1, ..., x_n, \bar{x}_1, ..., \bar{x}_n\}$ and where the set of arcs U is defined in the following way: each monomial $T_k = y_1^k\, y_2^k$ of the posiform φ generates the two arcs $(y_1^k, \bar{y}_2^k)$ and $(y_2^k, \bar{y}_1^k)$. The graph $G = (V, U)$ is called the implication graph associated with the posiform φ. We can interpret these arcs in the following way: if y_1^k is equal to 1, then, so that the function φ takes the value 0, $\bar{y}_2^k$ must necessarily take the value 1. In the same way, if y_2^k is equal to 1, then $\bar{y}_1^k$ must take the value 1. This definition of the implication graph associated with a posiform leads on very naturally to theorem 8.7.

THEOREM 8.7.– *The equation $\varphi(x, \bar{x}) = 0$, where $\varphi(x, \bar{x})$ is a quadratic posiform, allows a solution if and only if the implication graph associated with φ does not contain any circuits that go through x_i and $\bar{x}_i$ for an index i of {1,2,. . . ,n}.*

In an equivalent way, we can express the previous result as follows: the equation $\varphi(x, \bar{x}) = 0$ has a solution if and only if no strongly connected component of the implication graph associated with φ contains both the vertices x_i and $\bar{x}_i$ for an index i of $\{1, 2, ..., n\}$. Note that this condition can be checked by an algorithm in linear time as a function of the number of arcs in the graph. We therefore deduce that we can answer the question "Does the equation $\varphi(x, \bar{x}) = 0$ have a solution?" with a linear algorithm as a function of the number of terms appearing in the posiform. Note that in the case of posiforms of a degree greater than or equal to 3, the decision problem "Does a solution to the equation $\varphi(x, \bar{x}) = 0$ exist?" is equivalent to the SAT problem and is therefore **NP**-complete. Let us consider the function $\varphi(x, \bar{x}) = 2x_1\bar{x}_2 + 4x_1x_3 + 6x_2\bar{x}_3$. The implication graph does not contain any circuits going through x_i and $\bar{x}_i$. Therefore a solution to the equation $2x_1\bar{x}_2 + 4x_1x_3 + 6x_2\bar{x}_3 = 0$ exists. Note that the fact that there is a *path* from x_1 to $\bar{x}_1$ in this implication graph means that any solution of the equation satisfies $x_1 = 0$. Let us assume that we seek to minimize a qpBf $f(x)$. If we express this function in the form $c + \varphi(x, \bar{x})$, it is obvious that c is a lower bound of the minimum of f. Theorem 8.7 allows us to know whether c is the minimum of f. In fact, c is the minimum of f if and only if $x \in \{0, 1\}^n$ exists such that $\varphi(x, \bar{x}) = 0$. We can therefore answer the question "Is c the minimum of f?" in polynomial time. If c is not the minimum of f, theorem 8.7 can be used to obtain a better lower bound than c [BIL 94b]. For this, we look in $\varphi(x, \bar{x})$ for a sum of monomials for which the associated implication graph contains a circuit that goes through x_i and $\bar{x}_i$. This sum of monomials, which is a subfunction of $\varphi(x, \bar{x})$, cannot take the value 0. It is therefore greater than or equal to the smallest coefficient, $c_{\min}$, appearing in this sum. In this way we obtain a new lower bound of the minimum of f: $c + c_{\min}$. By applying this idea in an iterative way to the posiform $\varphi(x, \bar{x})$, we obtain a lower bound of the minimum of f that can be better than c. Moreover, we can use this implication graph to break down, in a specific way, the problem of optimizing a qpBf [BIL 89b].

8.8. Optimizing a qpBf: special cases and polynomial cases

8.8.1. *Maximizing negative–positive functions*

Let us consider the qpBf $f(x) = -\sum_{i=1}^{n} c_i x_i + \sum_{i=1}^{n-1} \sum_{j=i+1}^{n} c_{ij} x_i x_j$, where $c_i > 0$ for each $i \in \{1, ..., n\}$ and $c_{ij} \geq 0$ for each $(i, j) \in \{1, ..., n\}^2$ such that $i < j$. These functions are called *negative–positive* functions because the coefficients of the linear terms are negative, while those of the quadratic terms present in the function are positive. By replacing all the variables x_i with $1 - \bar{x}_i$ in the linear part of the function f, we obtain the quadratic posiform $\varphi(x, \bar{x}) = -\sum_{i=1}^{n} c_i + \sum_{i=1}^{n} c_i \bar{x}_i + \sum_{i=1}^{n-1} \sum_{j=i+1}^{n} c_{ij} x_i x_j$. According to theorem 8.6, the maximal value of this function is equal to the quantity $-\sum_{i=1}^{n} c_i$ added to the weight of the maximum weight stable set in the conflict graph associated with the posiform $\sum_{i=1}^{n} c_i \bar{x}_i + \sum_{i=1}^{n-1} \sum_{j=i+1}^{n} c_{ij} x_i x_j$. We can easily see that, in this special case, the conflict graph is bipartite. Moreover, we know that establishing a maximum weight stable set in a bipartite graph is a polynomial problem that can be solved by using, for example, a maximum flow algorithm in a transport network. We therefore deduce from this that the problem of maximizing negative–positive functions is a polynomial problem (see also [RHY 70, PIC 75]). This reasoning can easily be extended to negative–positive pseudo-Boolean functions of any degree, that is to the functions whose coefficients of the linear terms are negative, while the coefficients of the non-linear terms are positive. In an equivalent way, the problem of minimizing *positive–negative* functions is also polynomial. We can also show that negative–positive qpBfs correspond to supermodular qpBfs. Let us remember that a set function f on a finite set E is submodular if $\forall A \subseteq E, \forall B \subseteq E, f(A) + f(B) \geq f(A \cup B) + f(A \cap B)$. We say that the function f is supermodular if the function $(-f)$ is submodular. We can show that the search for the maximum of supermodular pseudo-Boolean functions of degree 3 can also be reduced to the search for a maximum weight stable set in a bipartite graph [BIL 85]. The method is based on a rewriting of the function f in which the non-linear terms make up a special posiform: in all the quadratic terms the variables are in their direct form, and in all the cubic terms they are either all in their direct form, or all in their complemented form. Note that if $P \neq NP$, the problem of recognizing supermodular functions of degree greater than 3 is difficult [CRA 89] (see [HAN 86] and [SIM 90] for unimodular functions).

EXAMPLE 8.4.– Let $f(x) = -2x_1 - 3x_2 - 2x_3 - 3x_4 + 2x_1x_2 + 3x_1x_3 + 6x_3x_4$. The associated posiform is $f(x) = -10 + 2\bar{x}_1 + 3\bar{x}_2 + 2\bar{x}_3 + 3\bar{x}_4 + 2x_1x_2 + 3x_1x_3 + 6x_3x_4$. The point that maximizes this function f is defined by $x_2 = 0, x_1 = x_3 = x_4 = 1$.

8.8.2. *Maximizing functions associated with k-trees*

Let us consider the qpBf $f(x) = \sum_{i=1}^{n} c_i x_i + \sum_{i=1}^{n-1} \sum_{j=i+1}^{n} c_{ij} x_i x_j$ and let us associate with it the graph $G = (V, E)$, defined in the following way: $V = \{v_1, ..., v_n\}$ is the set of vertices and $[v_i, v_j] \in E \Leftrightarrow c_{ij} \neq 0$. For example, the graph associated with the qpBf $6x_1 + 4x_2 + 5x_3 + 2x_5 - x_1x_2 + 2x_1x_5 + 3x_1x_6 + 2x_2x_3 + 6x_2x_4$ is a tree. Let us remember the following definitions that allow us to define a k-tree: a vertex v is simplicial if the neighbors of v make up a clique and a perfect elimination k-scheme is an order of vertices $v_1, ..., v_n$ such that $v_j, j = 1, 2, ..., n-k$ is simplicial and of degree k in the subgraph generated by $\{v_j, v_{j+1}, ..., v_n\}$. We say that a graph is a k-tree if it allows a perfect elimination k-scheme.

THEOREM 8.8.– *If the graph G associated with the qpBf $f(x_1, ..., x_n)$ is a partial k-tree, then the maximum of $f(x)$ can be established using an algorithm in $O(n)$ (for a fixed k) when we know a perfect elimination k-scheme [CRA 90].*

This algorithm is based on Hammer and Rudneau's fundamental (iterative) algorithm, which allows us to minimize non-linear functions in 0–1 variables by eliminating one variable with each iteration [HAM 68].

8.8.3. *Maximizing a quadratic posiform whose terms are associated with two consecutive arcs of a directed multigraph*

Let $G = (X, U)$ be a directed multigraph without loops. Zverovich [ZVE 01] has studied the maximum of quadratic posiforms whose terms are associated with the triplets (u, x, u') such that: $u \in U, x \in X, u' \in U$, x is an extremity of both u and u', and $u \neq u'$. He proposes a very simple polynomial algorithm, based on the degrees of the vertices of G, for calculating the maximum of such posiforms. He also proposes a polynomial characterization of these posiforms. Another case of qpBf optimization in polynomial time is introduced in [CHA 92].

8.8.4. *Quadratic pseudo-Boolean functions equal to the product of two linear functions*

Let us consider the problem of maximizing the qpBf:

$$f(x) = (a_0 + \sum_{i=1}^{n} a_i x_i)(b_0 + \sum_{i=1}^{n} b_i x_i)$$

This is an **NP**-hard problem whose continuous relaxation:

$$\max \left\{ (a_0 + \sum_{i=1}^{n} a_i x_i)(b_0 + \sum_{i=1}^{n} b_i x_i) : x \in [0,1]^n \right\}$$

can be solved by an algorithm of complexity $O(n \log n)$ [HAM 02]. This result is based on the following property: if the variables are numbered in such a way that: $b_1/a_1 \geq b_2/a_2 \geq ... \geq b_n/a_n$, then the continuous solution is of the form $(1, 1, ..., 1, \lambda, 0, ..., 0)$ with $0 \leq \lambda \leq 1$. This particularity allows us to solve instances of this problem containing several thousand variables using an implicit enumeration algorithm.

8.9. Reductions, relaxations, linearizations, bound calculation and persistence

8.9.1. *Complementation*

Let us consider the qpBf $f(x) = \sum_{i=1}^{n} q_i x_i + \sum_{i=1}^{n-1} \sum_{j=i+1}^{n} q_{ij} x_i x_j$ and let us express it in the form $f(x) = c + \varphi(x, \bar{x})$, where c is a constant and $\varphi(x, \bar{x})$ is a quadratic posiform without a constant term. This means that for each $x \in \{0,1\}^n$, $f(x) = c + \varphi(x, \bar{x})$. We see directly that the value of c is a lower bound of the minimum of f since $\varphi(x, \bar{x}) \geq 0$ for each $x \in \{0,1\}^n$. For example, $f(x) = 2x_1 - 5x_2 + 4x_3 - 2x_1x_2 + 4x_1x_3 - 6x_2x_3$ can be expressed by $-9 + 2x_1 + 7\bar{x}_2 + 2\bar{x}_3 + 2\bar{x}_1x_2 + 4x_1x_3 + 6\bar{x}_2x_3$ or $-11 + 11\bar{x}_2 + 4x_3 + 2x_1\bar{x}_2 + 4x_1x_3 + 6x_2\bar{x}_3$. From the point of view of the lower bound, the first form is the most interesting because it tells us that the minimum of f is greater than or equal to –9, while the second only tells us that it is greater than or equal to –11. This remark leads us to the following problem: given a qpBf, how can we write it in the form $c + \varphi(x, \bar{x})$ while maximizing the value of the constant c. We wish to express the qpBf $f(x) = \sum_{i=1}^{n} q_i x_i + \sum_{i=1}^{n-1} \sum_{j=i+1}^{n} q_{ij} x_i x_j$ in the form $c + \varphi(x, \bar{x})$, that is in the form:

$$c + \sum_{i=1}^{n} (a_i x_i + b_i \bar{x}_i) + \sum_{i=1}^{n-1} \sum_{j=i+1}^{n} (r_{ij} x_i x_j + s_{ij} x_i \bar{x}_j + t_{ij} \bar{x}_i x_j + u_{ij} \bar{x}_i \bar{x}_j)$$

while maximizing the constant c. Let us write the mathematical program $(P1)$ corresponding to this maximization, taking into account the fact that the pseudo-Boolean function defined by $c + \varphi(x, \bar{x})$ allows a unique polynomial form, the function $f(x)$:

$$\left\{ \begin{array}{lll} \max \ c & & \\ q_{ij} = r_{ij} - s_{ij} - t_{ij} + u_{ij} & i = 1, ..., n-1 \quad j = i+1, ..., n & [C1.1] \\ q_i = a_i - b_i + \sum\limits_{i<j} s_{ij} + \sum\limits_{j<i} t_{ji} & & [C1.2] \\ \quad - \sum\limits_{j<i} u_{ji} - \sum\limits_{i<j} u_{ij} & i = 1, ..., n & [C1.3] \\ 0 = c + \sum\limits_{i} b_i + \sum\limits_{i<j} u_{ij} & & [C1.4] \\ a_i \geq 0, b_i \geq 0 & i = 1, ..., n & [C1.5] \\ r_{ij}, s_{ij}, t_{ij}, u_{ij} \geq 0 & i = 1, ..., n-1 \quad j = i+1, ..., n & [C1.6] \end{array} \right.$$

THEOREM 8.9.– *Let $f(x) = \sum_{i=1}^{n} q_i x_i + \sum_{i=1}^{n-1} \sum_{j=i+1}^{n} q_{ij} x_i x_j$ be a qpBf and c^* be the value of the largest constant c such that a quadratic posiform $\varphi(x, \bar{x})$ exists* *that satisfies* $f(x) = c + \varphi(x, \bar{x})$ *for each* $x \in \{0,1\}^n$. *This largest constant* c^* *is equal to the value of an optimal solution of the linear program* $P1$.

The point of calculating $c*$ is that it is a lower bound of f and, in a certain sense, the best, and also that some extra work on the posiform may possibly improve this lower bound (see section 8.7.2). Experiments concerning this limit, sometimes called the *roof dual value*, are presented in [WIL 85] (see also [LU 87] for functions of degree greater than 2).

8.9.2. *Linearization*

Let us see how to formulate the problem of minimizing the qpBf $f(x) = \sum_{i=1}^{n} q_i x_i + \sum_{i=1}^{n-1} \sum_{j=i+1}^{n} q_{ij} x_i x_j$ using a linear program in bivalent variables. Let us replace the products $x_i x_j$ with the variable y_{ij} and let us add the linearization constraints $y_{ij} \leq x_i$, $y_{ij} \leq x_j$, $1 - x_i - x_j + y_{ij} \geq 0$ and $y_{ij} \geq 0$. We obtain the mathematical program ($P2$), a classical linearization of the problem of minimizing a qpBf. It is easy to check the validity of this formulation by inspecting the different possible values of the product $x_i x_j$.

$$\left\{\begin{array}{llll} \min \sum\limits_{i=1}^{n} q_i x_i + \sum\limits_{i=1}^{n-1} \sum\limits_{j=i+1}^{n} q_{ij} y_{ij} & & & \\ y_{ij} \leq x_i & i = 1, ..., n-1 & j = i+1, ..., n & [C2.1] \\ y_{ij} \leq x_j & i = 1, ..., n-1 & j = i+1, ..., n & [C2.2] \\ 1 - x_i - x_i + y_{ij} \geq 0 & i = 1, ..., n-1 & j = i+1, ..., n & [C2.3] \\ y_{ij} \geq 0 & i = 1, ..., n-1 & j = i+1, ..., n & [C2.4] \\ x_i \in \{0,1\} & i = 1, ..., n & & [C2.5] \end{array}\right.$$

The optimal value of $\underline{P2}$, the continuous relaxation of $P2$, is a lower bound of the minimum of $f(x)$, but we also have theorem 8.10 [HAM 84], which is shown by pointing out that the linear program $\underline{P2}$ is the dual of the linear program $P1$.

THEOREM 8.10.– *The optimal value of the linear program* $\underline{P2}$*, the continuous relaxation of the linear program in bivalent variables* $P2$*, is equal to* c^*.

Note that the constraints [C2.1] and [C2.2] can be limited to the pairs (i, j) such that $q_{ij} < 0$, the constraint [C2.3] to the pairs (i, j) such that $q_{ij} > 0$, and the constraint [C2.4] to the pairs (i, j) such that $q_{ij} \neq 0$ (see also [BOR 90]). Let us add to the five constraints of $P2$ a sixth constraint, [C2.6]:

$$y_{ij} \in \{0,1\}, \ i = 1, ..., n-1, j = i+1, ..., n$$

The quadratic Boolean polytope is defined by:

$$QP^n = conv \left\{ (x, y) \in \mathbb{R}^{n(n+1)/2} : C2.1 - C2.6 \right\}$$

There are many families of facets of QP^n (see, for example, [BOR 92, DES 89, FAY 03a, PAD 89]). The constraints [C2.1], [C2.2], [C2.3] and [C2.4] correspond to trivial facets. By considering three indices i, j, k such that $1 \leq i < j < k \leq n$, we obtain other valid inequalities that are facets. We call them *triangular inequalities*. The inequalities $x_i - y_{ij} - y_{ik} + y_{jk} \geq 0$, $x_j - y_{ij} + y_{ik} - y_{jk} \geq 0$, and $x_k + y_{ij} - y_{ik} - y_{jk} \geq 0$ are the "cut" inequalities, and $1 - x_i - x_j - x_k + y_{ij} + y_{ik} + y_{jk} \geq 0$ are the "clique" inequalities. The reader can refer to Chapter 10 for a general introduction to polyhedral methods.

8.9.3. *Lagrangian duality*

Let us consider the qpBf $f(x) = \sum_{i=1}^{n} q_i x_i + \sum_{i=1}^{n-1} \sum_{j=i+1}^{n} q_{ij} x_i x_j$ and let us state that

$$S = \left\{ (i,j) \in \{1, ..., n\}^2 : i < j, q_{ij} < 0 \right\}$$

$$T = \left\{ (i,j) \in \{1, ..., n\}^2 : i < j, q_{ij} > 0 \right\}$$

When separating the quadratic terms whose coefficients are positive from the quadratic terms whose coefficients are negative, the function $f(x)$ is expressed as $f(x) = \sum_{i=1}^{n} q_i x_i + \sum_{(i,j) \in S} q_{ij} x_i x_j + \sum_{(i,j) \in T} q_{ij} x_i x_j$. By replacing x_i with $1 - \bar{x}_i$ in the quadratic terms with positive coefficients and by grouping the linear terms obtained, we obtain an expression of f as a function of x_i and $\bar{x}_i$ for which all the quadratic terms have a negative coefficient,

$$f(x, \bar{x}) = \sum_{i=1}^{n} l_i x_i + \sum_{(i,j) \in S} q_{ij} x_i x_j - \sum_{(i,j) \in T} q_{ij} \bar{x}_i x_j$$

By renaming the variable $\bar{x}_i$ as y_i, we can formulate the problem of minimizing f as that of minimizing a positive–negative function with constraints; this is the following program ($P3$):

$$\begin{cases} \min f(x) = \sum_{i=1}^{n} l_i x_i + \sum_{(i,j) \in S} q_{ij} x_i x_j - \sum_{(i,j) \in T} q_{ij} y_i x_j \\ x_i + y_i = 1 \quad i = 1, ..., n \quad [C3.1] \\ x_i \in \{0, 1\} \quad i = 1, ..., n \quad [C3.2] \end{cases}$$

It is then natural, to obtain a bound to the minimum value of f, to study the dual Lagrangian problem of $P3$, since minimizing positive–negative pseudo-Boolean functions without constraints is an easy (polynomial) problem. By denoting by λ_i the Lagrange multiplier associated with the constraint $x_i + y_i = 1$, we obtain the Lagrange function:

$$L(x, y, \lambda) = \sum_{i=1}^{n} (l_i + \lambda_i) x_i + \sum_{i=1}^{n} \lambda_i y_i + \sum_{(i,j) \in S} q_{ij}\, x_i x_j - \sum_{(i,j) \in T} q_{ij}\, y_i x_j - \sum_{i=1}^{n} \lambda_i$$

the dual function $w(\lambda) = \min\{L(x, y, \lambda) : x \in \{0,1\}^n, y \in \{0,1\}^n\}$ and the dual problem $\max\{w(\lambda) : \lambda \in \mathbb{R}^n\}$. Theorem 8.11 shows that in fact the three approaches (complementation, linearization and Lagrangian duality) introduced above to obtain lower bounds to the minimum value of f all give the same bound [ADA 90, ADA 94a, HAN 90, MIC 93].

THEOREM 8.11.– *opt*($\underline{P2}$) $= c^* = \max\{w(\lambda) : \lambda \in \mathbb{R}^n\}$.

Let us make clear that the bound c^* and the corresponding posiform $\varphi(x, \bar{x})$, that is rewriting f in the form $c^* + \varphi(x, \bar{x})$, can be calculated efficiently by determining a maximum flow in a bipartite graph [HAM 84].

8.9.4. *Another linearization*

In this section we introduce a formulation of the problem of maximizing a qpBf using a mixed integer linear program that contains $O(n)$ variables and $O(n)$ constraints. Let the qpBf to be maximized be $f(x) = \sum_{i=1}^{n} q_i x_i + \sum_{i=1}^{n-1} \sum_{j=i+1}^{n} q_{ij} x_i x_j$. Let us denote by z_i the quantity $x_i \sum_{j=i+1}^{n} q_{ij} x_i x_j$. Maximizing f can now be stated as the following mixed integer linear program ($P4$):

$$\begin{cases} \max \sum_{i=1}^{n} q_i x_i + \sum_{i=1}^{n-1} z_i & \\ z_i \leq B_i^1 x_i & i = 1, ..., n-1 \quad [C4.1] \\ z_i \leq \sum_{j=i+1}^{n} q_{ij} x_j - B_i^2 (1 - x_i) & i = 1, ..., n-1 \quad [C4.2] \\ x_i \in \{0,1\} & i = 1, ..., n \quad [C4.3] \end{cases}$$

with $B_i^1 = \sum_{j|j>i, q_{ij}>0} q_{ij}$ and $B_i^2 = \sum_{j|j>i, q_{ij}<0} q_{ij} (i = 1, ..., n-1)$. We can easily check that the formulation is correct by successively inspecting the case where $x_i = 1$, then the case where $x_i = 0$. This linearization contains n variables x_i, n-1 variables z_i, and 2(n-1) constraints, without taking into account the integrity constraints on the variables x_i. This type of linearization has been proposed in [GLO 75] for the general problem of optimizing qpBfs with linear constraints. We see directly that there are several ways of using this linearization principle. In [ADA 04], the authors study the best practices, in a specific sense and in the slightly more general context of mixed integer quadratic programming.

8.9.5. *Convex quadratic relaxation*

Let us express the qpBf $f(x) = \sum_{i=1}^{n} c_i x_i + \sum_{i=1}^{n-1} \sum_{j=i+1}^{n} q_{ij} x_i x_j$ in the matrix form $f(x) = c^T x + x^T Q x$. The vector $c = (c_1, ..., c_n)$ corresponds to the coefficients of the linear terms and Q is a symmetric square matrix whose diagonal terms are zero and whose general term Q_{ij} is equal to $\frac{1}{2} q_{ij}$. It is easy to check that for each $x \in \{0,1\}^n$, $f(x) = c^T x + u^T x + x^T (Q - diag(u)) x$, where $diag(u)$ is a square matrix $n \times n$, all of whose terms are zero apart from the diagonal terms, which are equal to $u_1, ..., u_n$. We know that the function $c^T x + u^T x + x^T(Q - diag(u))x$ is convex if $Q - diag(u)$ is a positive matrix (see the definition in the following section). In this case, calculating the minimum of f with the constraint $x \in [0,1]^n$ is an easy problem. It is a relaxation of the initial problem that therefore gives a lower bound to the minimum value of f with the constraint $x \in \{0,1\}^n$. A simple way of obtaining a positive matrix $Q - diag(u)$ is to choose the vector $u = (u_1, ..., u_n)$ such that for each $i \in \{1, ..., n\}$, $u_i = \lambda_{\min}$, the smallest eigenvalue of Q [CAR 84, HAM 70, PLA 05]. We will see an algorithm for minimizing qpBfs based on this idea of convex quadratic relaxation in section 8.11.3. Let us note here that, for reasons of convexity, the maximum of f with the constraint $x \in [0,1]^n$ is a point of $\{0,1\}^n$. On the other hand, establishing this maximum is a hard problem. Recognizing the convexity of certain extensions of pseudo-Boolean functions is studied in [CRA 93].

EXAMPLE 8.5.– Let us consider the following qpBf to be minimized: $f(x) = 37x_1 + 18x_2 + 86x_3 + 69x_4 - 90x_1x_2 + 36x_1x_3 + 36x_1x_4 + 4x_2x_3 + 66x_2x_4 + 34x_3x_4$. The smallest eigenvalue of the associated matrix Q is $\lambda_{\min} = -65.46$. A relaxation of the problem therefore consists of minimizing the function $(37 - 65.46)x_1 + (18 - 65.46)x_2 + (86 - 65.46)x_3 + (69 - 65.46)x_4 + 65.46(x_1^2 + x_2^2 + x_3^2 + x_4^2) - 90x_1x_2 + 36x_1x_3 + 36x_1x_4 + 4x_2x_3 + 66x_2x_4 + 34x_3x_4$ with the constraint $(x_1, ..., x_4) \in [0,1]^4$. We find as the optimum of this relaxation $x = (0.88 \quad 0.97 \quad 0 \quad 0)$ and its value is -35.62. The integer optimum is (1 1 0 0) and its value is -35.

8.9.6. *Positive semi-definite relaxation*

Positive semi-definite programming is known for the spectacular results that it has allowed us to obtain in combinatorial optimization, in the domain of ε-approximation algorithms. Parallel to these studies, more and more effective interior point algorithms have been developed (see Chapter 7 for interior point methods in linear programming). In this way it has become possible to solve hard combinatorial optimization problems in an exact or approximate manner, through positive semi-definite relaxations (see, for example, [DEL 09, HEL 98, POL 95a, POL 95b, ROU 04]). Given a real, square and symmetric matrix A of dimension $n \times n$, we say that this matrix is positive, which we denote by $A \dot{\geq} 0$, if and only if $\forall y \in \mathbb{R}^n$, $y^T A y \geq 0$. We call the mathematical program $(P5)$ a (primal) positive semi-definite program:

$$\min \left\{ \sum_{i=1}^{m} c_i x_i : \sum_{i=1}^{m} x_i A_i - A_0 \dot{\geq} 0 \right\}$$

where $A_0, A_1, ..., A_n$ are real, square and symmetric matrices $n \times n$, where the coefficients $c_i (i = 1, ..., m)$ are real numbers and where the variables x_i $(i = 1, ..., m)$ are real numbers. The tensorial product of the vector x by itself, which we express by xx^T, is the square symmetric matrix whose general term (i,j) is equal to $x_i x_j$. Given A and B, two square symmetric matrices of dimensions $n \times n$, the trace of AB, denoted by $A \bullet B$, is equal to $\sum_{i=1}^{n} \sum_{j=1}^{n} A_{ij} B_{ij}$. Let us now consider the positive semi-definite program ($P6$), the dual of the program ($P5$):

$$\max \left\{ A_0 \bullet Z : A_i \bullet Z = c_i, \ i = 1, ..., m, Z \dot{\geq} 0 \right\}$$

where the unknown is the square and symmetric matrix Z, of dimension $n \times n$ and with the general term Z_{ij}. The problem:

$$\max \left\{ \sum_{i=1}^{n} q_i x_i + \sum_{i=1}^{n-1} \sum_{j=i+1}^{n} q_{ij} x_i x_j : x \in \{0,1\}^n \right\}$$

can be written $\max \left\{ q^T x + \frac{1}{2} Q \bullet x x^T : x \in \{0,1\}^n \right\}$, where Q is a square and symmetric matrix of dimensions $n \times n$. A relaxation of this problem is obtained in the following way: we replace the matrix xx^T with the matrix X and, instead of adding the constraint $X = xx^T$, we only add the constraint $X - xx^T \dot{\geq} 0$. Furthermore, we relax the integrity constraint on the variables x by only imposing that they be between 0 and 1. In this way we obtain the following mathematical program ($P7$), which is really a relaxation of the initial problem, since the matrix whose terms are all zero is a positive matrix:

$$\max \left\{ q^T x + \frac{1}{2} Q \bullet X : X - xx^T \dot{\geq} 0, x \in [0,1]^n \right\}$$

We can show that the constraints set ($X - xx^T \dot{\geq} 0$, $x \in [0,1]^n$) is equivalent to the constraints set ($\begin{pmatrix} 1 & x^T \\ x & X \end{pmatrix} \dot{\geq} 0$, $d(X) = x$), where $d(X)$ refers to the vector obtained from the diagonal elements of the matrix X. A positive semi-definite relaxation of the initial problem is thus given by the following mathematical program ($P8$):

$$\left\{ \begin{array}{ll} \max q^T x + \frac{1}{2} Q \bullet X & \\ \quad d(X) = x & [C8.1] \\ \begin{pmatrix} 1 & x^T \\ x & X \end{pmatrix} \dot{\geq} 0 & [C8.2] \\ x \in \mathbb{R}^n, X \in \mathbb{R}^{n \times n} & [C8.3] \end{array} \right.$$

that is the positive semi-definite program ($P9$):

$$\left\{ \begin{array}{lll} \max \ C_0 \bullet Z & & \\ Z \dot{\geq} 0 & & [C9.1] \\ z_{00} = 1 & & [C9.2] \\ z_{ii} = z_{0i} & i = 1, ..., n & [C9.3] \end{array} \right.$$

with:

$$C_0 = \begin{pmatrix} 0 & \frac{1}{2} q^T \\ \frac{1}{2} q & \frac{1}{2} Q \end{pmatrix}$$

8.9.7. *Persistence*

Let us again consider the problem of minimizing a qpBf $f(x)$. Let us establish the largest constant c^* and the posiform $\phi(x, \bar{x})$ such that for each $x \in \{0,1\}^n$,$f(x) = c^* + \phi(x, \bar{x})$. We can show that certain variables have a set value in any optimum x^* of $f(x)$. Algorithm 8.3 [BIL 92f], which is particularly easy to use, allows us to establish these variables and their values.

Algorithm 8.3

1: express $f(x)$ in the form $f(x) = c^* + \phi(x, \bar{x})$
2: **while** the posiform contains linear terms **do**
3: set the variables in such a way as to cancel the linear terms
4: simplify the posiform taking into account the preceding settings
5: **end while**

EXAMPLE 8.6.– Let the function $f(x) = -3x_1 - 2x_3 - 10x_4 + 2x_5 + x_6 - 6x_1x_2 + 2x_1x_3 + 10x_1x_4 - 2x_2x_3 + 8x_2x_4 - 3x_5x_6$. The largest constant is equal to -12 and the quadratic function $f(x)$ is expressed by $f(x) = -12 + \bar{x}_3 + x_1x_3 + 8x_1x_4 + \bar{x}_1\bar{x}_3 + 2\bar{x}_1\bar{x}_4 + 2x_2\bar{x}_3 + 8\bar{x}_2\bar{x}_4 + 2x_5\bar{x}_6 + 6\bar{x}_1x_2 + \bar{x}_5x_6$. Let us apply algorithm 8.3: $x_3 = 1$ at any point minimizing $f(x)$. We can therefore say:

$$\min f(x) = -12 + \min\{x_1 + 8x_1x_4 + 2\bar{x}_1\bar{x}_4 + 8\bar{x}_2\bar{x}_4 + 2x_5\bar{x}_6 + 6\bar{x}_1x_2 + \bar{x}_5x_6\}$$

We then see that in any optimum of $f(x)$, $x_1 = 0$. We deduce from this the new equality $\min f(x) = -12 + \min\{2\bar{x}_4 + 8\bar{x}_2\bar{x}_4 + 2x_5\bar{x}_6 + 6x_2 + \bar{x}_5x_6\}$. We then see that in any optimum of $f(x)$, $x_4 = 1$ and $x_2 = 0$. Finally, by setting the variables x_5 and x_6 to 0, we obtain $\min f(x) = -12 + \min\{2x_5\bar{x}_6 + \bar{x}_5x_6\} = -12$. We can find other results concerning persistence in [ADA 98] and graph theory applications in [HAM 81].

8.10. Local optimum

We say that $\tilde{x}$ is a local minimum of a pseudo-Boolean function $f(x)$ if and only if, for each $i \in \{1, ..., n\}$, $f(\tilde{x}_1, ..., \tilde{x}_i, ..., \tilde{x}_n) \leq f(\tilde{x}^i) = f(\tilde{x}_1, ..., 1 - \tilde{x}_i, ..., \tilde{x}_n)$. How do we find a local optimum? What is the complexity of the problem? Algorithm 8.4 enables us to find a local optimum. Note that this algorithm can be used as a heuristic to find a good solution to the problem $\min\{f(x) : x \in \{0,1\}^n\}$. We simply need to apply it a great many times with different initial points. For example, it often gives us the optimal solution of the problem or a solution that is very close to the optimal solution for qpBfs generated randomly with integer coefficients and uniformly distributed over an interval $[a, b]$.

We do not know if this algorithm is polynomial. We only know that, for a qpBf with integer coefficients, the number of iterations is bounded by the absolute value of

Algorithm 8.4

```
1: choose x̃ ∈ {0,1}^n
2: establish i such that f(x̃) − f(x̃^i) = max{ f(x̃) − f(x̃^k) : k = 1,...,n }
3: if f(x̃) − f(x̃^i) ≤ 0 then
4:     stop: x̃ is a local optimum
5: else
6:     x̃ ← x̃^i
7: end if
8: go back to 2.
```

the minimum of $f(x)$ if we take the vector $(0, ..., 0)$ as the initial point. Indeed, this point gives the value 0 to the function and, with each iteration of the algorithm, the improvement in the value of the function is at least one unit. Many experiments have been done with this algorithm. They have allowed us to observe that the number of iterations was, in general, less than n, the number of variables. The reader can refer to Chapter 14 of Volume 2 for a general introduction to local searching.

Let us now consider a pBf with a single local optimum $f(x)$ (this local optimum is therefore a global optimum) and a constant K. The decision problem (Π): "Does $x \in \{0,1\}^n$ exist such that $f(x) \leq K$?" belongs to **NP**. Let us also consider the complementary problem $(\overline{\Pi})$: "Is it true that there is no $x \in \{0,1\}^n$ such that $f(x) \leq K$?". By definition, this last problem belongs to **co-NP**. It is, however, easy to show that it also belongs to **NP**. In fact, the polynomial certificate of the reply "yes" is given by showing the local (global) minimum x^*. We prove polynomially that this is a global optimum by checking that modifying any component of x^* does not improve the value of the function. We therefore deduce from this that the decision problem (Π) belongs to the intersection of the two classes **NP** and **co-NP** and we can therefore assume that it is a polynomial problem. We can also, therefore, assume that there is a polynomial algorithm for optimizing pBfs with a single local minimum. On the other hand, we do not know if knowing such an algorithm would allow us to construct an algorithm for establishing the local minimum of any pBf. Let us note here that recognizing a pBf with a single local minimum is an **NP**-complete problem, even for cubic functions [CRA 89]. It is an open problem for qpBfs. Let us also point out that, for any qpBf $f(x)$, the problem consisting of establishing a local minimum $(\tilde{x}_1, ..., \tilde{x}_n)$ of f, such that $\tilde{x}_{n-1} = \tilde{x}_n = 0$, is **NP**-hard [PAR 92]. Let us now give some experimental results concerning calculating local optima using algorithm 8.4. We consider a qpBf with 200 variables, generated randomly. The coefficients of the linear and quadratic terms of this function are integers uniformly distributed over the interval $[-100, 100]$. The previous algorithm has been applied 200 times, choosing a new initial point at random each time. This point is generated in the following way: for each $i \in \{1, ..., 200\}$, the variable x_i is set to 1 with probability $1/2$ and, therefore, to 0 with the same probability. All the initial points chosen at random have given to

$f(x)$ a value approximately between $-10\ 000$ and $+10\ 000$, and all the local optima obtained by algorithm 8.4 have given to $f(x)$ a value relatively close to $-45\ 000$. In this instance, we have obtained $-56\ 000$ as the lower bound using the convex quadratic relaxation method based on the smallest eigenvalue method introduced in section 8.9.5.

Let us point out here that the decision problem concerning the unicity of a qpBf $f(x)$ minimum, "Does $f(x)$ allow a single (global) minimum?", is **NP**-complete [PAR 92].

8.11. Exact algorithms and heuristic methods for optimizing qpBfs

8.11.1. *Different approaches*

Three syntheses for the solution of the problem of optimizing qpBfs are presented in [BEA 98, HAN 00, HEL 98]. In reference [HAN 00] the authors give a quick classification of the different exact solution algorithms that have been proposed in the literature: algebraic and dynamic programming methods [CRA 90, HAM 68, ROS 74]; linearization methods followed by solution of a 0–1 variables or mixed variables linear program [FOR 59, FOR 60, GLO 73, GLO 74, GLO 75, GUE 02, ORA 92]; cut methods [ADA 86, BAR 89, GRA 76, HEL 98]; methods consisting of reformulating the problem as the minimization of a continuous concave function [KAL 90, KON 80, THO 98]; and, lastly, enumeration methods [BIL 94b, BIL 07, CHA 95, GUL 84, HAM 72, HAN 00, HAN 70, HAN 71, HEN 83, LU 84, PAR 90]. Note that many studies combine several approaches. The reader can refer to Chapter 3 for a general presentation of branch-and-bound methods. More recently, new solution methods, based on second-order cone constraint programming, have been proposed [KIM 01, MUR 03]. This concerns non-linear convex programs whose linear programming and quadratic (convex) programming are special cases. They can be expressed as:

$$\min \left\{ e^t x : \| A_i x + b_i \| \leq c_i^t x + d_i \ (i = 1, ..., N) \right\}$$

with $e \in \mathbb{R}^n$, $A_i \in \mathbb{R}^{(n_i-1)\times n}$, $b_i \in \mathbb{R}^{n_i-1} c_i \in \mathbb{R}^n$ and $d_i \in \mathbb{R}$. The vector $x \in \mathbb{R}^n$ is the vector of the variables and the considered norm is the standard Euclidean norm, that is $\| u \| = (u^t u)^{1/2}$.

Note that it is generally difficult to find the optimum of quadratic pseudo-Boolean functions containing about a hundred variables and that this difficulty strongly depends on the density of the function (number of non-zero coefficients divided by $n(n+1)/2$, where n refers to the number of variables). We introduce two algorithms below. The first allows us to calculate very good quality lower bounds; it is based on a decomposition method inspired by Lagrangian decomposition. The second consists of finding a good reformulation of the problem in the form of a minimization of a convex quadratic function. This algorithm then uses general optimization software. It gives very good results and, using general software, is relatively easy to put to work. Very many heuristic methods have been proposed for optimizing a qpBf. We will not describe them here for lack of space. The method described in section 8.10 is an efficient method and it is particularly easy to set up. Beasley [BEA 98] describes the application of classical metaheuristics (tabu search, simulated annealing, genetic algorithm) to the problem of optimizing a qpBf and gives experimental results. The reader can also refer to the article by Glover *et al.* [GLO 00], which describes a method that is applicable to very large functions, and which gives a great number of references. We also find in [MER 05] a brief synthesis of these methods and very many references. The references [ALK 98, GLO 98, LOD 99, PAL 95] also deal with heuristic methods.

8.11.2. *An algorithm based on Lagrangian decomposition*

Let us now present the outline of an algorithm that is very effective for establishing a "good" lower bound of the minimum of a qpBf (see [CHA 95]):

$$f(x) = \sum_{i=1}^{n} c_i x_i + \sum_{i=1}^{n} \sum_{j \neq i} c_{ij} x_i x_j$$

Let us express the function $f(x)$ to be minimized as the sum of p qpBf:

$$f(x) = \sum_{k=1}^{p} f_k(x)$$

We will assume that minimizing the functions f_k is not too difficult, that is that it can be done in a "reasonable" CPU time. A lower bound of the minimum of $f(x)$ is given, obviously, by the quantity:

$$\sum_{k=1}^{p} \min \{ f_k(x) : x \in \{0,1\}^n \}$$

The problem is therefore to establish these functions $f_k(x)$ in such a way as to have a "good" bound for the minimum of $f(x)$. In general, this problem is expressed in terms of the following program $(P10)$:

$$\max \left\{ \sum_{k=1}^{p} \min_{x \in \{0,1\}^n} f_k(x) : f_1, ..., f_p \text{qpBf and } f(x) = \sum_{k=1}^{p} f_k(x) \right\}$$

In fact, the authors limit themselves to the functions $f_k, of \{0,1\}^n$ in $\mathbb{R}$ made up of a quadratic function plus a bilinear function. Let us consider a partition of $X = \{x_1, ..., x_n\}$ in p clusters $X_1, X_2, ..., X_p$. For $k = 1, ..., p$ let us state that $Y_k = X - X_k$, I_k = the set of indices of the variables of X_k, and J_k = the set of indices of the variables of Y_k. Let us now define, for $k = 1, ..., p$, the functions:

$$f_k(x) = \sum_{i=1}^{n} \alpha_i^k x_i + \sum_{i \in I_k} \sum_{j \in I_k - \{i\}} c_{ij} x_i x_j + \sum_{i \in I_k} \sum_{j \in J_k} \alpha_{ij}^k \; x_i x_j$$

where the different coefficients satisfy the following conditions in order to satisfy the equality $f(x) = \sum_{k=1}^{p} f_k(x)$ for each $x \in \{0,1\}^n$:

– [C1]: $\sum_{k=1}^{p} \alpha_i^k = c_i \ (i = 1, ..., n)$;

– [C2]: $\alpha_{ij}^{cl(i)} + \alpha_{ji}^{cl(j)} = c_{ij} + c_{ji} \ ((i,j) \in \{1, ..., n\}^2 \ ; cl(i) \neq cl(j))$, where $cl(i)$ is the index of the cluster containing the variable x_i.

Note that when the variables of I_k are fixed, the reduced function $f_k(x)$ is a linear function. Given a partition $X_1, ..., X_p$, finding the best decomposition comes down to solving the problem:

$$\max \left\{ \sum_{k=1}^{p} \min_x f_k(x) : \alpha \text{ satisfying } C1 \text{ and } C2 \right\}$$

The authors show that solving this problem comes down to solving a Lagrangian dual problem. They solve it by a subgradient algorithm in which, at each stage, the value of the dual function is established heuristically by the simulated annealing method [KIR 83]. At the last iteration, its value is established exactly by a *branch-and-bound* algorithm. This Lagrangian decomposition type approach enables us to establish, using clusters of 25 variables, lower bounds at less than 3% of the optimum on instances of 100 randomly generated variables, whatever the density of the function, on an HP-720 workstation. This decomposition method has been successfully applied to other 0–1 optimization problems in [BIL 99, BIL 04b, FAY 94].

8.11.3. *An algorithm based on convex quadratic programming*

Let us now consider the problem of minimizing a qpBf expressed in its matrix form: $\min \{ q(x) = x^t Q x + c^t x : x \in \{0,1\}^n \}$, where Q is a real matrix and $c \in \mathbb{R}^n$. Without loss of generality, we can assume that the matrix Q is symmetric and that all the diagonal elements are zero. The method [BIL 07] consists of considering the perturbed function $q_u(x) = x^t(Q - diag(u))x + (c + u)^t x$ instead of $q(x)$, choosing the vector u in such a way that the matrix $(Q - diag(u))$ is positive, in order to obtain a convex function ($diag(u)$ refers to the diagonal matrix constructed from the vector u). Let us also note that the two problems $\min \{ q(x) : x \in \{0,1\}^n \}$ and $\min \{ q_u(x) : x \in \{0,1\}^n \}$ are equivalent, since, in the case of bivalent variables, $x_i^2 = x_i$. In order to obtain a "good" vector u, we choose it in such a way as to maximize the quantity $\min \{ q(x) : x \in [0,1]^n \}$, while keeping the matrix $(Q - diag(u))$ positive. In this way we seek to obtain a "good" lower bound (the best in a certain sense) of the root of the *branch-and-bound* procedure which will be used to solve the problem, because it is well known that the effectiveness of the procedure strongly depends on the quality of this bound. Let u^* be such a vector. In practice, it is obtained by solving the following positive semi-definite program ($P11$):

$$\begin{cases} \max r & \\ \begin{pmatrix} -r & \frac{1}{2}(c+u)^T \\ \frac{1}{2}(c+u) & Q - diag(u) \end{pmatrix} \dot{\succeq} 0 & [C11.1] \\ r \in \mathbb{R}, u \in \mathbb{R}^n & [C11.2] \end{cases}$$

The method next consists of solving the problem $\min \{ q_{u^*}(x) : x \in \{0,1\}^n \}$ by using a solver that allows us to minimize convex quadratic functions with integer variables. In each node of the search tree, the lower bound is then given by the minimum of a convex quadratic function whose variables must take values between 0 and 1, which is calculated effectively. The implementation is therefore particularly easy if we have such a solver and experiments show that the method is effective for solving instances containing about a hundred variables, whatever the density. Let us note, however, that, in this method, a non-trivial but quick pretreatment is necessary: the calculation of u^*. This pretreatment can be simplified by replacing u^* with $-\lambda_{\min}(Q)$ but the continuous relaxation is then not as good.

8.12. Approximation algorithms

Given the problem of maximizing $\max\{f(x) : x \in S\}$ with positive values and a constant $\varepsilon \geq 1$, we will define as the ε-approximation algorithm for this problem, an algorithm that, for any instance I, provides an feasible solution $x^a(I)$ such that: $f(x^*(I))/f(x^a(I)) \leq \varepsilon$, $x^*(I)$, referring to the optimal solution associated with the instance I. For a minimization problem, an ε-approximation algorithm satisfies $f(x^a(I))/f(x^*(I)) \leq \varepsilon$. From an optimization point of view, both the problems $\{max\ f(x) : x \in S\}$ and $\{max\ f(x) + K : x \in S\}$, where K refers to a positive or negative constant, are equivalent. It is easy to verify that this is no longer the case from an approximation point of view. We introduce below a few results concerning posiform optimization and also the MAX-SAT problem, introduced in section 8.4, which can easily be given, from an optimization point of view, as a posiform minimization problem. The reader can refer to Chapter 11, Volume 2 for a general introduction to polynomial approximation and to [BAD 96] for pseudo-Boolean functions.

8.12.1. *A 2-approximation algorithm for maximizing a quadratic posiform*

Let the quadratic posiform to be maximized be defined by:

$$f(x,\bar{x}) = \sum_{i=1}^{n} (a_i x_i + b_i \bar{x}_i) + \sum_{i=1}^{n-1} \sum_{j=i+1}^{n} (r_{ij} x_i x_j + s_{ij} x_i \bar{x}_j + t_{ij} \bar{x}_i x_j + u_{ij} \bar{x}_i \bar{x}_j)$$

All the coefficients $a_i, b_i, r_{ij}, s_{ij}, t_{ij}$ and u_{ij} are positive. Let us state this problem as a linear program $(P12)$ in bivalent variables:

$$\left\{\begin{array}{llll} \max f_L(x,\bar{x},y,z,v,w) = \sum\limits_{i=1}^{n} (a_i x_i + b_i \bar{x}_i) & & & \\ + \sum\limits_{i=1}^{n-1} \sum\limits_{j=i+1}^{n} (r_{ij}y_{ij} + s_{ij}z_{ij} + t_{ij}v_{ij} + u_{ij}w_{ij}) + M \sum\limits_{i=1}^{n} (x_i + \bar{x}_i - 1) & & & \\ \quad x_i + \bar{x}_i \leq 1 & i = 1,...,n & & [C12.1] \\ \quad y_{ij} \leq 1 - \bar{x}_i; y_{ij} \leq 1 - \bar{x}_j & i = 1,...,n-1 & j = i+1,...,n & [C12.2] \\ \quad z_{ij} \leq 1 - \bar{x}_i; z_{ij} \leq 1 - x_j & i = 1,...,n-1 & j = i+1,...,n & [C12.3] \\ \quad v_{ij} \leq 1 - x_i; v_{ij} \leq 1 - \bar{x}_j & i = 1,...,n-1 & j = i+1,...,n & [C12.4] \\ \quad w_{ij} \leq 1 - x_i; w_{ij} \leq 1 - x_j & i = 1,...,n-1 & j = i+1,...,n & [C12.5] \\ \quad x_i, \bar{x}_i \in \{0,1\} & i = 1,...,n & & [C12.6] \\ \quad y_{ij}, z_{ij}, v_{ij}, w_{ij} \in \{0,1\} & i = 1,...,n-1 & j = i+1,...,n & [C12.7] \end{array}\right.$$

where M is a sufficiently large constant. Let us note that the continuous relaxation of $P12$, which we express by $\overline{P12}$, can be considered as the continuous relaxation of a natural linearization of the maximum weight stable set problem in the graph made up from the vertices x_i, $\bar{x}_i (i = 1, ..., n)$, $y_{ij}, z_{ij}, v_{ij}, w_{ij} (i = 1, ..., n-1; j = i+1, ..., n)$, and in which there is an edge between two vertices a and b if the constraint $a + b \leq 1$ is present in the program $P12$. We can therefore establish an optimal solution of $\overline{P12}$ whose components are 0, $1/2$ or 1 using a maximum flow algorithm [PIC 77]. Let $(x, \bar{x}, y, z, v, w)$ be such a solution. We can easily show that $f(x, \bar{x}) \geq f_L(x, \bar{x}, y, z, v, w)/2$. Moreover, using algorithm 8.5 from section 8.5, we can calculate an (integer) approximate solution x^a in polynomial time such that $f(x^a) \geq f(x, \bar{x})$. From this we finally deduce theorem 8.12.

THEOREM 8.12.– *The problem of maximizing a quadratic posiform allows a 2-approximation polynomial algorithm* $(f(x^*)/f(x^a) \leq 2)$.

Using positive semi-definite programming, it is possible to obtain a better guarantee: 1.164 instead of 2 [FEI 95]. Negative results concerning the approximation of the maximum of quadratic posiforms can be deduced from known results for the MAX-CUT problem introduced in section 8.4. The interested reader can refer to [HOC 97]. Let us mention here only that the MAX-CUT problem allows a 1.138-approximation polynomial algorithm [GOE 95]. Note that if $P \neq NP$, then an ε-approximate polynomial algorithm, ε being a constant, for a posiform of any degree does not exist. There are many negative results concerning the approximation of the cardinal of the maximum cardinal clique in a graph [HOC 97]. These results can be used to study the approximation of the maximum of a posiform of any degree, since the cardinal of the maximum cardinal clique of a graph can be interpreted as the maximum value of a certain posiform.

8.12.2. MAX-SAT *approximation*

Let us now consider the MAX-SAT problem, which was introduced in section 8.4, and its weighted version WEIGHTED MAX-SAT. In this version, a positive or zero weight is assigned to each clause and we seek to satisfy a set of clauses whose sum associated weight is maximal. Although the MAX-SAT problem is easily stated as the minimization of a posiform, there are ε-approximate polynomial algorithms for MAX-SAT, while none exist for the minimization of a posiform, at least if $P \neq NP$. By relying on all known techniques for approximating the solution of the WEIGHTED MAX-SAT problem, it is possible to construct a 1.29-approximation polynomial algorithm for this problem [ASA 97]. We obtain the same guarantee for the MAX-SAT problem. For the WEIGHTED MAX-2-SAT problem, there is a 1.074-approximation polynomial algorithm based on positive semi-definite programming [CRE 96, FEI 95]. The MAX-k-SAT problem, where each clause contains exactly k literals, allows a $(1/(1-2^{-k}))$-approximation polynomial algorithm [JOH 74]. Lastly, the MIN-k-SAT problem, which consists of minimizing the number of clauses satisfied in a problem where the number of literals per clause is less than or equal to k, allows a $2(1-1/2^k)$-approximation algorithm [BER 96].

8.13. Optimizing a quadratic pseudo-Boolean function with linear constraints

In this section and in the following two sections, we tackle the optimization of a function of bivalent variables subject to some constraints. This optimization model is extremely general and we limit ourselves here to the problem of optimizing a quadratic pseudo-Boolean function with linear constraints, expressed as the following program ($P13$):

$$\begin{cases} \min f = \sum_{i=1}^{n} c_i x_i + \sum_{i=1}^{n-1} \sum_{j=i+1}^{n} c_{ij} x_i x_j & & \\ \sum_{j=1}^{n} a_{kj} x_j \leq b_k & k = 1, ..., p & [C13.1] \\ \sum_{j=1}^{n} a_{kj} x_j = b_k & k = p+1, ..., p+m & [C13.2] \\ x_i \in \{0, 1\} & i = 1, ..., n & [C13.3] \end{cases}$$

Very many combinatorial optimization problems are formulated as for $P13$. Due to lack of space, we will only tackle certain aspects of this model. Some examples are presented below. The reader interested in the problem of optimizing a non-linear function with bivalent variables subject to some *non-linear* constraints can refer to [HAN 93], where four approaches are introduced: linearizations, algebraic methods, enumeration methods, and cut methods. The reference [CHA 00a] introduces a linearization technique for a general problem of non-linear programming in mixed variables (binary variables and real number variables). The program is expressed in a polynomial form, where each monomial appearing in the program is a product of bivalent variables and of a real number variable. The objective function to optimize is a sum of monomials and the constraints specify that a certain sum of monomials must be non-negative (see also [CHA 00b]).

The constrained problem:

$$\min\left\{f(x):\sum\nolimits_{i=1}^{n}a_{ki}x_i=b_k(k=1,...,m),x\in\{0,1\}^n\right\}$$

where $f(x)$ is a qpBf, is easily formulated as a non-constrained problem:

$$\min\left\{f(x)+M\sum\nolimits_{k=1}^{m}(\sum\nolimits_{i=1}^{n}a_{ki}x_i-b_k)^2:x\in\{0,1\}^n\right\}$$

where M is a sufficiently large positive constant. If the constraints are given in the form of inequalities, that is $\sum_{i=1}^{n}a_{ki}x_i\leq b_k$, with integer coefficients, we can transform them into equality constraints:

$$\sum\nolimits_{i=1}^{n}a_{ki}x_i+\sum\nolimits_{i=1}^{i(k)}2^{i-1}y_i=b_k$$

where the variables y_i are bivalent variables and where $i(k)$ is a sufficiently large integer. This classical transformation of a constrained problem to a non-constrained problem is rarely of interest from a practical point of view; it is principally of theoretical interest.

8.13.1. *Examples of formulations*

Let us look at a few classical combinatorial optimization problems that can be formulated in this way. The quadratic 0–1 knapsack problem, introduced in [GAL 80] and already studied in [LAU 70], is one of the fundamental integer non-linear optimization problems. It consists of maximizing a qpBf with positive coefficients with a capacity constraint:

$$\max\left\{\sum_{i=1}^{n}c_ix_i+\sum_{i=1}^{n-1}\sum_{j=i+1}^{n}c_{ij}x_ix_j:\sum_{i=1}^{n}a_ix_i\leq b,\ x\in\{0,1\}^n\right\}$$

We currently know how to solve instances of this problem containing about 300 variables [CAP 99, BIL 04b] (see also [BIL 96, CHA 86, HAM 97, MIC 96, SOU 00]). The problem of bipartitioning a graph into two subsets of fixed cardinality is easily formulated as $P13$ (see [KAR 00] for dealing with this problem using positive semi-definite programming). Let $G=(V,E)$ be a non-directed graph. For each edge $[i,j]$ of E, let us express its value as c_{ij}. The problem consists of establishing a partition (V_1,V_2) of V such that $|V_1|=K$, $|V_2|=n-K$ and the sum of the weights of the edges having one extremity in V_1 and the other in V_2 is minimal. We can easily express this partition problem in the form (see [BRU 97] for the equipartition problem):

$$\min\left\{\sum_{i=1}^{n}\sum_{j|[v_i,v_j]\in E}c_{ij}x_i-2\sum_{(i,j)|i<j,[v_i,v_j]\in E}c_{ij}x_ix_j,x\in\{0,1\}^n\right\}$$

Let us now look at the quadratic semi-assignment problem, which can be considered as a general graph partitioning problem. It consists of minimizing the 0–1 quadratic function:

$$\sum_{i=1}^{m}\sum_{k=1}^{n} e_{ik}x_{ik} + \sum_{i=1}^{m-1}\sum_{j=i+1}^{m}\sum_{k=1}^{n}\sum_{l=1}^{n} c_{ijkl}x_{ik}x_{jl}$$

with the constraints $\sum_{k=1}^{m} x_{ik} = 1(i = 1, ..., m)$. Note that it is easy, for this problem, to construct, from any $x \in [0,1]^n$, a solution $\tilde{x} \in \{0,1\}^n$ of less than or equal value [CAR 66]. The method consists of considering the constraints one after the other and setting the most interesting variable of this constraint to 1, taking into account the values of the other variables. At each stage, establishing the most interesting variable consists of minimizing a linear function of the variables x_{ik} with the constraint $\sum_{k=1}^{n} x_{ik} = 1$. The quadratic assignment problem is a classical but difficult combinatorial optimization problem, which can be formulated as $P13$. We obtain its formulation by stating that $n = m$ in the formulation of the quadratic semi-assignment and by adding the series of constraints $\sum_{i=1}^{n} x_{ik} = 1(k = 1, ..., n)$. [KAI 99] introduces a polyhedral approach to this problem. The k-cluster problem consists of establishing in a graph $G = (V, E)$ a subset of V of k vertices having the greatest possible number of edges. By stating that $IJ = \left\{ (i,j) \in \{1, ..., n\}^2 : i < j, [v_i, v_j] \in E \right\}$, we can formulate it in the following way: $max \left\{ \sum_{(i,j) \in IJ} x_i x_j : \sum_{i=1}^{n} x_i = k, x \in \{0,1\}^n \right\}$.

Another interesting quadratic optimization problem in bivalent variables is the uniform cost task assignment problem. This concerns a special case of the quadratic semi-assignment problem. We consider a set of communicating tasks $\{T_1, T_2, ..., T_n\}$ and a network of processors $\{P_1, P_2, ..., P_m\}$. The coefficient e_{ik} $(i = 1, ..., n; k = 1, ..., m)$ refers to the task execution cost T_i on the processor P_k, and c_{ij} refers to the cost of communication between the tasks T_i and T_j if they are placed on two different processors. We assume that, if they are placed on the same processor, the communication costs are zero. The problem consists of finding an assignment of the tasks to the processors that minimizes the sum of the execution and communication costs. The associated 0–1 quadratic program is $P14$:

$$\begin{cases} \min \sum_{i=1}^{n-1}\sum_{j=i+1}^{n} c_{ij} + \sum_{i=1}^{n}\sum_{k=1}^{m} e_{ik}x_{ik} - \sum_{i=1}^{n-1}\sum_{j=i+1}^{n} c_{ij} \sum_{k=1}^{m} x_{ik}x_{jk} \\ \sum_{k=1}^{m} x_{ik} = 1 \qquad i = 1, ..., n \qquad [C14.1] \\ x_{ik} \in \{0,1\} \quad i = 1, ..., n;\ k = 1, ..., m \quad [C14.2] \end{cases}$$

The reader interested in the task assignment problem (or very similar problems) can refer to the references [BIL 89a, BIL 92b, BIL 92c, BIL 92d, BIL 93, BIL 94a, BIL 94c, BIL 95, ELL 91, ELL 03, FAY 94, GAL 91, ROU 96, SUT 89].

Let us now introduce a classical optimal investment problem: the optimal selection of projects in an uncertain world. Let $\{P_1, P_2, ..., P_n\}$ be a set of projects. The return provided by the project P_i is a random variable expressed as R_i. E_i is the mathematical expectation of the return provided by the project P_i, Var_i is the variance of the return provided by P_i, Cov_{ij} is the covariance of the returns provided by P_i and P_j, c_i is the cost of P_i, and B is the available capital. The first problem consists of retaining a certain number of projects from $\{P_1, P_2, ..., P_n\}$ in such a way as to optimize an objective function taking into account both the return and the risk. The return is measured by its mathematical expectation and the risk by the variance of this return. By stating that $x_i = 1 \Leftrightarrow P_i$ is retained, we obtain the following program ($P15$):

$$\begin{cases} \min -\alpha \sum_{i=1}^{n} E_i x_i \\ +\ (1-\alpha)(\sum_{i=1}^{n} Var_i\ x_i + 2\sum_{i=1}^{n-1}\sum_{j=i+1}^{n} Cov_{i,j}\, x_i x_j) \\ \sum_{i=1}^{n} c_i\ x_i \leq B \quad [C15.1] \\ x \in \{0,1\}^n \quad [C15.2] \end{cases}$$

The first part of the objective function represents the expectation of the return weighted by the coefficient α, and the second part represents the risk weighted by the coefficient $(1-\alpha)$. The constraint shows that the cost of the set of projects retained must not exceed the available capital B. Let us note here that the objective function is convex if we replace the term $\sum_{i=1}^{n} Var_i\ x_i$ with the equivalent term in binary variables $\sum_{i=1}^{n} Var_i\ x_i^2$. We can therefore solve $P15$ directly by using a standard mathematical programming software package.

We can express this project selection problem in a slightly different way by establishing the subset of projects that maximizes the probability that the return is greater than a certain constant t, that is:

$$\max\ \{\ P(R^T x \geq t) : cx \leq B, x \in \{0,1\}^n\}$$

Assuming that the random variables $R_1, ..., R_n$ follow a multivariate normal distribution, the problem is formulated as the following maximization problem:

$$\max\Big\{\ (E^T x - t)/\sqrt{x^T V x} : cx \leq B, x \in \{0,1\}^n\Big\}$$

where V is the variance–covariance matrix of $\mathbb{R}$. This maximal probability is thus equal to $(1/\sqrt{2\pi})\int_s^\infty e^{\frac{-t^2}{2}}dt$, with $s = (t - E^T x^*)/\sqrt{x^* V^T x^*}$, where x^* refers to the optimal vector x. Let us consider the case where t is less than or equal to the mathematical expectation of the feasible solution of maximal expectation. (The reasoning would be similar in the opposite case.) The problem that we wish to solve can be stated in the following way: $\max\big\{\ (E^T x - t)^2/(x^T V x) : cx \leq b, E^T x \geq t, x \in \{0,1\}^n\big\}$. To solve this fractional optimization problem we can then use Dinkelbach's method, presented in section 8.4, which can be generalized to constrained problems. This method calls, in each iteration, for the solution of a parametric auxiliary problem that consists of maximizing a qpBf subject to two linear constraints:

$$\max\big\{(E^T x - t)^2 - \lambda(x^T V x) : cx \leq b, E^T x \geq t, x \in \{0,1\}^n\big\}$$

To sum up, this second approach to the project selection problem shows that a deterministic version associated with the linear knapsack problem in which the objective function coefficients follow a multivariate normal distribution consists of optimizing a series of qpBfs subject to linear constraints.

8.13.2. *Some polynomial and pseudo-polynomial cases*

Let $\max\left\{\sum_{i=1}^{n} c_i x_i + \sum_{i=1}^{n-1}\sum_{j=i+1}^{n} c_{ij}x_i x_j : \sum_{i=1}^{n} x_i \leq k, x \in \{0,1\}^n\right\}$. We associate the graph $G = (V, E)$, where $V = \{v_1, v_2, ..., v_n\}$ and $[v_i, v_j] \in E \Leftrightarrow c_{ij} > 0$, with this problem. For example, the graph associated with $f = 6x_1 + 4x_2 + 5x_3 + 2x_5 - x_1x_2 + 2x_1x_5 + 3x_1x_6 + 2x_2x_3 + 6x_2x_4$ is a tree. Theorem 8.13 [BIL 92a] shows that the optimum of f, with a cardinality constraint, can be obtained by a polynomial algorithm if the graph associated with f is a tree.

THEOREM 8.13.– *The problem* $\max\{f(x) : \sum_{i=1}^{n} x_i \leq k, x \in \{0,1\}^n\}$ *can be solved by an algorithm based on dynamic programming, of complexity* $O(n^2)$*, if the function* $f(x)$ *is a qpBf whose associated graph is a tree.*

Let us consider a variant of the uniform cost task assignment problem introduced in section 8.13.1. In this variant, we assume that the communication graph is a tree and that the number of tasks that can be assigned to each processor is limited, which is expressed by the constraint:

$$\sum_{i=1}^{n} x_{ik} \leq C_k (k = 1, ..., n)$$

This problem can be solved by an algorithm of complexity $O(m^{2n-1})$, based on a sophisticated technique of dynamic programming that exploits the tree structure of the communication graph [BIL 92b]. Therefore, for a fixed number of processors, the problem is polynomial but, as we can see, the complexity of the algorithm increases rapidly with the number of processors.

The quadratic knapsack problem introduced in section 8.13.1 can be solved by a pseudo-polynomial algorithm based on dynamic programming in the case where the graph associated with the function is an edge-series-parallel graph [RAD 02]. It is also shown in this reference that the problem associated with a vertex-series-parallel graph is, conversely, strongly **NP**-complete.

8.13.3. *Complementation*

In section 8.9.1 we studied complementation in the non-constrained case. We will see here how to generalize this result in certain constrained cases. Given a qpBf $f(x)$, let us express by $H[f(x)]$ the greatest constant c such that $f(x)$ can be expressed as:

$$f(x) = c + \varphi(x, \bar{x}),$$

$\varphi(x, \bar{x})$ being a quadratic posiform. Let us consider the problem:

$$\min \{ f(x) : x \in X \subset \{0,1\}^n \}$$

where $f(x)$ is a qpBf: $\sum_{i=1}^n c_i x_i + \sum_{i=1}^{n-1} \sum_{j=i+1}^n c_{ij} x_i x_j$. We wish to find the greatest possible constant c and the quadratic posiform $\varphi(x, \bar{x})$ such that $f(x) = c + \varphi(x, \bar{x}), \forall x \in X$.

In the non-constrained case, that is in the case where $X = \{0,1\}^n$, it was sufficient to use the relationship $x_i = 1 - \bar{x}_i$. In the constrained case, things are more complicated because we can modify the initial function by adding to it qpBfs that are zero on all the feasible domain points, then use the relationship $x_i = 1 - \bar{x}_i$. For example, let the qpBf to be minimized be:

$$f(x) = 4x_1 - x_2 + 2x_3 - 3x_1x_2 - 4x_1x_3 - x_2x_3$$

with the constraint $2x_1 + x_2 - x_3 = 0$. $H[f(x)] = -3$ and, therefore, not taking into account the constraint, the function f can be expressed as:

$$f(x) = \ -3 \ + 2\bar{x}_1 + \bar{x}_2 + 3x_1\bar{x}_2 + 3x_1\bar{x}_3 + \bar{x}_1x_3 + x_2\bar{x}_3$$

for each $x \in \{0,1\}^3$. Let us now consider the function $f'(x) = f(x) + 2x_1 + x_2 - x_3$; $H[f'(x)] = -1$ so this function can now be expressed by $f'(x) = -1 + \bar{x}_1 + 3x_1\bar{x}_2 + 4x_1\bar{x}_3 + x_2\bar{x}_3$ for each $x \in \{0,1\}^3$. We therefore deduce from this that $f(x) = -1 + \bar{x}_1 + 3x_1\bar{x}_2 + 4x_1\bar{x}_3 + x_2\bar{x}_3$ for each $x \in \{0,1\}^3$ satisfying the constraint $2x_1 + x_2 - x_3 = 0$.

Generalizing this idea gives the following theorem [BIL 97a].

THEOREM 8.14.– *Let $f(x)$ be a qpBf and c^* the greatest constant c such that $f(x) = c + \varphi(x, \bar{x}), \forall x \in X \subseteq \{0,1\}^n$, $\varphi(x, \bar{x})$ being a quadratic posiform.*

$$c^* = \max \{ H[f(x) + g(x)] : g(x) \ \textit{a qpBf such that } g(x) = 0, \ \forall x \in X \}$$

.

Let us consider, for example, the set of feasible solutions defined in the following way: $X = \{ x : x \in \{0,1\}^n, \sum_{i=1}^n x_i = k \}$. In order to be able to use theorem 8.14, we must be able to characterize the qpBf that vanishes at any point of X. This is given by theorem 8.15 [BIL 97a].

THEOREM 8.15.– *The qpBf $g(x)$ is equal to 0 for each $x \in X$ if and only if real numbers $a_0, a_1, ..., a_n$ exist such that:*

$$g(x) = (x_1 + x_2 + ... + x_n - k)(a_0 + a_1x_1 + a_2x_2 + ... + a_nx_n)$$

for each $x \in \{0,1\}^n$.

Using theorems 8.14 and 8.15, we can establish theorem 8.16 [BIL 97a], which shows how to calculate c^* using a (continuous) linear program.

THEOREM 8.16.– *A qpBf* $f(x) = \sum_{i=1}^{n} c_i x_i + \sum_{i=1}^{n-1} \sum_{j=i+1}^{n} c_{ij} x_i x_j$ *is to be minimized on the domain* $X = \{ x : x \in \{0,1\}^n, \sum_{i=1}^{n} x_i = k \}$. *The greatest constant* c^* *such that* $f(x)$ *can be expressed by* $f(x) = c + \varphi(x, \bar{x})$ *for each* $x \in X$, $\varphi(x, \bar{x})$ *being a quadratic posiform, is given by the optimal value of the following linear program* $(P16)$*:*

$$\begin{cases} c^* = \min \sum\limits_{i=1}^{n} q_i x_i + \sum\limits_{i=1}^{n-1} \sum\limits_{j=i+1}^{n} q_{ij} y_{ij} & & & \\ \sum\limits_{i=1}^{n} x_i = k & & & [C16.1] \\ \sum\limits_{j=1}^{i-1} y_{ji} + \sum\limits_{j=i+1}^{n} y_{ij} = (k-1) x_i & i = 1, ..., n & & [C16.2] \\ y_{ij} \leq x_i & i = 1, ..., n-1 & j = i+1, ..., n & [C16.3] \\ y_{ij} \leq x_j & i = 1, ..., n-1 & j = i+1, ..., n & [C16.4] \\ 1 - x_i - x_j + y_{ij} \geq 0 & i = 1, ..., n-1 & j = i+1, ..., n & [C16.5] \\ y_{ij} \geq 0 & i = 1, ..., n-1 & j = i+1, ..., n & [C16.6] \end{cases}$$

Let us remember that for a function without constraints, the constant c^* is calculated by the program $(P16)$ without the first two constraints.

In the case of the quadratic semi-assignment problem introduced in section 8.13.1, we can prove theorem 8.17 [BIL 01]:

THEOREM 8.17.– *The qpBf:*

$$\sum_{i=1}^{m} \sum_{k=1}^{n} e_{ik} x_{ik} + \sum_{i=1}^{m-1} \sum_{j=i+1}^{m} \sum_{k=1}^{n} \sum_{l=1}^{n} c_{ikjl} x_{ik} x_{jl}$$

is to be minimized on the domain

$$X = \left\{ x : x \in \{0,1\}^{m \times n}, \sum_{k=1}^{n} x_{ik} = 1 (i = 1, ..., m) \right\}$$

The greatest constant c^* *such that* $f(x)$ *can be expressed by* $f(x) = c + \varphi(x, \bar{x})$ *for each* $x \in X$, $\varphi(x, \bar{x})$ *being a quadratic posiform, is given by the optimal value of the* *linear program (P17):*

$$\begin{cases} c^* = \min \sum_{i=1}^{m} \sum_{k=1}^{n} e_{ik} x_{ik} + \sum_{i=1}^{m-1} \sum_{j=i+1}^{m} \sum_{k=1}^{n} \sum_{l=1}^{n} c_{ikjl} y_{ikjl} \\ \sum_{k=1}^{n} x_{ik} = 1 \qquad 1 \leq i \leq m & [C17.1] \\ \sum_{k=1}^{n} y_{ikjl} = x_{jl} \quad 1 \leq i < j \leq m \quad l = 1, ..., n & [C17.2] \\ \sum_{k=1}^{n} y_{jlik} = x_{jl} \quad 1 \leq j < i \leq m \quad l = 1, ..., n & [C17.3] \\ y_{ikjl} \geq 0 \qquad 1 \leq i < j \leq m \quad k = 1, ..., n \quad l = 1, ..., n & [C17.4] \end{cases}$$

Note that, in the case of quadratic semi-assignment, a complemented variable can still be expressed uniquely as a function of the problem variables in their direct form, thanks to the semi-assignment constraints. Expressing the function f in the form $f(x) = c^* + \varphi(x, \bar{x})$ enables us to transform the initial problem into a new quadratic semi-assignment problem, concerning the same variables, but with a new objective function:

$$c^* + \sum_{i=1}^{m} \sum_{k=1}^{n} \hat{e}_{ik} x_{ik} + \sum_{i=1}^{m-1} \sum_{j=i+1}^{m} \sum_{k=1}^{n} \sum_{l=1}^{n} \hat{c}_{ikjl} x_{ik} x_{jl}$$

all of whose coefficients are positive or zero. This concerns a *reduction* of the problem and this reduction is the best possible in the sense that the constant c^* is the largest possible. Let us mention here that a $P17$ type linearization is used in [FRI 83] for the quadratic assignment problem.

This idea of complementation can be generalized to posiforms of a degree d greater than 2. We can, for example, for a qpBf f and a subset X of $\{0, 1\}^n$, define the bound $B_d(f, X)$ equal to the largest constant c such that the function f can be expressed by $f(x) = c + \varphi_d(x, \bar{x}), \forall x \in X$, $\varphi_d(x, \bar{x})$ being a posiform of degree d which, expressed in its unique polynomial form, is a quadratic function. By relaxing the condition on the quadratic aspect of the unique polynomial form of the posiform, we can define the constant $C_d(f, X)$ equal to the largest constant c such that the function f can be expressed by $f(x) = c + \varphi_d(x, \bar{x}), \forall x \in X$, $\varphi_d(x, \bar{x})$ being a posiform of degree d. We can see directly that for any X included in $\{0, 1\}^n$:

$$B_d(f, \{0, 1\}^n) = C_d(f, \{0, 1\}^n) \leq B_d(f, X) \leq C_d(f, X)$$

A detailed study of these bounds is presented in [BIL 98, DJA 98, FAY 03a], notably for the graph bipartitioning problem.

8.14. Linearization, convexification and Lagrangian relaxation for optimizing a qpBf with linear constraints

Linearization consists of reformulating the initial quadratic problem as a linear program of bivalent variables or of mixed variables. We can then use one of the many solvers available to solve the program obtained and take advantage of the many known results in linear 0–1 programming. For example, valid inequality generation methods are presented in Chapter 3, Volume 3 (see also [FAY 05]). In general, there are several ways of linearizing a problem which is stated, in a natural way, such as optimizing a qpBf with linear constraints. For example, [BIL 02] introduces six bivalent or mixed variable linear programs formulating the k-CLUSTER problem introduced in section 8.13.1. We present below the two important linearization techniques that we have already introduced for the non-constrained case. The reader can refer to [ELL 00, ELL 02] for theoretical and experimental comparisons between linearization and Lagrangian decomposition applied to the general problem of optimizing a qpBf subject to linear constraints (see also [ADA 86, ADA 94b, BAL 84a, BAL 84b]). The convexification technique consists of reformulating the initial objective as a convex quadratic function by using the fact that, if x_i is a bivalent variable, then $x_i^2 = x_i$. In the case of equality constraints, an effective technique, from the practical solution point of view, is to add to the objective function a quadratic function that is zero on all points of the domain and then to convexify the function obtained in this way.

8.14.1. *Linearization*

Let us consider problem $P13$ (section 8.13). Classical linearization of this problem consists of replacing, as in the non-constrained case, each product $x_i x_j$ appearing in the objective function by a variable y_{ij}, and adding three linearization constraints which mean that $y_{ij} = x_i x_j$, that is $y_{ij} \leq x_i$, $y_{ij} \leq x_j$, and $1 - x_i - x_j + y_{ij} \geq 0$. Note that this linearization is of interest when many coefficients c_{ij} are zero. Let us now inspect "compact" linearization, with $O(n)$ variables and $O(n)$ constraints, of the problem in the case where all the objective function coefficients are positive or zero. For $i = 1, ..., n-1$, let us define a new variable $z_i = x_i \sum_{j=i+1}^{n} c_{ij} x_j$. The problem can then be formulated as the following mixed variable linear program ($P18$):

$$\left\{\begin{array}{lll} \max \sum_{i=1}^{n} c_i x_i + \sum_{i=1}^{n-1} z_i & & \\ \sum_{i=1}^{n} a_{ki} x_i \leq b_k & k = 1, ..., m & [C18.1] \\ \sum_{i=1}^{n} a_{ki} x_i = b_k & k = p+1, ..., p+m & [C18.2] \\ z_i \leq (\sum_{j=i+1}^{n} c_{ij}) x_i & i \in \{1, ..., n-1\} & [C18.3] \\ z_i \leq \sum_{j=i+1}^{n} c_{ij} x_j & i \in \{1, ..., n-1\} & [C18.4] \\ x \in \{0,1\}^n & & [C18.5] \end{array}\right.$$

This type of linearization was introduced in [GLO 75]. It can be of interest to reduce the feasible domain of the continuous relaxation of $P18$ by more precisely adjusting the coefficient $\sum_{j=i+1}^{n} c_{ij}$ in the constraint [C18.3]. We can, for example, replace it by

$$\bar{\varphi}_i = \max \left\{ \sum_{j=i+1}^{n} c_{ij} x_j : x \in S, x_i = 1 \right\}$$

where S refers to the set of feasible solutions. If $\bar{\varphi}_i$ is too difficult to calculate, we can relax the constraint $x \in \{0,1\}^n$ by replacing it with $x \in [0,1]^n$. From a practical solution point of view, this technique is often effective. It has been used successfully to solve the quadratic knapsack problem [BIL 04a] and for the problem of task assignment with capacities on processors [ELL 03]. A study concerning the links between classical linearization, to which we add a certain type of cut, and compact linearization is presented in [ADA 04].

8.14.2. *Convexification*

The qpBf $q(x) = \sum_{i=1}^{n} c_i x_i + \sum_{i=1}^{n-1} \sum_{j=i+1}^{n} c_{ij} x_i x_j$ is to be minimized with linear equality constraints $\sum_{i=1}^{n} a_{ki} x_i = b_k$ $(k = 1, ..., m)$. Let $\alpha \in \mathbb{R}^{n\times m}$, $\lambda \in \mathbb{R}^n$, and let us consider the perturbed function:

$$q_{\alpha,\lambda}(x) = q(x) + \sum_{j=1}^{n} x_j \sum_{k=1}^{m} \alpha_{jk} (\sum_{i=1}^{n} a_{ki} x_i - b_k) + \sum_{i=1}^{n} \lambda_i (x_i^2 - x_i)$$

Minimizing $q_{\alpha,\lambda}(x)$, with the constraints $\sum_{i=1}^{n} a_{ki} x_i = b_k$ $(k = 1, ..., m)$, is equivalent to the initial problem but, for certain vectors α and λ, the objective is convex and the problem can then be solved directly using a standard solver. It is well known that the effectiveness of the approach depends on the value of the continuous relaxation of the problem. We will therefore try to establish α and λ in such a way that $\min\{\, q_{\alpha,\lambda}(x) : \sum^{a_{ki}x_i} = b_k \ (k = 1, ..., m), x \in [0, 1]^n\}$ is as large as possible. We can show that these optimal vectors, α^* and λ^*, correspond to the values of the dual variables associated with the constraints [C19.1] and [C19.2] to the optimum of the positive semi-definite program ($P19$) below:

$$\left\{ \begin{array}{lll} \min \sum\limits_{i=1}^{n} c_i x_i + \frac{1}{2} \sum\limits_{i=1}^{n-1} \sum\limits_{j=i+1}^{n} c_{ij} X_{ij} + \frac{1}{2} \sum\limits_{i=2}^{n} \sum\limits_{j=1}^{n-1} c_{ji} X_{ij} & & \\ X_{ii} = x_i & i = 1, ..., n & [C19.1] \\ -b_k x_i + \sum\limits_{j=1}^{n} a_{kj} X_{ij} = 0 & i = 1, ..., n; k = 1, ..., m & [C19.2] \\ \sum\limits_{j=1}^{n} a_{kj} x_j = b_k & k = 1, ..., m & [C19.3] \\ \begin{pmatrix} 1 & x^t \\ x & X \end{pmatrix} \dot{\geq} 0 & & [C19.4] \\ x \in \mathbb{R}^n, X \in \mathbb{R}^{n\times n} & & [C19.5] \end{array} \right.$$

We will therefore submit the perturbed problem to a mixed variable quadratic program solver:

$$\min\left\{ q_{\alpha^*,\lambda^*}(x) : \sum\nolimits_{i=1}^{n} a_{ki} x_i = b_k (k = 1, ..., m), x \in \{0, 1\}^n \right\}$$

Early calculation experiments, on instances of the k-cluster problem (see section 8.13.1) containing about a hundred vertices, have shown the great effectiveness of this approach in relation to other methods such as mixed variable linear programming (for a discussion of a special case of this approach see [PLA 05], and see [FAY 07, ROU 04] for a discussion of some positive semi-definite relaxations).

8.14.3. *Lagrangian duality*

Let us consider the problem of minimizing a qpBf subject to linear constraints:

$$\min\left\{f(x) = \sum_{i=1}^{n} c_i x_i + \sum_{i=1}^{n-1}\sum_{j=i+1}^{n} c_{ij} x_i x_j : Ax \leq b, x \in \{0,1\}^n\right\}$$

where all the coefficients c_{ij} are negative or zero. The function $f(x)$ is therefore submodular and approaching its minimization using Lagrangian duality seems natural, since minimizing a submodular function, without constraints, is a polynomial problem. In fact, we can easily show that the optimal value of the dual Lagrangian problem is equal to the optimal value of the continuous relaxation of the classical linearization of the initial problem (see section 8.14.1). This property comes from the fact that the minimal value of a submodular qpBf is equal to the optimal value of the continuous relaxation of the associated classical linearization. The link between Lagrangian duality and linearization is studied in [BIL 92d] for the uniform cost task assignment problem with capacity constraints. Applying Lagrangian methods to the general problem $P13$ is studied in [MIC 92] (see also [FAY 07]).

8.15. ε-Approximation algorithms for optimizing a qpBf with linear constraints

We introduce below, by way of example, some approximation results concerning classical qpBf optimization problems with linear constraints. The MAX-k-CUT problem consists of finding in a graph $G = (V, E)$ a partition of the vertices in K parts in such a way as to maximize the sum of the weights of the edges incident to these parts. This problem can be formulated by the following 0–1 variable quadratic program ($P20$) with linear constraints:

$$\begin{cases} \max f(x) = \sum_{i=1}^{n-1}\sum_{j=i+1}^{n}\sum_{p=1}^{K} c_{ij} x_{ip}(1 - x_{jp}) & \\ \sum_{p=1}^{K} x_{ip} = 1 \quad i = 1,...,n & [C20.1] \\ x \in \{0,1\}^{n\times K} & [C20.2] \end{cases}$$

Let us show how to easily obtain a polynomial approximation algorithm for this problem. Let us consider the feasible solution of the continuous relaxation of $P20$: $x_{ip} = 1/K(i = 1,...,n,\ p = 1,...,K)$. This gives $\sum_{i=1}^{n-1}\sum_{j=i+1}^{n} c_{ij}(1 - 1/K)$ to the objective function. Moreover, the value of the optimal solution is less than or equal to $\sum_{i=1}^{n-1}\sum_{j=i+1}^{n} c_{ij}$. From this we therefore deduce the following relationship: $f(x) \geq (1 - \frac{1}{K})f(x^*)$. The method introduced in section 8.13.1 for the quadratic semi-assignment problem allows us to construct an integer solution x', at least as good, from the non-integer solution x. x' is therefore an approximate solution of the problem MAX-k-CUT such that $f(x') \geq (1 - 1/K)f(x^*)$.

Positive semi-definite programming enables us to obtain a ratio equal to $1/(1 - 1/K + 2\log k/k^2)$ [FRI 97]. The MIN-k-CUT problem allows a (2-2/K)-approximation algorithm [SAR 91].

The k-CLUSTER problem described in section 8.13.1 consists of establishing, in a graph, a subset of k vertices having the greatest possible number of edges. Using linear programming for this problem enables us to obtain theorem 8.18 [BIL 08].

THEOREM 8.18.– *A polynomial approximation algorithm for the k-CLUSTER problem exists such that $f(x^*)/f(x^a) \leq 9n/8k$, where x^* refers to an optimal solution and x^a, the approximate solution.*

Establishing the approximate solution is done in three stages. Let us state:

$$IJ = \left\{(i,j) \in \{1,...,n\}^2 : i < j, [v_i, v_j] \in E\right\}$$

The first stage consists of calculating an optimal solution (x^1, y^1) of the continuous relaxation of the classical linearization of the problem:

$$\max\left\{ f_L(x,y) = \sum\nolimits_{(i,j)\in IJ} y_{ij} : y_{ij} \leq x_i, y_{ij} \leq x_j, \sum\nolimits_{i=1}^{n} x_i = k, x \in [0,1]^n \right\}$$

The second stage consists of transforming this solution into an equivalent solution (x^2, y^2) for which all the non-integer components are equal. Obviously, x^2 is an feasible solution of the relaxed quadratic problem ($x \in [0,1]^n$). We can show that the value of this solution, $\sum_{(i,j)\in IJ} x_i^2 x_j^2$, is greater than or equal to $k\, f_L(x^1, y^1)/n$.

The last stage consists of constructing an approximate solution x^a, from x^2 (see also [BIL 97b] for approximating 0–1 variable quadratic programs). Note that the weighted version of the k-cluster problem, where each edge of the graph carries a positive or zero integer weight, allows a polynomial 2-approximation algorithm if these weights satisfy the triangular inequality [HAS 97].

For the quadratic knapsack problem described in section 8.13.1, there is a completely polynomial approximation scheme, based on dynamic programming, if the graph associated with the objective function is edge-series-parallel [RAD 02].

8.16. Bibliography

[ADA 86] ADAMS W.P., SHERALI H.D., "A tight linearization and an algorithm for zero-one quadratic programming problems", *Management Science*, vol. 32, no. 10, p. 1274–1290, 1986.

[ADA 90] ADAMS W.P., BILLIONNET A., SUTTER A., "Unconstrained 0-1 optimization and Lagrangian relaxation", *Discrete Applied Mathematics*, vol. 29, no. 2–3, p. 131–142, 1990.

[ADA 94a] ADAMS W.P, DEARING P.M., "On the equivalence between roof duality and Lagrangian for unconstrained 0-1 quadratic programming problems", *Discrete Applied Mathematics*, vol. 48, no. 1, p. 1–20, 1994.

[ADA 94b] ADAMS W.P., JOHNSON T.A., "Improved linear programming-based lower bounds for the quadratic assignment problem", *Proceedings of the DIMACS Workshop on Quadratic Assignment Problems, DIMACS Series in Discrete Mathematics and Theoretical Computer Science, American Mathematical Society*, vol. 16, p. 43–75, 1994.

[ADA 98] ADAMS W.P., LASSITER J.B., SHERALI H.D., "Persistency in 0-1 polynomial programming", *Mathematics of Operations Research*, vol. 23, p. 267–283, 1998.

[ADA 04] ADAMS W.P, FORRESTER R.J., GLOVER F.W., "Comparisons and enhancement strategies for linearizing mixed 0-1 quadratic programs", *Discrete Optimization*, vol. 1, p. 99–120, 2004.

[ALK 98] ALKHAMIS T.M., HASAN M., AHMED M.A., "Simulated annealing for the unconstrained quadratic pseudo-Boolean function", *European Journal of Operational Research*, vol. 108, no. 3, p. 641–652, 1998.

[ASA 97] ASANO T., HORI K., ONO T., HIRATA T., "A theoretical framework of hybrid approaches to MAX SAT", *Proc. Int. Symp. on Algorithms and Computation, Lecture Notes in Comput. Sci.*, 1350, p. 153–162, Springer-Verlag, 1997.

[BAD 96] BADICS T., Approximation of some nonlinear binary optimization problems, PhD thesis, RUTCOR, Rutgers University, USA, 1996.

[BAL 84a] BALAS E., MAZZOLA J.B., "Nonlinear 0-1 programming: I. Linearization techniques", *Mathematical Programming*, vol. 30, p. 1–21, 1984.

[BAL 84b] BALAS E., MAZZOLA J.B., "Nonlinear 0-1 programming: II. Dominance relations and algorithms", *Mathematical Programming*, vol. 30, p. 22–45, 1984.

[BAR 86] BARAHONA F., "A solvable case of quadratic 0-1 programming", *Discrete Applied Mathematics*, vol. 13, p. 23–26, 1986.

[BAR 89] BARAHONA F., JÜNGER M., REINELT G., "Experiments in quadratic 0-1 programming", *Mathematical Programming*, vol. 44, p. 127–137, 1989.

[BEA 98] BEASLEY J.E., Heuristic algorithms for the unconstrained binary quadratic programming problem, Technical Report, Management School, Imperial College, London, UK, 1998.

[BER 96] BERTSIMAS D., TEO C-P., VOHRA R., "On dependent randomized rounding algorithms", *Proc. Int. Conf. on Integer Prog. and Combinatorial Optimization, Lecture Notes in Comput. Sci.*, 1084, p. 330–344, Springer-Verlag, 1996.

[BIL 85] BILLIONNET A., MINOUX M., "Maximizing a supermodular pseudoboolean function: a polynomial algorithm for cubic functions", *Discrete Applied Mathematics*, vol. 12, p.1–11, 1985.

[BIL 89a] BILLIONNET A., COSTA M.-C., SUTTER A., "Les problèmes de placement dans les systèmes distribués", *Technique et Science Informatiques*, vol. 8, no. 4, p. 307–337, 1989.

[BIL 89b] BILLIONNET A., JAUMARD B., "A decomposition method for minimizing quadratic pseudo-boolean functions", *Operations Research Letters*, vol. 8, p. 161–163, 1989.

[BIL 92a] BILLIONNET A., "Maximizing a tree-structured pseudo-Boolean function with a cardinality constraint", *International Colloquium on Graphs and Optimization*, Grimentz, Switzerland, 1992.

[BIL 92b] BILLIONNET A., "Un algorithme polynomial pour le placement de tâches à structure arborescente dans un système distribué avec contraintes de charge", *Technique et Science Informatiques*, vol. 11, no. 1, p. 117–137, 1992.

[BIL 92c] BILLIONNET A., COSTA M.-C., SUTTER A., "An efficient algorithm for a task allocation problem", *Journal of the Association for Computing Machinery*, vol. 39, no. 3, p. 502–518, 1992.

[BIL 92d] BILLIONNET A., ELLOUMI S., "Placement de tâches dans un système distribué et dualité lagrangienne", *Revue d'Automatique, d'Informatique et de Recherche Opérationnelle (R.A.I.R.O.), série verte*, vol. 26, no. 1, p. 83–97, 1992.

[BIL 92e] BILLIONNET A., SUTTER A., "An efficient algorithm for the 3-satisfiability problem", *Operations Research Letters*, vol. 12, p. 29–36, 1992.

[BIL 92f] BILLIONNET A., SUTTER A., "Persistency in quadratic 0-1 optimization", *Mathematical Programming*, vol. 54, p. 115–119, 1992.

[BIL 93] BILLIONNET A., "Partitioning multiple chains-like task across a host-satellite system", *Information Processing Letters*, vol. 48, p. 261–266, 1993.

[BIL 94a] BILLIONNET A., "Allocating tree structured programs in a distributed system with uniform communication costs", *IEEE Transactions on Parallel and Distributed Systems*, vol. 5, no. 4, p. 445–448, 1994.

[BIL 94b] BILLIONNET A., SUTTER A., "Minimization of a quadratic pseudo-Boolean function", *European Journal of Operational Research*, vol. 78, p. 106–115, 1994.

[BIL 94c] BILLIONNET A., ELLOUMI S., "Placement des tâches d'un programme à structure arborescente sur un réseau de processeurs: synthèse de résultats récents", *Information Systems and Operational Research*, vol. 32, no. 2, p. 65–86, 1994.

[BIL 95] BILLIONNET A., ELLOUMI S., "An algorithm for finding the k-best allocations of a tree-structured program", *Journal of Parallel and Distributed Computing*, vol. 26, no. 2, p. 225–232, 1995.

[BIL 96] BILLIONNET A., CALMELS F., "Linear programming for the 0-1 quadratic knapsack problem", *European Journal of Operational Research*, vol. 92, no. 2, p. 310–325, 1996.

[BIL 97a] BILLIONNET A., FAYE A., "A lower bound for a constrained quadratic 0-1 minimization problem", *Discrete Applied Mathematics*, vol. 74, p. 135–146, 1997.

[BIL 97b] BILLIONNET A., ROUPIN F., "Linear programming to approximate quadratic 0-1 maximization problems", *Southeast ACM Conf.*, Murfreesboro, TN, United States, 2–4 April 1997.

[BIL 98] BILLIONNET A., DJABALI R., FAYE A., "Lower bounds for the graph bipartitioning problem", *Optimisation et Décision, Actes de FRANCORO II*, Sousse, Tunisia, p. 11–23, 1998.

[BIL 99] BILLIONNET A., FAYE A., SOUTIF E., "A new upper bound for the 0-1 quadratic knapsack problem", *European Journal of Operational Research*, vol. 112, p. 664–672, 1999.

[BIL 01] BILLIONNET A., ELLOUMI S., "Best reduction of the quadratic semi-assignment problem", *Discrete Applied Mathematics*, vol. 109, p. 197–213, 2001.

[BIL 02] BILLIONNET A., "Different formulations for the heaviest k-subgraph problem", *Information Systems and Operational Research*, 43, 2005, 171–186.

[BIL 04a] BILLIONNET A., SOUTIF E., "Using a mixed integer programming tool for solving the 0-1 quadratic knapsack problem", *INFORMS Journal on Computing*, vol. 16, p. 188–197, 2004.

[BIL 04b] BILLIONNET A., SOUTIF E., "An exact method based on lagrangian decomposition for the 0-1 quadratic knapsack problem", *European Journal of Operational Research*, vol. 157, p. 565–575, 2004.

[BIL 07] BILLIONNET A., ELLOUMI S., "Using a mixed integer quadratic programming solver for the unconstrained quadratic 0-1 problem", *Mathematical Programming*, 109, 55–68, 2007.

[BIL 08] BILLIONNET A., ROUPIN F., "A deterministic approximation algorithm for the densest k-subgraph problem", *International Journal of Operational Research*, vol. 3, 301–314, 2008

[BOI 94] BOISSIN N., Optimisation des fonctions quadratiques en variables bivalentes, PhD thesis, CNAM, Paris, 1994.

[BOR 90] BOROS E., CRAMA Y., HAMMER P.L., "Upper bounds for quadratic 0-1 maximization", *Operations Research Letters*, vol. 9, p. 73–79, 1990.

[BOR 92] BOROS E., CRAMA Y., HAMMER P.L., "Chvátal cuts and odd cycle inequalities in quadratic 0-1 optimization", *SIAM J. Disc. Math.*, vol. 5, no. 2, p. 163–177, 1992.

[BOR 02] BOROS E., HAMMER P.L., "Pseudo-Boolean optimization", *Discrete Applied Mathematics*, vol. 123, p. 155–225, 2002.

[BRU 97] BRUNETA L., CONFORTI M., RINALDI G., "A branch and cut algorithm for the equicut problem", *Mathematical Programming*, vol. 78, no. 2, p. 243–263, 1997.

[CAP 99] CAPRARA A., PISINGER D., TOTH P., "Exact Solution of the Quadratic Knapsack Problem", *INFORMS Journal on Computing*, vol. 11, p. 125–137, 1999.

[CAR 66] CARLSON R., NEMHAUSER G., "Clustering to minimize interaction costs", *Operations Research*, vol. 14, p. 52–58, 1966.

[CAR 99] CARRARESI P., FARINACCIO F., MALUCELLI F., "Testing optimality for quadratic 0-1 problems", *Mathematical Programming*, vol. 85, no. 2, p. 407–421, 1999.

[CAR 84] CARTER M.W., "The indefinite zero-one quadratic problem", *Discrete Applied Mathematics*, vol. 7, p. 23–44, 1984.

[CHA 86] CHAILLOU P., HANSEN P., MAHIEU Y., "Best network flow bound for the quadratic knapsack problem", *Lecture Notes in Mathematics*, 1403, p. 226–235, 1986.

[CHA 92] CHAKRADAR S.T., BUSHNELL M.L., "A solvable class of quadratic 0-1 programming", *Discrete Applied Mathematics*, vol. 36, p. 233–251, 1992.

[CHA 00a] CHANG C.-T., "An efficient linearization approach for mixed integer programs", *European Journal of Operational Research*, vol. 123, p. 652–659, 2000.

[CHA 00b] CHANG C.-T., CHANG C.-C., "A linearization method for mixed 0-1 polynomial programs", *Computers and Operations Research*, vol. 27, p. 1005–1016, 2000.

[CHA 95] CHARDAIRE P., SUTTER A., "A decomposition method for quadratic zero-one programming", *Management Science*, vol. 41, no. 4, p. 704–712, 1995.

[CRA 89] CRAMA Y., "Recognition problems for special classes of polynomials in 0-1 variables", *Mathematical Programming*, vol. 44, p. 139–155, 1989.

[CRA 90] CRAMA Y., HANSEN P., JAUMARD B., "The basic algorithm for pseudo-Boolean programming revisited", *Discrete Applied Mathematics*, vol. 29, no. 2–3, p. 171–185, 1990.

[CRA 93] CRAMA Y., "Concave extensions for nonlinear 0-1 maximization problems", *Mathematical Programming*, vol. 61, p. 53–60, 1993.

[CRE 96] CRESCENZI P., SILVESTRI R., TREVISAN L., "To weight or not to weight: Where is the question?", *Proc. 4th Israel Symp. on Theory of Computing and Systems*, IEEE Comp. Society, p. 68–77, 1996.

[DEL 09] DELAPORTE G., JOUTEAU S., ROUPIN F., Sdp_s: A tool to formulate and solve semidefinite relaxations for bivalent quadratic problems, http://semidef.free.fr, 2009.

[DES 89] DE SIMONE C., "The cut polytope and the Boolean quadric polytope", *Discrete Mathematics*, vol. 79, p.71–75, 1989.

[DIN 67] Dinkelbach W., "On nonlinear fractional programming", *Management Science*, vol. 13, p. 492–498, 1967.

[DJA 98] DJABALI R., Optimisation non linéaire en variables bivalentes et applications, PhD thesis, CNAM, Paris, 1998.

[ELL 91] ELLOUMI S., Contribution à la résolution des programmes non linéaires en variables 0-1, application aux problèmes de placement de tâches dans les systèmes distribués, PhD theisi, CNAM, Paris, 1991.

[ELL 00] ELLOUMI S., FAYE A., SOUTIF E., "Decomposition and Linearization for 0-1 Quadratic Programming", *Annals of Operations Research*, vol. 99, p. 79–93, 2000.

[ELL 02] ELLOUMI S., Contributions à l'optimisation combinatoire, Mémoire d'habilitation à diriger des recherches, University of Paris 13, 2002.

[ELL 03] ELLOUMI S., ROUPIN F., SOUTIF E., Comparison of different lower bounds for the constrained module allocation problem, Technical report CEDRIC no. 473, CNAM, Paris, 2003.

[FAY 94] FAYE A., Programmation quadratique en variables bivalentes sous contraintes linéaires. Application au placement de tâches dans les systèmes distribués et à la partition de graphes, PhD thesis, CNAM, Paris, 1994.

[FAY 03a] FAYE A., Programmation quadratique en variables bivalentes sous contraintes linéaires, Mémoire d'habilitation à diriger des recherches. University of Paris 13, 2003.

[FAY 05] FAYE A., TRINH Q.-A., "A polyhedral approach for a constrained quadratic 0-1 problem", *Discrete Applied Mathematics*, 149(1), 87–100, 2005.

[FAY 07] FAYE A., ROUPIN F., "Partial lagrangian relaxation for general quadratic programming", *4'OR*, 5(1), 75–88, 2007.

[FEI 95] FEIGE U., GOEMANS M. X., "Approximating the value of two prover proof systems, with applications to MAX 2-SAT and MAX DICUT", *Proc. Israel Symp. on Theory of Computing and Systems, IEEE Computer Society*, p. 182–189, 1995.

[FOR 59] FORTET R., "L'algèbre de Boole et ses applications en recherche opérationnelle", *Cahiers du Centre d'Études de Recherche Opérationnelle*, vol. 4, p. 5–36, 1959.

[FOR 60] FORTET R., "Applications de l'algèbre de Boole en recherche opérationnelle", *RAIRO-Recherche Opérationnelle*, vol. 4, p. 17–26, 1960.

[FRI 83] FRIEZE A., YADEGAR J., "On the Quadratic Assignment Problem", *Discrete Applied Mathematics*, vol. 5, p. 89–98, 1983.

[FRI 97] FRIEZE A., JERRUM M., "Improved approximation algorithms for MAX K-CUT and MAX BISECTION", *Algorithmica*, vol. 18, p. 67–81, 1997.

[GAL 80] GALLO G., HAMMER P.L., SIMEONE B., "Quadratic knapsack problems", *Mathematical Programming Study*, vol. 12, p. 132–149, 1980.

[GAL 91] GALLO G., SIMEONE B., "Optimal grouping of researchers into departments", *Ricerca Operativa*, no. 57, p. 45–69, 1991.

[GOE 95] GOEMANS M.X., WILLIAMSON D.P., "Improved approximation algorithms for maximum cut and satisfiability problems using semidefinite programming", *J. ACM*, vol. 42, p. 1115–1145, 1995.

[GLO 73] GLOVER F., WOLSEY E., "Further reduction of zero-one polynomial programs to zero-one linear programming problems", *Operations Research*, vol. 21, p.156–161, 1973.

[GLO 74] GLOVER F., WOLSEY E., "Converting a 0-1 polynomial programming problem to a 0-1 linear program", *Operations Research*, vol. 22, p. 180–182, 1974.

[GLO 75] GLOVER F., "Improved linear integer programming formulations of nonlinear integer problems", *Management Science*, vol. 22, p. 455–460, 1975.

[GLO 98] GLOVER F., KOCHENBERGER G.A., ALIDAEE B., "Adaptative memory tabu search for binary quadratic programs", *Management Science*, vol. 44, no. 3, p. 336–345, 1998.

[GLO 00] GLOVER F., ALIDAEE B., REGO C., KOCHENBERGER G.A., One-pass heuristics for large-scale unconstrained binary quadratic problems, Report, Hearing Center for Enterprise Science, HCES-09-00, 2000.

[GLO 02] GLOVER F., KOCHENBERGER G.A., "A royal road to combinatorial optimization ? – The 0-1 programming problems", *Cumberland Conference on Combinatorics, Graph Theory and Computing*, University of Mississippi, 2002.

[GRA 76] GRANOT D., GRANOT F., "On solving fractional (0,1) programs by implicit enumeration", *INFOR*, vol. 14, p. 241–249, 1976.

[GUE 02] GUEYE S.A., Linéarisation et relaxation lagrangienne pour problèmes quadratiques en variables binaires, PhD thesis, University of Avignon, 2002.

[GUL 84] GULATI V.P., GUPTA S.K., MITTAL A.K., "Unconstrained quadratic bivalent programming problem", *European Journal of Operational Research*, vol. 15, p. 121–155, 1984.

[HAM 65] HAMMER P.L., "Some network flow problems solved with pseudo-Boolean programming", *Operations Research*, vol. 13, p. 388–399, 1965.

[HAM 68] HAMMER P.L., RUDEANU S., *Boolean Methods in Operations Research*, Springer-Verlag, Berlin, 1968.

[HAM 70] HAMMER P.L., RUBIN A.A., "Some remarks on quadratic programming with 0-1 variables", *RAIRO-Recherche Opérationnelle*, vol. 3, p. 67–79, 1970.

[HAM 72] HAMMER P.L., PELED U.N., "On the maximization of a pseudo-Boolean function", *Journal of the Association for Computing Machinery*, vol. 19 , no. 2, p. 265–282, 1972.

[HAM 81] HAMMER P.L., HANSEN P., SIMEONE B., "Upper planes of quadratic 0-1 functions and stability in graphs", MANGASARIAN O, MEYER R.R., ROBINSON S.M., Eds., *Nonlinear programming 4*, p. 395–414, Academic Press, New York, 1981.

[HAM 84] HAMMER P.L., HANSEN P., SIMEONE B., "Roof duality, complementation and persistency in quadratic 0-1 optimization", *Mathematical Programming*, vol. 28, p. 121–195, 1984.

[HAM 97] HAMMER P.L., RADER D.J. JR, "Efficient methods for solving Quadratic 0-1 Knapsack Problems", *INFOR*, vol. 35, no. 3, p. 170–182, 1997.

[HAM 02] HAMMER P.L., HANSEN P., PARDALOS P.M., RADER D.J. JR, "Maximizing the product of two linear functions in 0-1 variables", *Optimization*, vol. 51, p. 511–537, 2002.

[HAN 70] HANSEN P., "Un Algorithme pour les programmes non linéaires en variables zéro-un", *C.R. Acad. Sc. Paris*, vol. 270, p. 1700–1702, 1970.

[HAN 71] HANSEN P., "Pénalités additives pour les programmes en variables zéro-un", *C.R. Acad. Sc. Paris*, vol. 273, p. 175–177, 1971.

[HAN 79] HANSEN P., "Methods of nonlinear 0-1 programming", *Annals of Discrete Mathematics*, vol. 5, p. 53–70, 1979.

[HAN 86] HANSEN P., SIMEONE B., "Unimodular functions", *Discrete Applied Mathematics*, vol. 14, p. 269–281, 1986.

[HAN 90] HANSEN P., LU S.H., SIMEONE B., "On the equivalence of paved-duality and standard linearization in nonlinear optimization", *Discrete Applied Mathematics*, vol. 29, p. 187–193, 1990.

[HAN 93] HANSEN P., JAUMARD B., MATHON V., "Constrained nonlinear 0-1 programming", *ORSA Journal on Computing*, vol. 5, no. 2, p. 87–118, 1993.

[HAN 00] HANSEN P., JAUMARD B., MEYER C., A simple enumerative algorithm for unconstrained 0-1 quadratic programming, *Les cahiers du GERAD*, G-2000-59, Montreal, Canada, 2000.

[HAS 97] HASSIN R., RUBINSTEIN S., TAMIR A., "Approximation algorithms for maximum dispersion", *Operations Research Letters*, vol. 21, p. 133–137, 1997.

[HEL 98] HELMBERG C., RENDL F., "Solving quadratic 0-1 problems by semidefinite programs and cutting planes", *Mathematical Programming*, vol. 82, p. 291–315, 1998.

[HEN 83] HENNET J.C., "Comparaison de deux méthodes de résolution d'un problème combinatoire quadratique", *RAIRO-Recherche Opérationnelle*, vol. 17, no. 3, p. 285–295, 1983.

[HOC 97] HOCHBAUM D.S. (Ed.), *Approximation algorithms for NP-hard problems*, PWS Publishing Company, Boston, 1997.

[IAS 01] IASEMIDIS L.D., PARDALOS P., SACKELLARES J.C., SHIAU D.-S., "Quadratic binary programming and dynamical system approach to determine the predictability of epileptic seizures", *Journal of Combinatorial Optimization*, vol. 5, p. 9–26, 2001.

[JOH 74] JOHNSON D.S., "Approximation algorithms for combinatorial problems", *J. Comput. System Sci.*, vol. 9, p. 256–278, 1974.

[KAI 99] KAIBEL V., "Polyhedral Methods for the QAP", PARDALOS P.M., PITSOULIS L., Eds., *Nonlinear Assignment Problems*, p. 1–34, Kluwer Academic Publishers, 1999.

[KAL 90] KALANTARI B., BAGCHI A., "An algorithm for quadratic 0-1 programs", *Naval Research Logistics*, vol. 37, no. 4, p. 527–538, 1990.

[KAR 00] KARISCH S.E., RENDL F., CLAUSEN J., "Solving graph bisection problems with semidefinite programming", *INFORMS Journal on Computing*, vol. 12, no. 3, p. 177–191, 2000.

[KIM 01] KIM S., KOJIMA M., "Second order cone programming relaxation of nonconvex quadratic optimization problems", *Optimization Methods and Software*, vol. 15, p. 201–224, 2001.

[KIR 83] KIRKPATRICK S., GELATT C.D., VECCHI M.P., "Optimization by simulated annealing", *Science*, vol. 220, p. 671–680, 1983.

[KON 80] KONNO H., "Maximizing a convex quadratic function over a hypercube", *Journal of the Operations Research Society of Japan*, vol. 23, no. 2, p. 171–189, 1980.

[LAU 70] LAUGHUNN D.J., "Quadratic binary programming with application to capital budgeting problems", *Operations Research*, vol. 18, p. 454–461, 1970.

[LI 94] LI H.-L., "A global approach for general 0-1 fractional programming", *European Journal of Operational Research*, vol. 73, p. 590–596, 1994.

[LOD 99] LODI A., ALLEMAND K., LIEBLING T.M., "An evolutionary heuristic for quadratic 0-1 programming", *European Journal of Operational Research*, vol. 119, p. 662–670, 1999.

[LU 84] LU S.H., "An improved enumerative algorithm for solving quadratic zero-one programming", *European Journal of Operational Research*, vol. 15, p. 110–120, 1984.

[LU 87] LU S.H., WILLIAMS A.C., "Roof duality for polynomial 0-1 optimization", *Mathematical Programming*, vol. 37, p. 357–360, 1987.

[MER 05] MERZ P., KATAYAMA K., "Memetic algorithms for the unconstrained binary quadratic programming problem", *BioSystems*, vol. 78, p. 99–118, 2004.

[MIC 92] MICHELON P., "Unconstrained nonlinear programming: a non differentiable approach", *Journal of Global Optimization*, vol. 2, p. 155–165, 1992.

[MIC 93] MICHELON P., MACULAN N., "Lagrangean methods for 0-1 quadratic programming", *Discrete Applied Mathematics*, vol. 42, p. 257–269, 1993.

[MIC 96] MICHELON P., VEUILLEUX L., "Lagrangean methods for the 0-1 quadratic knapsack problem", *European Journal of Operational Research*, vol. 92, p. 326–341, 1996.

[MUR 03] MURAMATSU M., SUZUKI T., "A new second order cone programming relaxation for max-cut problem", *J. of Operations Research Society of Japan*, vol. 46, p. 164–177, 2003.

[ORA 92] ORAL M., KETTANI O., "A linearization procedure for quadratic and cubic mixed integer problem", *Operations Research*, vol. 40, p. 109–116, 1992.

[PAD 89] PADBERG M., "The boolean quadric polytope: some characteristics, facets and relatives", *Mathematical Programming*, vol. 45, p. 139–172, 1989.

[PAL 95] PALUBEKIS G., "A heuristic-based branch and bound algorithm for unconstrained quadratic 0-1 programming", *Computing*, vol. 54, no. 4, p. 283–301, 1995.

[PAR 90] PARDALOS P., RODGERS G.P., "Computational aspect of a branch and bound algorithm for quadratic 0-1 programming", *Computing*, vol. 45, p. 131–144, 1990.

[PAR 92] PARDALOS P., JHA S., "Complexity of uniqueness and local search in quadratic 0-1 programming", *Operations Research Letters*, vol. 11, p. 119–123, 1992.

[PIC 75] PICARD J.C., RATLIFF H.D., "Minimum cuts and related problems", *Networks*, vol. 5, p. 357–370, 1975.

[PIC 77] PICARD J.C., QUEYRANNE M., "On the integer-valued variables in the linear vertex packing problem", *Mathematical Programming*, vol. 12, p. 97–101, 1977.

[PLA 05] PLATEAU M.-C., BILLIONNET A., ELLOUMI S., "Eigen value methods for linearly-constrained quadratic 0-1 problems with application to the densest k-subgraph problem", *Congrès ROADEF' 05*, Tours, 14–16 February, 2005.

[POL 95a] POLJAK S., RENDL F., WOLKOWICZ H., "A recipe for semidefinite relaxation for 0-1 quadratic programming", *Journal of Global Optimization*, vol. 7, p. 51–73, 1995.

[POL 95b] POLJAK S., WOLKOWICZ H., "Convex relaxations of 0-1 quadratic programming", *Mathematics of Operations Research*, vol. 20, no. 3, p. 550–561, 1995.

[RAD 98] RADZIK T., "Fractional combinatorial optimization", In *Handbook of Combinatorial Optimization*, p. 429–478, Z.-Z. DU and P.M. PARDALOS (Eds.), Kluwer Academic Publishers, 1998.

[RAD 02] RADER D.J. JR., WOEGINGER G.J., "The quadratic 0-1 knapsack problem with series-parallel support", *Operations Research Letters*, vol. 30, no. 3, p. 159–166, 2002.

[RHY 70] RHYS J., "A selection problem of shared fixed costs and networks", *Management Science*, vol. 17, p. 200–207, 1970.

[ROB 76] ROBERT P.D., TERRELL M.P., "An approximation technique for pseudo-boolean maximization problems", *AIIE Transactions*, vol. 8, no. 3, p. 365–368, 1976.

[ROS 72] ROSENBERG I.G., "0-1 optimization and nonlinear programming", *RAIRO* (série bleue), vol. 2, p. 95–97, 1972.

[ROS 74] ROSENBERG I.G., "Minimisation of pseudo-Boolean functions by binary developpement", *Discrete Mathematics*, vol. 7, p. 151–165, 1974.

[ROS 75] ROSENBERG I.G., "Reduction of bivalent maximization to the quadratic case", *Cahier du Centre d'Études de Recherche Opérationnelle*, vol. 17, p. 71–74, 1975.

[ROU 96] ROUPIN F., Approximation de programmes quadratiques en 0-1 soumis à des contraintes linéaires, Application aux problèmes de placement et de partition de graphes, PhD thesis, CNAM, Paris, 1996.

[ROU 04] ROUPIN F., "From linear to semidefinite programming: an algorithm to obtain semidefinite relaxations for bivalent quadratic problems", *Journal of Combinatorial Optimization*, vol. 8, no. 4, p. 469–493, 2004.

[SAR 91] SARAN H., VAZIRANI V., "Finding k-cuts within twice the optimal", *Proc. IEEE Symp. on Foundations of Comput. Sci.*, p. 743–751, IEEE Computer Society, 1991.

[SIM 90] SIMEONE B., DE WERRA D., COCHAND M., "Recognition of a class of unimodular functions", *Discrete Applied Mathematics*, vol. 29, p. 243–250, 1990.

[SOU 00] SOUTIF E., Résolution du problème du sac-à-dos quadratique en variables bivalentes, PhD thesis, CNAM, Paris, 2000.

[SUT 89] SUTTER A., Programmation non linéaire en variables 0-1, application à des problèmes de placement de tâches dans des systèmes distribués, PhD thesis, CNAM, Paris, 1989.

[THO 98] THOAI N.V., "Global optimization technique for solving the general quadratic integer programming problem", *Computational Optimization and Applications*, vol. 10, no. 2, p. 149–163, 1998.

[WIL 85] WILLIAMS A.C., Quadratic 0-1 programming using the roof dual with computational results, RUTCOR Research Report , no. 8-85, Rutgers University, 1985.

[ZVE 01] ZVEROVICH I., Maximization of quadratic posiforms corresponding to 2-paths of a directed multigraph, RUTCOR Research Report, no. 30-2001, 2001.

Chapter 9

Column Generation in Integer Linear Programming

In this chapter we introduce exact methods for solving integer linear programming problems with a large number of variables. These methods are known as *branch-and-price* methods. In the case where these problems all have columns whose components are 0 or 1, we develop geometrical cuts for eliminating columns generated beforehand. This schema can be applied to problems without a specific structure. We also make a comparison between the linear relaxations of different formulations of integer linear programming (ILP) problems. We discuss the difficulties in choosing separation rules (*branching*) inside column generation algorithms. For partitioning problems, we present Ryan and Foster's separation rule. We present two applications for column generation algorithms in detail.

9.1. Introduction

Early techniques for automatic column generation were presented in the 1960s as methods for solving large linear programming (LP) problems. When the data for the problem to be solved do not fit into the main memory of the computer, the columns that are the basis of the simplex method are generated by an auxiliary optimization problem. Dantzig and Wolfe [DAN 60] introduced these techniques in the context of their decomposition algorithm, originally developed for computers whose main memory capacity was, at the time, very limited.

Chapter written by Irène LOISEAU, Alberto CESELLI, Nelson MACULAN and Matteo SALANI.

Gilmore and Gomory [GIL 61, GIL 63] used a new approach to model and solve the one-dimensional cutting problem, which becomes an integer linear programming (ILP) problem with a huge number of variables. Column generation is used to solve linear relaxation. They generate columns using the knapsack problem. This relaxation provides excellent lower bounds in practice. Two decades later, Marcotte [MAR 85] showed that the optimal value of this (ILP) is frequently the rounding up of the optimal value of its linear relaxation.

We can therefore think of two classes of problem for which the solution is obtained using automatic column generation. The first is associated with the method proposed by Dantzig and Wolfe. The second is made up of problems for which we naturally have a formulation of the type introduced by Gilmore and Gomory.

First of all, we introduce, in the first section, a bounded variable linear programming column generation method, because this will be used in the problems that are dealt with in this chapter.

In section 9.3, we introduce an inequality for eliminating 0–1 column generation and in section 9.4, we introduce two integer linear programming problem models and a comparison between their linear relaxations. In section 9.5, we present a schema of a column generation method for solving an ILP and the difficulties that can appear at the time of implementation. In section 9.6, two column generation algorithms are introduced: for the p-medians problem and for vehicle routing problems.

9.2. A column generation method for a bounded variable linear programming problem

In order to be able to solve integer linear programming (ILP) problems using classical "branch-and-bound" methods, we must consider a linear programming problem at each node of the enumeration tree. In our case, this problem will be solved by column generation techniques. Let us consider the following problem:

$$(BLP): \text{ minimize } z = \sum_{j=1}^{p} c_j x_j \qquad [9.1]$$

with the constraints

$$\sum_{j=1}^{p} a_j x_j = b \qquad [9.2]$$

$$0 \leqslant x_j \leqslant d_j, \ j = 1, 2, \ldots, p \qquad [9.3]$$

where $a_j \in \mathcal{R}^m$, $b \in \mathcal{R}^m$, $b \geqslant 0$, $d = (d_1 \ \ldots \ d_p)^T \in \mathcal{R}^p_+$, $c_j \in \mathcal{R}$. When all the elements c_j, a_j and d_j, on the face of it, are known, this problem can be solved using the simplex methods proposed by Dantzig [CHV 83, DAN 95, DAN 63, LAS 70].

We assume that a_j belongs to a finite set $\mathcal{K}$, and $|\mathcal{K}| = p$. It is unthinkable to write this (BLP) problem explicitly when p has a huge value. For example, let us consider $\mathcal{K}$, the set of 0–1 vectors associated with the spanning tree of a complete graph with n vertices; we have $|\mathcal{K}| = n^{n-2}$, for $n = 10$, then $|\mathcal{K}| = p = 100\,000\,000$.

In this case, we could solve the (BLP) using column generation methods associated with the simplex method. For this, we must assume that $c_j = f(a_j)$ and $d_j = g(a_j)$, where $f : \mathcal{K} \to \mathcal{R}$ and $g : \mathcal{K} \to \mathcal{Z}_+$. Using the usual notation (see [CHV 83]), let $B = (a_{B(1)}\ a_{B(2)}\ \cdots\ a_{B(m)})$ be a square matrix $m \times m$ whose $det(B) \neq 0$, and let N be a matrix $m \times (p - m)$ whose columns a_j are not in B. Furthermore let $x_B = (x_{B(1)}\ x_{B(2)}\ \cdots\ x_{B(m)})^T$ and x_N be a vector associated with the columns of N.. Thus [9.2] can be expressed in the following way: $Bx_B + Nx_N = b$, and then we will have:

$$x_B = B^{-1}b - B^{-1}Nx_N \qquad [9.4]$$

Let N_2 be the matrix whose columns a_j of N are associated with $x_j = d_j$ and N_1 the matrix whose columns of N are associated with $x_j = 0$. We can also express [9.4] in the following way:

$$x_B = B^{-1}b - B^{-1}N_1x_{N_1} - B^{-1}N_2x_{N_2} \qquad [9.5]$$

Let $\bar{x}_B = B^{-1}b$, and $\hat{x}_B = \bar{x}_B - B^{-1}N_2\bar{x}_{N_2}$, where $\bar{x}_{N_2} = d^T_{N_2}$. If $0 \leqslant \hat{x}_{B(i)} \leqslant d_{B(i)}$, $i = 1, 2, \ldots, m$, then B is a feasible primal base of (BLP).

Let $u = c_B^T B^{-1}$, where $c_B = (c_{B(1)}\ c_{B(2)}\ \cdots\ c_{B(m)})^T \in \mathcal{R}^m$, and $\bar{c}_j = c_j - ua_j$. We assume that $\bar{c}_j \leqslant 0$, for each j associated with $x_j = d_j$. It is always possible to have this if (BLP) is not empty: if we had $\bar{c}_j > 0$ for some $x_j = d_j$, we could solve (BLP) without considering the columns of N_1, and we would obtain a new matrix N_2 for which $\bar{c}_j \leqslant 0$ for every $x_j = d_j$.

At each iteration of the simplex method, we must solve an auxiliary subproblem (*pricing problem* or *oracle*) in order to know whether the base B is optimal for (BLP) or to generate a column that will be part of the new base. This auxiliary problem will be:

$$(AP) : \text{ minimize } f(a) - ua$$

with the constraint

$$a \in \mathcal{K}$$

Using this schema, we may need to calculate the k best solutions of (AP), because the $k - 1$ solutions found may be columns of N_2, or may have already been generated.

Let a_k be an optimal solution of (AP), that is $val(AP) = f(a_k) - ua_k$, where $val(\cdot)$ is the optimal value of the problem $(\cdot)$. We need to obtain the s-best solutions of (AP) for $s = 1, 2, \ldots$ such that: $val(s - best) \leqslant val([s+1] - best)$. Thus $val(1 - best) = val(AP)$.

Let $I = \{1, 2, \ldots, p\}$, $I_{N_1} = \{j \mid a_j \in N_1\}$, $I_{N_2} = \{j \mid a_j \in N_2\}$, and $I_B = \{j \mid a_j \in B\}$. To solve (BLP), we propose the following method. Let:

$$(LP, B, N_2) : \text{ minimize } ub + \sum_{j \in N_2} \bar{c}_j x_j$$

with the constraints

$$x_B = \bar{x}_B - \textstyle\sum_{j \in N_2} y_j x_j$$
$$0 \leqslant x_{B(i)} \leqslant d_j,\ i = 1, 2, \ldots, m$$
$$0 \leqslant x_j \leqslant d_j,\ j \in N_2$$

where $u = c_B^T B^{-1}$, $\bar{c}_j = f(a_j) - ua_j$, $d_j = g(a_j)$, and $y_j = B^{-1} a_j$. Given a feasible primal base B of (PLB), we present the COLGEN algorithm:

```
BEGIN
        SOLVE (LP, B, I_N2);
        s := 0;
1       s := s + 1;
        t := val(s − best) := f(ā) − uā; (oracle)
        if ā ∈ N2, goto 1;
        if t ⩾ 0, an optimal solution STOP;
        ā ∈ I_N1 put ā into B or into N2;
        SOLVE (LP, B, I_N2);
        s := 0;
        goto 1;
END.
```

9.3. An inequality to eliminate the generation of a 0–1 column

When the set of all the columns $a \in \mathcal{K}$ can be represented by:

$$Ka \leqslant h \quad [9.6]$$

$$a \in \{0, 1\}^m \quad [9.7]$$

where $K \in \mathcal{R}^{q \times m}$ is a known matrix and $h \in \mathcal{R}^q$ is a given vector, we propose a way of stopping the generation of a column $\bar{a}$ already generated in previous iterations.

We use a linear inequality to eliminate a point of $\{0,1\}^m$. Let $\bar{a} \in \{0,1\}^m$; we wish to construct an inequality that only eliminates $\bar{a}$ in [9.6] and [9.7]. That is, given $\mathcal{B} = \{0,1\}^m - \{\bar{a}\}$, how do we find a facet of the convex envelope of $\mathcal{B}$, such that $\bar{a}$ does not belong to the inequality associated with this facet? All the points of the sphere centered on $\bar{a}$ with a radius equal to 1 satisfy the following equation:

$$\sum_{j=1}^{m} (a_j - \bar{a}_j)^2 = 1 \qquad [9.8]$$

All the neighbors of $\bar{a}$ in the cube whose vertices are the points of $\{0,1\}^m$ belong to the surface of this sphere. If we wish to consider all the points of $\mathcal{B}$, we write:

$$\sum_{j=1}^{m} (a_j - \bar{a}_j)^2 \geqslant 1 \qquad [9.9]$$

which implies:

$$\sum_{j=1}^{m} (a_j^2 - 2\bar{a}_j a_j + \bar{a}_j^2) \geqslant 1$$

As we have $a_j^2 = a_j$, because $a_j \in \{0,1\}$, then:

$$\sum_{j=1}^{m} (1 - 2\bar{a}_j) a_j \geqslant 1 - \sum_{j=1}^{m} \bar{a}_j \qquad [9.10]$$

It is not difficult to verify that [9.10] is associated with a facet of the convex envelope of $\mathcal{B}$, which is not satisfied by $\bar{a}$. (For more details on this type of inequality see [MAC 01]). We add this inequality to COLGEN which becomes:

BEGIN
1 SOLVE(LP, B, I_{N_2});
add to (AP) $\sum_{j=1}^{m}(1 - 2\bar{a}_j)a_j \geqslant 1 - \sum_{j=1}^{m} \bar{a}_j, \forall \bar{a} \in N_2$;
SOLVE(AP),
$val(AP) := f(\hat{a}) - u\hat{a}$; (pricing problem)
if $val(AP) \geqslant 0$, an optimal solution STOP;
$\hat{a}$ will go into B or N_2; simplex algorithm
goto1;
END.

9.4. Formulations for an integer linear program

Let there be an integer linear programming problem:

$$(P): \ \min cx$$

with the constraints

$$Ax = b \quad [9.11]$$

$$Dx = d \quad [9.12]$$

$$x \in \mathcal{Z}_+^n$$

where $c^T \in \mathcal{R}^n, \ A \in \mathcal{R}^{m \times n}, \ D \in \mathcal{R}^{q \times n}, \ b \in \mathcal{R}^m, \ d \in \mathcal{R}^q$.

Using the ideas conceived by Dantzig and Wolfe [DAN 60], we can consider:

$$X = \{x \in \mathcal{Z}_+^n \mid Dx = d\} = \{v^1, \ v^2, \ \ldots, v^p\}$$

This means that the set X possesses p elements, or even that $|X| = p$. It is obvious that if $x \in X$ then we can write:

$$x = \sum_{j=1}^{p} v^j \lambda_j$$

$$\sum_{j=1}^{p} \lambda_j = 1$$

$$\lambda_j \in \{0, 1\}, \ j = 1, 2, \ldots, p$$

and therefore (P) can be stated in the following way:

$$(P1): \ \min \sum_{j=1}^{p} (cv^j)\lambda_j$$

with the constraints:

$$\sum_{j=1}^{p} (Av^j)\lambda_j = b$$

$$\sum_{j=1}^{p} \lambda_j = 1$$

$$\lambda_j \in \{0,1\},\ j = 1,2,\ldots,p$$

$(P1)$ is equivalent to (P). Given that to solve an ILP we resort to its linear relaxation, it is useful to compare the linear relaxation of $(P1)$ with that of (P).

The linear relaxation of $(P1)$, which is given by a lower bound provided by the solution of $(P1)$ is:

$$(LP1):\ \min \sum_{j=1}^{p} (cv^j)\lambda_j$$

with the constraints

$$\sum_{j=1}^{p} (Av^j)\lambda_j = b$$

$$\sum_{j=1}^{p} \lambda_j = 1$$

$$\lambda_j \geqslant 0,\ j = 1,2,\ldots,p$$

We will draw a comparison between the dual Lagrangian problems of (P) and $(LP1)$. If we associate a vector $u = (u_1\ u_2\ \ldots\ u_m)^T$ with the constraints made by $Ax = b$, we can express the Lagrangian by $\mathcal{L}(x,u) = cx + u(Ax - b)$. The dual function will be $l(u) = \min\{\mathcal{L}(x,u) \mid x \in X\}$.

Therefore we can express the dual Lagrangian problem of (P) in the following way:

$$(DLP):\ \max\{l(u) \mid u^T \in \mathcal{R}^m\}$$

(DLP) can also be expressed:

$$(DLP):\ \max\ t$$

with the constraints

$$t \leqslant cv^k + u(Av^k - b),\ k = 1,2,\ldots,p$$

$$u^T \in \mathcal{R}^m$$

or even:

$$(DLP):\ \max\ t$$

with the constraints

$$t - u(Av^k - b) \leqslant cv^k,\ k = 1,2,\ldots,p$$

$$u^T \in \mathcal{R}^m$$

Let us associate the variable $\lambda_k \geqslant 0$ with each constraint $t - u(Av^k - b) \leqslant cv^k$, $k = 1, 2, \ldots, p$. The dual of (DLP) is expressed:

$$(DDLP): \min \sum_{k=1}^{p} (cv^k)\lambda_k$$

with the constraints:

$$\sum_{k=1}^{p} (Av^k - b)\lambda_k = 0 \qquad [9.13]$$

$$\sum_{k=1}^{p} \lambda_k = 1 \qquad [9.14]$$

$$\lambda_k \geqslant 0, \; k = 1, 2, \ldots, p$$

The constraints [9.13] can be rewritten in the following way:

$$\sum_{k=1}^{p} (Av^k)\lambda_k = b \sum_{k=1}^{p} \lambda_k$$

But, according to [9.14], $\sum_{k=1}^{p} \lambda_k = 1$, so actually we can write:

$$(DDLP): \min \sum_{k=1}^{p} (cv^k)\lambda_k$$

with the constraints:

$$\sum_{k=1}^{p} (Av^k)\lambda_k = b$$

$$\sum_{k=1}^{p} \lambda_k = 1$$

$$\lambda_k \geqslant 0, \; k = 1, 2, \ldots, p$$

Note that $(DDLP)$ is identical to $(LP1)$. In this way we see that $val(DLP) = val(DDLP) = val(LP1)$. If $(\bar{P})$ represents the linear relaxation of (P), that is the constraint $x \in \mathcal{Z}_+^n$ is replaced with $x \in \mathcal{R}_+^n$, then $val(\bar{P}) \leqslant val(DLP)$ (see [NEM 88]). This implies:

$$val(\bar{P}) \leqslant val(LP1)$$

and we can conclude that the linear relaxation of $(P1)$ is stronger than that of (P).

9.5. Solving an integer linear program using column generation

There are problems that can be naturally formulated as integer linear programming problems with a very large number of columns, and the formulation will be close to that of $(P1)$, for example the one-dimensional cutting problem proposed by Gilmory and Gomory [GIL 61, GIL 63]. In this case we will not have any constraints:

$$\sum_{j=1}^{p} \lambda_j = 1$$

We therefore have a formulation of the following type:

$$(PE): \text{ minimize } \sum_{j=1}^{p} c_j x_j$$

with the constraints:

$$\sum_{j=1}^{p} a_j x_j = b$$

$$0 \leqslant x_j \leqslant d_j, \; j = 1, 2, \ldots, p$$

$$x_j \in \mathcal{Z}, \; j = 1, 2, \ldots, p$$

where c_j, a_j, b and d_j are defined in (BLP).

We solve (PE) (or $(P1)$) with branch-and-bound techniques (see [DAK 65, NEM 88]).

9.5.1. *Auxiliary problem* (pricing problem)

Let (PEL) be the linear relaxation of (PE) (or of $(P1)$), that is the constraints $x_j \in \mathcal{Z}, \; j = 1, 2, \ldots, p$ will not be considered in the problem (PEL).

At the initial node, we solve (PEL), and at each node of the enumeration tree we must also solve a linear problem. These linear relaxations can be solved by the COL-GEN method, using an *ad hoc* algorithm in each case to solve the auxiliary *pricing* problem (AP). The problem (AP) is a combinatorial optimization problem that can also be **NP**-hard.

In some applications, we can find a good heuristic solution for the auxiliary problem (AP) whose objective function for this solution is negative. This solution will be associated with a column that will be put into the base. When we can find no more good solutions that give a negative value to the objective function, we will need to solve the auxiliary problems exactly.

9.5.2. *Branching*

If the optimal solution of the linear relaxation is integer, then we have also obtained an optimal solution of (PE) (or $(P1)$). If not, we have a k such that $\hat{x}_{B(k)}$ is not integer. In this case, we must branch off, for which we have known integer linear programming rules (see [NEM 88]). To start with, we will mention the traditional *branching* rule. We will then introduce a rule for special cases.

With the traditional rule, we will define the two successors of this node by solving the two linear programming problems:

– $(PEL) \cap \{x_{B(k)} \leqslant \lfloor \hat{x}_{B(k)} \rfloor\}$

– $(PEL) \cap \{x_{B(k)} \geqslant \lfloor \hat{x}_{B(k)} \rfloor + 1\}$

where $\lfloor r \rfloor$ is the largest integer $\leqslant r$.

We assume that the following linear programming problem is associated with the node i of the branch-and-bound tree:

$$(PEL_i) : \text{minimize} \sum_{j=1}^{p} c_j x_j$$

with the constraints:

$$\sum_{j=1}^{p} a_j x_j = b$$

$$\alpha_j \leqslant x_j \leqslant \beta_j, \; j \in S_i$$

$$0 \leqslant x_j \leqslant d_j, \; j \in I - S_i$$

where S_i is the set of indices associated with the columns generated during the procedure to get to the node i, including the columns in B. We can have $\alpha_j = 0$ or $\beta_j = d_j$ for some $j \in S_i$. A feasible primal initial solution for (PEL_i) is associated with the matrix B for which $\alpha_{B(i)} \leqslant \bar{x}_{B(i)} \leqslant \beta_{B(i)}$, $i = 1, 2, \ldots, m$, $\bar{x}_j = \alpha_j$ or $\bar{x}_j = \beta_j$, $j \in S_i - I_B$, and $\bar{x}_j = 0$, $j \in I - S_i$.

Let us now consider a feasible base solution for (PEL_i), found using the solution of:

$$(PELX_i) : \text{minimize} \sum_{j \in S_i} c_j x_j$$

with the constraints:

$$\sum_{j \in S_i} a_j x_j = b$$

$$\alpha_j \leqslant x_j \leqslant \beta_j, \; j \in S_i$$

where $\bar{c}_j = 0$, $j \in I_B$; $\bar{c}_j \geqslant 0$ for $\bar{x}_j = \alpha_j$, and $j \in S_i - I_B$; $\bar{c}_j \leqslant 0$ for $\bar{x}_j = \beta_j$ and $j \in S_i - I_B$.

If (PEL_i) is a leaf of the branch-and-bound tree, then we either *backtrack*, or choose a $k \in I_B$ such that $\hat{x}_{B(k)}$ is not integer, and we develop the two successors of the node i :

– $(PEL_{i+1}) := (PEL_i) \cap \{x_{B(k)} \leqslant \lfloor \hat{x}_{B(k)} \rfloor\}$;
– $(PEL_{i+2}) := (PEL_i) \cap \{x_{B(k)} \geqslant \lfloor \hat{x}_{B(k)} \rfloor + 1\}$.

In this case we can solve the linear programming problems by only considering all the columns generated to reach the node i :

– $(PELX_{i+1}) := (PELX_i) \cap \{x_{B(k)} \leqslant \lfloor \hat{x}_{B(k)} \rfloor\}$;
– $(PELX_{i+2}) := (PELX_i) \cap \{x_{B(k)} \geqslant \lfloor \hat{x}_{B(k)} \rfloor + 1\}$.

Note that the base solution of $(PELX_i)$ is again a feasible dual for $(PELX_{i+1})$ and $(PELX_{i+2})$.

Unfortunately, $(PELX_{i+1})$ may be empty and (PEL_{i+1}) is not. This same reasoning can be applied to $(PELX_{i+2})$ in relation to (PEL_{i+2}). When $(PELX_{i+1})$ is not empty, its optimal base solution will be the feasible primal initial solution for (PEL_{i+1}). So the COLGEN algorithm is used to solve (PEL_{i+1}). The same development can be used to solve (PEL_{i+2}) from an optimal base solution of $(PELX_{i+2})$.

If $(PELX_{i+1})$ is empty, we need to solve (PEL_{i+1}) using the two-phase simplex algorithm. An artificial variable r is placed in row k in the following way:

$$x_{B(k)} + r = \bar{x}_{B(k)} - \sum_{j \in I_{N_1}} y_{kj} x_j - \sum_{j \in I_{N_2}} y_{kj} x_j$$

When $x_{B(k)} = \lfloor \hat{x}_{B(k)} \rfloor$, then:

$$\bar{r} = \bar{x}_{B(k)} - \sum_{j \in I_{N_1}} y_{kj} \alpha_j - \sum_{j \in I_{N_2}} y_{kj} \beta_j - \lfloor \hat{x}_{B(k)} \rfloor > 0$$

Since $a_{B(k)}$ is in the base B, we have $y_{B(k)} = B^{-1} a_{B(k)} = e_k$. If we consider $I_{N_2} := I_{N_2} \cup \{B(k)\}$, and $\beta_{B(k)} = \lfloor \hat{x}_{B(k)} \rfloor$, then we can solve the following linear programming problem using the COLGEN algorithm:

$$(PLB_1) : \text{ minimize } r$$

with the constraints:

$$\sum_{j=1}^{p} a_j x_j + a_{B(k)} r = b$$

$$\alpha_j \leqslant x_j \leqslant \beta_j, \; j \in S_i$$

$$0 \leqslant x_j \leqslant d_j, \; j \in I - S_i, \text{ and } r \geqslant 0$$

A feasible primal base for (PLB_1) is the same as the one associated with the optimal solution of $(PELX_i)$, where we replace $x_{B(k)}$ by r in the variables associated with the base.

If $val(PLB_1) = 0$, we have obtained a feasible primal base solution for (PEL_{i+1}). If not, (PEL_{i+1}) is empty and the node $(i+1)$ is closed and the *backtracking* operation is carried out.

To solve (PEL_{i+2}) when $(PELX_{i+2})$ is empty, we proceed in a similar way.

The traditional *branching* rule that we have just presented, at each node of the branch-and-bound tree, adds a new constraint to the auxiliary problems (AP), which generate new columns. This can hugely change the initial structure of the auxiliary problem, but it is possible that the solution conditions are not easy.

In the various published applications, we can see different *ad hoc branching rules*, defined to be compatible with the structure of the auxiliary (*pricing*) problem (AP). In particular, many of the applications for which this column generation method has been successful are problems that can be modeled as partitioning and covering problems. For this type of problem, Ryan and Foster [RYA 81] have proposed a *branching* strategy. A partitioning problem can be defined in the following way:

$$(PP) : \text{ minimize} \sum_{j=1}^{p} c(a_j)x_j \qquad [9.15]$$

with the constraints:

$$\sum_{j=1}^{p} a_j x_j = e \qquad [9.16]$$

$$x_j \in \{0, 1\}, \ j = 1, 2, \ldots, p \qquad [9.17]$$

where $a_j \in \{0,1\}^m, \ e = (1\,1\,\ldots\,1)^T \in \mathcal{R}^m, \ c(a_j) \in \mathcal{R}$.

Let (LPP) be the linear relaxation of (PP), that is:

$$(LPP) : \text{ minimize} \sum_{j=1}^{p} c(a_j)x_j \qquad [9.18]$$

with the constraints:

$$\sum_{j=1}^{p} a_j x_j = e \qquad [9.19]$$

$$x_j \geqslant 0, \ j = 1, 2, \ldots, p \qquad [9.20]$$

If the optimal solution provided by the solution of (LPP) is not integer, we must use *branching*.

Ryan and Foster [RYA 81] have shown that if a fractional solution $\bar{\lambda} \geqslant 0$ of $A\lambda = e$, exists, where $A \in \{0,1\}^{m \times p}$, two rows r and s of A exist such that:

$$0 < \sum_{k:a_{rk}=1,a_{sk}=1} \bar{\lambda}_k < 1$$

To eliminate this fractional solution $\bar{\lambda}$ we develop two branches. The left-hand branch will have the added constraint:

$$\sum_{k:a_{rk}=1,a_{sk}=1} \lambda_k = 0$$

and the right-hand branch will have the added constraint:

$$\sum_{k:a_{rk}=1,a_{sk}=1} \lambda_k = 1$$

To the column generation problem associated with the left-hand branch we must add the constraint $a_{rl} + a_{sl} \leqslant 1$, where l is the index of the selected column. For the right-hand branch, we must add the constraint $a_{rl} = a_{sl}$ to the column generation problem. These two constraints can be considered without changing the structures of the auxiliary problems (see [BAR 00, BAR 98, VAN 94b, WOL 98]). In [VAN 94b], we find other general branching rules of this kind.

9.6. Applications

Exact column generation methods for integer linear programming (ILP) are more recent than those for general linear programming, which appeared in the 1960s; see [AKK 99, BAR 00, BAR 98, BOU 97, BRA 97, CAR 98, CES 02, DES 84, DES 89, GAM 92, HAN 91, JOH 93, LOB 98, MAC 02, MAC 03, MEH 96, MIN 86, RIB 89, RIB 94, RYA 81, SAV 97, SIM 03, TAI 99, VAN 94b, VAN 98, VAN 99, VAN 00, VAN 94a, VAN 97, VAN 96], amongst others. In what follows, we will introduce two of these examples.

9.6.1. *The p-medians problem*

The p-medians problem (*PMP*) consists of partitioning the set $\mathcal{N}$ of vertices of a graph in p *clusters* (disjoint pair sets). We must select, from a set of candidates $\mathcal{M}$, a *median* vertex for each *cluster* by minimizing the sum of distances c_{ij} between each vertex and the median of its *cluster*. The *PMP* has been widely studied in all the multiresource localization problems (see, for example, [LAB 95, MIR 90]).

The vertices of the graph can represent places in which there will be clients: p service centers must be built in p different places, and each client assigned to a service center. The *PMP* can be generalized by associating a coefficient w_i with each vertex $i \in \mathcal{N}$, which will represent its weight, and associating a coefficient Q_j with each candidate for the median $j \in \mathcal{M}$, which will represent its capacity. This formulation is known as the *capacitated p-median problem* (*CPMP*). The *CPMP* can be modeled as an integer linear programming problem:

$$(\textit{CPMP}) \min \quad v = \sum_{i \in \mathcal{N}} \sum_{j \in \mathcal{M}} c_{ij} x_{ij}$$

with the constraints:

$$\sum_{j \in \mathcal{M}} x_{ij} = 1 \qquad \forall i \in \mathcal{N} \qquad [9.21]$$

$$\sum_{j \in \mathcal{M}} y_j = p \qquad [9.22]$$

$$\sum_{i \in \mathcal{N}} w_i x_{ij} \leqslant Q_j y_j \qquad \forall j \in \mathcal{M} \qquad [9.23]$$

$$x_{ij} \in \{0,\ 1\} \qquad \forall i \in \mathcal{N},\ \forall j \in \mathcal{M} \qquad [9.24]$$

$$y_j \in \{0,\ 1\} \qquad \forall j \in \mathcal{M}$$

The variables x_{ij} are assigned as $x_{ij} = 1$ if the vertex i is assigned to the median j, $x_{ij} = 0$ if it is not. The variables y_j are of localization $y_j = 1$ if the vertex j is chosen as median, $y_j = 0$ if it is not. The partitioning constraints [9.21] with the integrality constraints [9.24] impose the condittion that each vertex is assigned to exactly one median. The constraints [9.23] play a double role: ensuring that the sum of the weights of the vertices assigned to a median do not exceed the capacity, and that no vertex is assigned to a median associated with $y_j = 0$. Constraint [9.22] is satisfied if p vertices are selected for the medians. The objective is to minimize the sum of the distances between each vertex and the median with which it is associated. Henceforth, we refer to this model as the *natural formulation* of *CPMP*.

We will reformulate the *CPMP* problem using the ideas of Dantzig and Wolfe, which have already been presented. Let $(\mathbf{1}, p)$ be the vector of the second member associated with constraints [9.21] and [9.22], let $\boldsymbol{c^j} = (c_{1j} \dots c_{Nj})$ be the vector of the distances between each vertex and the candidate for the median j, let $\boldsymbol{x^j} =$

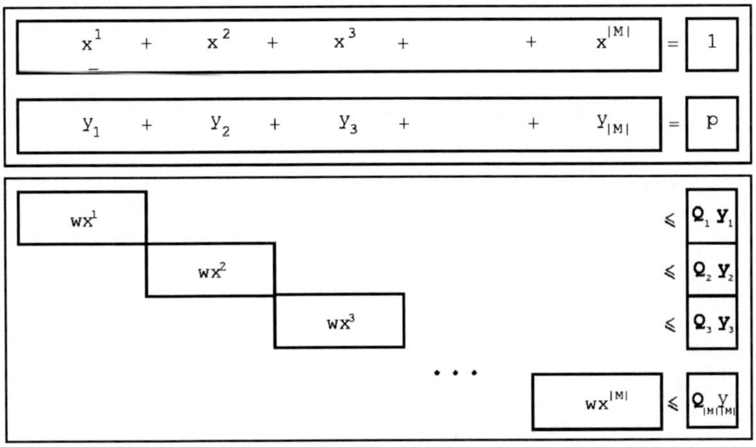

Figure 9.1. *Matrix of the constraints of the CPMP*

$(x_{1j},\ldots,x_{Nj})$ be the vector of the assignment variables relative to the median j, and let $\boldsymbol{w} = (w_1,\ldots,w_N)$ be the vector of the weights of the vertices. The matrix of the constraints of *CPMP* has the form represented by Figure 9.1. The first block corresponds to the constraints [9.11] and the second block to constraints [9.12]. The *CPMP* can be expressed in the following way:

$$\min \sum_{j\in\mathcal{M}} (\boldsymbol{c^j},0)(\boldsymbol{x^j},y_j)$$

with the constraints:

$$\begin{aligned} &\textstyle\sum_{j\in\mathcal{M}}(\boldsymbol{x^j},y_j) = (\mathbf{1},p) \\ &\quad(\boldsymbol{x^j},y_j) \in X^j \qquad\qquad \forall j \in \mathcal{M} \end{aligned}$$

where $X^j = \{(\boldsymbol{x^j},y_j) \in \{0,1\}^{N+1} | \boldsymbol{wx^j} \leqslant Q_j y_j\}$. Each X^j contains a finite number l_j+1 of points: $X^j = \{(\mathbf{0},0)),(\boldsymbol{\bar{x}^1},1),\ldots,(\boldsymbol{\bar{x}^{l_j}},1)\}$. For $(\boldsymbol{x^j},y_j) \in X^j$ we can write:

$$(\boldsymbol{x^j},y_j) = \sum_{k\in Z^j} (\boldsymbol{\bar{x}^k},\bar{y}^k)\lambda_k^j,\ \sum_{k\in Z^j} \lambda_k^j = 1,\ \lambda_k^j \in \{0,1\}\ \forall k \in Z^j \qquad [9.25]$$

where each $k \in Z^j$ is the index of a point in X^j. By replacing expression [9.25], the natural formulation becomes:

$$\min \sum_{j\in\mathcal{M}} (\boldsymbol{c^j},0) \sum_{k\in Z^j} (\boldsymbol{\bar{x}^k},\bar{y}^k)\lambda_k^j$$

with the constraints:

$$\sum_{j\in\mathcal{M}}\sum_{k\in Z^j}(\bar{\boldsymbol{x}}^{\boldsymbol{k}},\bar{y}^k)\lambda_k^j=(\mathbf{1},p)$$

$$\sum_{k\in Z^j}\lambda_k^j=1,\quad \forall j\in\mathcal{M}$$

$$\lambda_k^j\in\{0,1\},\quad \forall j\in\mathcal{M}\ \forall k\in Z^j$$

An alternative formulation for *CPMP* is therefore the following:

$$\min\sum_{j\in\mathcal{M}}\sum_{k\in Z^j}(\boldsymbol{c}^j\bar{\boldsymbol{x}}^{\boldsymbol{k}})\lambda_k^j$$

with the constraints

$$\sum_{j\in\mathcal{M}}\sum_{k\in Z^j}\bar{\boldsymbol{x}}^{\boldsymbol{k}}\lambda_k^j=\mathbf{1} \qquad [9.26]$$

$$\sum_{j\in\mathcal{M}}\sum_{k\in Z^j}\bar{y}^k\lambda_k^j=p \qquad [9.27]$$

$$\sum_{k\in Z^j}\lambda_k^j=1,\quad \forall j\in\mathcal{M} \qquad [9.28]$$

$$\lambda_k^j\in\{0,1\},\quad \forall j\in\mathcal{M}\ \forall k\in Z^j$$

The *CPMP* is formulated here as a partitioning problem, with a few more constraints. This new formulation is known as the *master problem.*

Each column is associated with a *feasible cluster*, that is a set of vertices assigned to the same median without exceeding the capacity of this median. The Z^j are the sets of indices of these clusters. Each cluster of index k is described by the coefficients $\bar{\boldsymbol{x}}^{\boldsymbol{k}}=(\bar{x}_1^k,\ldots,\bar{x}_N^k)$. Each $\bar{x}_i^k$ takes the value 1 if the vertex i belongs to the cluster k, 0 if it does not; the columns associated with $\bar{y}^k=1$ represent the active clusters. The cost of the cluster k is equal to the sum of the distances c_{ij} between each vertex i of the cluster and the median j. The binary variable λ_k^j shows whether or not k is selected for the median j. Constraints [9.26] and the integrality conditions ensure that each client is assigned to a median, constraints [9.28] ensure that one cluster is selected (possibly inactive) for each median, and constraint [9.27] requires p clusters to be put into play.

As we have already seen, the lower bound obtained by solving the linear relaxation of this new formulation is the same as that obtained by the Lagrangian relaxation of constraints [9.21] and [9.22], and can be stronger than that produced by the linear relaxation of the natural formulation. The number of clusters (the columns of the new formulation) is exponential in $|\mathcal{N}|$. However, we can use column generation techniques; let $\boldsymbol{u^1}$ be the vector of the dual variables associated with constraints [9.26], let u^2 be the dual variable associated with [9.27], and let $\boldsymbol{u^3}$ be the vector of the dual variables associated with constraints [9.28]. For each median $j \in \mathcal{M}$, the *pricing* problem (AP) to be solved is the following problem:

$$\min \quad \boldsymbol{c^j \bar{x}^k} - \boldsymbol{u^1}\bar{x}^k - u^2\bar{y}^k - u^3_j$$

with the constraints:

$$\boldsymbol{w\bar{x}^k} \leqslant Q_j\bar{y}^k$$
$$\boldsymbol{\bar{x}^k} \in \{0,1\}^N, \ \bar{y}^k \in \{0,1\}$$

which is a *knapsack* problem for each median (KP_j):

$$(KP_j) \ \max \sum_{i \in \mathcal{N}} (\boldsymbol{u^1} - \boldsymbol{c^j})\boldsymbol{\bar{x}^k}$$

with the constraints:

$$\boldsymbol{w\bar{x}^k} \leqslant Q_j$$
$$\boldsymbol{\bar{x}^k} \in \{0,1\}^N$$

There are very effective algorithms for solving the (**NP**-hard) knapsack problem (see [MAR 90] for a general study of the problem).

Since *CPMP* belongs to the partitioning problems category, once the optimal solution $(\widehat{\lambda}^j_k)$ of the linear relaxation of the master problem has been obtained, it is possible to use Ryan and Foster's branching procedure [RYA 81] as previously described. As we have seen, there will be two rows r and s such that:

$$0 < \sum_{j \in \mathcal{M}, k \in Z^j | \widehat{x}^k_r = 1, \widehat{x}^k_s = 1} \widehat{\lambda}^j_k < 1$$

that is two vertices r and s are assigned to the same median for some clusters, and to different medians for other clusters. Therefore we can separate the problem into two subproblems:

– P1, in which we add the constraint $\widehat{x}_r^k = \widehat{x}_s^k$ (the vertices r and s must be inserted in the same cluster);

– P2, in which we add the constraint $\widehat{x}_r^k + \widehat{x}_s^k \leqslant 1$ (the vertices r and s must be inserted in different clusters).

Nevertheless, the constraint added in P2 changes the structure of the pricing problem, which can no longer be solved as effectively. On the other hand, it is possible to reconstruct a fractional solution $(\widehat{x}_{ij})$ for the natural formulation by substituting for each $i \in \mathcal{N}$ and for each $j \in \mathcal{M}$: $\widehat{x}_{ij} = \sum_{k \in Z^j} \bar{x}_i^k \widehat{\lambda}_k^j$. The value $\widehat{x}_{ij}$ represents a fraction of the demand at the vertex i satisfied by the median j. If all the $\widehat{x}_{ij}$ have integer values, the solution found provides an upper bound whose value is equal to the lower bound, so evaluation of subproblems is eliminated. Otherwise, it is possible to apply separation techniques (*branching*) used in traditional separation and progressive evaluation (*branch-and-bound*) algorithms [NEM 88]. For example, it is possible to select a variable $x_{i^b j^b}$ with a value closer to 0.5 and separate the problem into two subproblems:

– P1, in which the variable $x_{i^b j^b}$ is set to 1 (that is the vertex i is assigned to the median j);

– P2, in which the variable $x_{i^b j^b}$ is set to 0 (that is the vertex i cannot be assigned to the median j).

In this way, the two subproblems do not change the structure of the pricing problem: there is a reduction in the number of variables considered. It is also possible to use more sophisticated separation techniques (*branching*; see [CES 02, SAV 97]).

9.6.2. *Vehicle routing*

One of the first applications of the column generation method in ILP has been to *vehicle routing problems (VRP)* [DES 84]. For an up-to-date and complete reference on vehicle routing problems, see Toth and Vigo [TOT 02], and for a detailed description of column generation methods for *VRP*s, see [DES 95, DES 98].

The *VRP* consists of a set $\mathcal{N}$ of vertices of a graph that must be visited by K vehicles with constraints (*side-constraints*, associated with the vehicles' capacities, with time windows, etc.) that will not be dealt with here because we want to focus our attention on the application of the method to routing problems in general.

Let $\mathcal{V} = \mathcal{N} \cup \{0\}$ and let $\mathcal{A}$ be the set of arcs. We define as c_{ij} the cost function associated with each arc $(i, j) \in \mathcal{A}$. One possible ILP formulation for routing problems is the following:

$$(VRPs) \quad \min \sum_{(i,j) \in \mathcal{A}} c_{ij}\, x_{ij}$$

with the constraints:

$$\sum_{i \in \mathcal{V}} x_{ij} = 1, \quad \forall i \in \mathcal{N} \qquad [9.29]$$

$$\sum_{j \in \mathcal{V}} x_{ji} = 1, \quad \forall j \in \mathcal{N} \qquad [9.30]$$

$$\sum_{j \in \mathcal{V}} x_{0j} = K \qquad [9.31]$$

$$\sum_{i \in \mathcal{V}} x_{i0} = K \qquad [9.32]$$

GSECs [9.33]

side-constraints [9.34]

$$x_{ij} \in \{0,1\}, \quad \forall (i,j) \in \mathcal{A}$$

A binary variable x_{ij} is associated with each arc $(i,j) \in \mathcal{A}$: if $x_{ij} = 1$, the arc is part of the solution (that means that a vehicle travels from vertex i to vertex j), otherwise $x_{ij} = 0$. Constraints [9.29], [9.30], [9.31] and [9.32] are associated with the flows: only one vehicle enters and exits each vertex, while K vehicles enter and exit the vertex that represents the depot. Constraints [9.33] eliminate subcycles and constraints [9.34] are associated with the specific *routing* problem. Many equivalent formulations and details of the elimination constraints can be found in [TOT 02].

We can reformulate the *VRP* as a *set partitioning* problem (see [DES 92]). Let $\mathcal{R}$ be the set of all the elementary paths (without cycles) on $\mathcal{V}$ that satisfy constraints [9.34] (*feasible paths*). The cost of each path, in relation to the original formulation, is $c_j = \sum_{(i,j) \in \mathcal{A}} c_{ij} x_{ij}$. Therefore the *VRP* can be expressed:

$$(MP) \ \min \sum_{j \in \mathcal{R}} c_j \, x_j$$

with the constraints:

$$\sum_{j \in \mathcal{R}} a_{ij} x_j = 1, \quad \forall i \in \mathcal{N} \qquad [9.35]$$

$$\sum_{j \in \mathcal{R}} x_j = K \qquad [9.36]$$

$$x_j \in \{0,1\}, \quad \forall j \in \mathcal{R}$$

The binary variable $x_j \in \{0,1\}$ shows whether or not path j^{th} is used, and the coefficients $a_{ij} \in \{0,1\}$ show whether or not vertex i is visited by route j. The linear relaxation of *MP* provides a *lower bound* valid for the original *VRP*.

The number of feasible paths in $\mathcal{R}$ is exponential. To solve *MP*, we will use automatic column generation techniques: we define as $\mathcal{R}'$ a reduced set of feasible paths; we solve the linear relaxation of the reduced *MP* and we use the dual information to generate new paths or to verify the end of the column generation procedure.

We have already seen that this formulation theoretically provides a lower bound equal to that of the Lagrangian relaxation and in [BRA 02] we can see the very good quality of the lower bounds for different examples.

Let $\boldsymbol{u^1}$ be the vector of the dual variables associated with the partitioning constraints [9.35] and let u^2 be the dual variable associated with constraint [9.36]. The *pricing* problem is to find a feasible path that is the optimal solution of the following problem (see [DES 84, DES 92, DES 98, FEI 03]):

$$(PA) \;\; \min \;\; \tilde{c} = \sum_{(i,j)\in\mathcal{A}} c_{ij}x_{ij} - \sum_{i\in\mathcal{N}} \boldsymbol{u^1}_i y_i - u^2$$

with the constraints:

$$\sum_{j\in\mathcal{V}} x_{ij} = y_i, \;\; \forall i \in \mathcal{N} \qquad [9.37]$$

$$\sum_{j\in\mathcal{V}} x_{ji} = y_i, \;\; \forall i \in \mathcal{N} \qquad [9.38]$$

$$\sum_{j\in\mathcal{V}} x_{0j} = 1 \qquad [9.39]$$

$$\sum_{i\in\mathcal{V}} x_{i0} = 1 \qquad [9.40]$$

The specific constraints [9.41]

$$x_{ij} \in \{0,1\}, \;\; \forall (i,j) \in \mathcal{A}$$

$$y_i \in \{0,1\}, \;\; i \in \mathcal{N}$$

Each variable y_i shows whether or not a vertex i is visited by the path; the variables x_{ij} have already been defined. constraints [9.37], [9.38], [9.39] and [9.40] are associated with a path, and constraints [9.41] ensure that the path found is feasible under the specific conditions.

If $\tilde{c}_{min} \leqslant 0$, we use the variables x_{ij} to define a new path to add to *MP*, otherwise we end the column generation method. If the optimal vector of the linear relaxation is integer, we have found an optimal solution of the original problem; if not, we carry out a separation of the fractional solution, and an enumeration tree is created.

Dror [DRO 94] has shown that this shortest path problem for generating columns in the *VRP pricing* problem is **NP**-hard in the strong sense. There are many algorithms for solving this *pricing* problem. In [DES 98] a general framework for several column generation algorithms for the *VRP* is described. In the case where the formulation of the *VRP* is a partitioning problem, one technique used for dealing with the associated *pricing* problem is dynamic programming. The exact solution methods for the *pricing* problem that use dynamic programming have an exponential complexity and are used only to solve small problems.

The *state-space* relaxation technique introduced by Christofides *et al.* [CHR 81] can be useful for reducing calculation time. This relaxed problem can be solved using a pseudo-polynomial algorithm. The paths obtained by dynamic programming with the relaxation of the *pricing* problem are superoptimal, that is the lower bound found is still valid. The schema to use in this case is to try to eliminate columns that are non-feasible for *MP*.

There are also many branching strategies. Two of them are separation on the arcs, and separation on the resources (see for example [DEL 03, DES 95, DES 98, GEL 92]). Separation on the arcs consists of selecting a fractional path with cycles and generating a child node for each arc of the cycle: the first k arcs are fixed to each node, the $k+1$ arc is forbidden and the others are free. When the path has a cycle, the succession of fixed arcs is interrupted before closing the cycle (the path becomes inadmissible). Branch-offs on the resources can be applied when a node is visited several times in a *fractional* way by a cycle of a path or by several different paths. In this case, we add a new bound to the quantity of resources used so that it is possible to visit a node with only one path (see [GEL 92]).

9.7. Bibliography

[AKK 99] DEN AKKER J.V., HOOGEVEEN J., VAN DE VELDE S., "Parallel machine scheduling by column generation", *Operations Research*, vol. 47, num. 6, p. 862–872, 1999.

[BAR 98] BARNHART C., JOHNSON E.L., NEMHAUSER G., SAVELSBERGH M.W., VANCE P.H., "Branch-and-price: Column generation for solving huge integer programs", *Operations Research*, vol. 46, p. 316–329, 1998.

[BAR 00] BARNHART C., HANNE C., VANCE P.H., "Using branch-and-price and cut to solve origin destination integer multicommodity flow problems", *Operations Research*, vol. 48, p. 318–326, 2000.

[BOU 97] BOURJOLLY J., LAPORTE G., MERCURE H., "A combinatorial column generation algorithm for the maximum stable set problem", *Operations Research Letters*, vol. 20, p. 21–29, 1997.

[BRA 97] BRAMEL J., SIMCHI-LEVI D., "On the effectiveness of set covering formulations for the vehicle routing problem with time windows", *Operations Research*, vol. 45, p. 295–301, 1997.

[BRA 02] BRAMEL J., SIMCHI-LEVI D., "Set-covering-based algorithms for the capacitated VRP", *SIAM Monographs on Discrete Mathematics and Applications*, SIAM, p. 85–108, 2002.

[CAR 98] DE CARVALHO J., "Exact solution of cutting stock problems using column generation and branch-and-bound", *International Transactions in Operational Research*, vol. 5, num. 1, p. 35–43, 1998.

[CES 02] CESELLI A., Algoritmi Branch & Bound e Branch & Price per il problema delle p-mediane con capacità, Università degli Studi di Milano, Milan, Italy, 2002.

[CHR 81] CHRISTOFIDES N., MINGOZZI A., TOTH P., "State-space relaxation procedures for the computation of bounds to routing problems", *Networks*, vol. 11, p. 145–164, 1981.

[CHV 83] CHVÁTAL V., *Linear Programming*, W. H. Freeman & Co., New York/San Francisco, 1983.

[DAK 65] DAKIN R., "A tree search algorithm for mixed integer programming problems", *Computer Journal*, vol. 8, p. 250–255, 1965.

[DAN 60] DANTZIG G., WOLFE P., "Decomposition principle for linear programming", *Operations Research*, vol. 8, p. 101–111, 1960.

[DAN 63] DANTZIG G., *Linear Programming and Extensions*, Princeton University Press, Princeton, NJ, 1963.

[DAN 95] DANTZIG G., "Upper bounds, secondary constraints and block triangularity in linear programming", *Econometrica*, vol. 23, p. 174–183, 1995.

[DEL 03] DELL'AMICO M., RIGHINI G., SALANI M., A branch-and-price algorithm for the vehicle routing problem with simultaneous pick-up and delivery, Report, Università degli Studi di Milano, Milan, Italy, 2003.

[DES 84] DESROSIERS J., SOUMIS F., DESROCHERS M., "Routing with time-windows by column generation", *Networks*, vol. 14, p. 545–565, 1984.

[DES 89] DESROCHERS J., SOUMIS F., "A column-generation approach to the urban transit crew scheduling problem", *Transportation Science*, vol. 23, p. 1–13, 1989.

[DES 92] DESROCHERS M., DESROSIERS J., SOLOMON M., "A new optimization algorithm for the vehicle routing problem with time windows", *Operations Research*, vol. 40, num. 2, p. 342–353, 1992.

[DES 95] DESROSIERS J., DUMAS Y., SOLOMON M., SOUMIS F., *Time Constrained Routing and Scheduling*, INFORMS, 1995.

[DES 98] DESAULNIERS G., DESROSIERS J., IOACHIM I., SOLOMON M., SOUMIS F., VILLENEUVE D., *A Unified Framework for Deterministic Time Constrained Vehicle Routing and Crew Scheduling Problems*, p. 57–93, Kluwer, Boston, MA, 1998.

[DRO 94] DROR M., "Note on the complexity of the shortest path models for column generation in VRPTW", *Operations Research*, vol. 42, num. 5, p. 977–978, 1994.

[FEI 03] FEILLT D., DEJAX P., GENDREAU M., GUEGUEN C., An exact algorithm for the elementary shortest path problem with resource constraints: application to some vehicle routing problems, Report , CRT, 2003.

[GAM 92] GAMACHE M., SOUMIS F., MARQUIS G., DESROSIERS J., "A column generation approach for large-scale aircrew rostering problems", *Operations Research*, vol. 48, num. 2, p. 247–263, 1992.

[GEL 92] GÉLINAS S., DESROCHERS M., DESROSIERS J., SOLOMON M., A new branching strategy for time constrained routing problems with application to backhauling, Report, GERAD, 1992.

[GIL 61] GILMORE P., GOMORY R., "A linear programming approach to the cutting stock problem", *Operations Research*, vol. 9, p. 849–859, 1961.

[GIL 63] GILMORE P., GOMORY R., "A linear programming approach to the cutting stock problem – part II", *Operations Research*, vol. 11, p. 863–888, 1963.

[HAN 91] HANSEN P., JAUMARD B., DE ARAGÃO M.P., Un algorithme primal de programmation linéaire généralisée pour les programmes mixtes, Report, C. R. Acad. Sci. Paris, 1991.

[JOH 93] JOHNSON E., MEHROTAL A., NEMHAUSER G., "Min-cut clustering", *Mathematical Programming*, vol. 62, p. 133–152, 1993.

[LAB 95] LABBÉ M., PEETERS D., THISSE J., *Location on Networks*, vol. 8, p. 551–624, Elsevier Science, 1995.

[LAS 70] LASDON L.S., *Optimization Theory for Large Systems*, Macmillan, New York, 1970.

[LOB 98] LOBEL A., "Vehicle scheduling in public transit and lagrangean pricing", *Management Science*, vol. 12, num. 44, p. 1637–1649, 1998.

[MAC 01] MACULAN N., MACAMBIRA E., DE SOUZA C., Geometric cuts for 0-1 integer programming, Report, Engenharia de Sistemas, COPPE, Universidade Federal do Rio de Janeiro, 2001.

[MAC 02] MACULAN N., PASSINI M., BRITO J., LISSER A., *Column Generation Method for Network Design*, Kluwer Academic Press, p. 165–179, 2002.

[MAC 03] MACULAN N., PASSINI M., BRITO J., LOISEAU I., "Column-generation in integer linear programming", *RAIRO-Recherche Opérationnelle*, vol. 37, num. 2, p. 67–83, 2003.

[MAR 85] MARCOTTE O., "The cutting stock problem and integer rounding", *Mathematical Programming*, vol. 13, p. 82–92, 1985.

[MAR 90] MARTELLO S., TOTH P., *Knapsack Problems – Algorithms and Computer Implementations*, John Wiley & Sons, New York, 1990.

[MEH 96] MEHROTRA A., TRICK M., "A column generation approach for graph coloring", *INFORMS Journal on Computing*, vol. 8, num. 4, p. 344–353, 1996.

[MIN 86] MINOUX M., "Optimal traffic assignment in a SS/TDMA frame: A new approach by set covering and column generation", *RAIRO-Recherche Operationnelle*, vol. 20, num. 4, p. 273–286, 1986.

[MIR 90] MIRCHANDANI P., "The p-median problem and generalizations", *Discrete Location Theory*, Wiley Interscience Series, John Wiley & Sons, New York, 1990.

[NEM 88] NEMHAUSER G.L., WOLSEY L.A., *Integer and Combinatorial Optimization*, John Wiley & Sons, New York, 1988.

[RIB 89] RIBEIRO C., MINOUX M., PENNA M., "An optimal column-generation-with-ranking algorithm for very large scale partitioning problems in traffic assignment", *European Journal of Operational Research*, vol. 41, p. 232–239, 1989.

[RIB 94] RIBEIRO C., SOUMIS F., "A column generation approach to the multiple-depot vehicle scheduling problem", *Operations Research*, vol. 42, p. 41–52, 1994.

[RYA 81] RYAN D., FOSTER B., "An integer programming approach to scheduling", *Computer Scheduling of Public Transport Urban Passenger Vehicle and Crew Scheduling*, North Holland, 1981.

[SAV 97] SAVELSBERGH M., "A branch-and-price algorithm for the generalized assignment problem", *Operations Research*, vol. 46, p. 831–841, 1997.

[SIM 03] SIMONETTI L.G., Geração de colunas para o problema de empacotamento de árvores de Steiner, Universidade Federal do Río de Janeiro, 2003.

[TAI 99] TAILLARD E., "A heuristic generation method for the heterogeneous fleet VRP", *RAIRO Recherche Operationnelle*, vol. 33, num. 1, p. 1–14, 1999.

[TOT 02] TOTH P., VIGO D., "The vehicle routing problem", *SIAM Monographs on Discrete Mathematics and Applications*, SIAM, 2002.

[VAN 94a] VANCE P.H., BARNHART C., JOHNSON E.L., NEMHAUSER G.L., "Solving binary cutting stock problems by column generation and branch-and-bound", *Computational Optimization and Applications*, vol. 3, p. 111–130, 1994.

[VAN 94b] VANDERBECK F., Decomposition and Column Generation for Integer Programming, Thèse de doctorat, PhD thesis, Catholic University of Louvain, Louvain, Belgium, 1994.

[VAN 96] VANDERBECK F., WOLSEY L., "An exact algorithm for IP column generation", *Operations Research Letters*, vol. 19, p. 151–159, 1996.

[VAN 97] VANCE P.H., BARNHART C., JOHNSON E.L., NEMHAUSER G.L., "Airline crew scheduling: a new formulation and decomposition algorithm", *Operations Research*, vol. 45, num. 2, p. 188–200, 1997.

[VAN 98] VANCE P.H., "Branch-and-price algorithms for the one-dimensional cutting stock problem", *Computational Optimization and Applications*, vol. 9, p. 211–228, 1998.

[VAN 99] VANDERBECK F., "Computational study of a column generation algorithm for bin packing and cut stocking problems", *Mathematical Programming*, vol. 86, p. 565–594, 1999.

[VAN 00] VANDERBECK F., "On Dantzig–Wolfe decomposition in integer programming and ways to perform branching in a branch-and-price algorithm", *Operations Research*, vol. 48, num. 1, p. 111–128, 2000.

[WOL 98] WOLSEY L., *Integer Programming*, John Wiley & Sons, New York, 1998.

Chapter 10

Polyhedral Approaches

Several problems from various areas reduce to maximizing (or minimizing) a linear function with linear constraints with bivalent variables. These problems, said to be combinatorial optimization problems, are generally **NP**-hard. Effective methods have therefore been developed to formulate and solve this type of problem. In particular, polyhedral approaches have proved to be powerful for optimally solving these problems. These consist of transforming the problem into a linear program by describing the convex hull of its solutions by a system of linear inequalities. The equivalence established between separation and optimization on a polyhedron, and the evolution of computational tools, have given an important boost to these methods. In fact, using these techniques we can develop polynomial algorithms and min–max relationships between combinatorial structures. These approaches have been successfully applied to several combinatorial optimization problems in the last few years. In this chapter, we will discuss these methods and introduce some applications to the maximum cut and network design problems.

10.1. Introduction

During the last three decades, combinatorial optimization has developed considerably, as much on the theoretical side as on the application side. Several techniques have been developed, and have been shown to be effective in formulating and solving hard combinatorial problems from areas as diverse as transport and telecommunications, biology, VLSI circuits, and statistical physics [GRA 95].

Chapter written by Ali Ridha MAHJOUB.

A *combinatorial optimization problem* is a problem of the following form: given a family $\mathcal{F}$ of subsets of a basic finite set $E = \{e_1, \ldots, e_n\}$ and a system of weights $\omega = (\omega(e_1), \ldots, \omega(e_n))$ associated with the elements of E, find a set $F \in \mathcal{F}$ of maximum (or minimum) weight $\omega(F) = \sum_{e \in F} \omega(e)$, that is:

$$\max(\text{or min}) \, \{\omega(F) \; : \; F \in \mathcal{F}\} \qquad [10.1]$$

Here the family $\mathcal{F}$ is the set of solutions of the problem; it may represent various combinatorial structures, such as paths, cycles, trees, etc., in graphs.

One of the basic problems in combinatorial optimization is *the traveling salesman problem*. Given a graph where each edge is given a certain weight, this consists of finding a cycle that goes through each vertex of the graph only once and whose weight is minimum. Such a cycle is called a *tour* or a *Hamiltonian cycle*. Here the basic set E is the set of edges of the graph and the family $\mathcal{F}$ is the set of tours. This problem has many applications, in particular in routing models [LAW 85].

Other problems, forming a large and important class of combinatorial optimization models, are those of set covering, set packing, and set partitioning problems [BAL 76, COR 01]. Let $M = \{e_1, \ldots, e_n\}$ be a finite set of elements and let $\{M_1, \ldots, M_m\}$ be a family of subsets of M. We say that $F \subseteq \{M_1, \ldots, M_m\}$ is a *covering* of M (resp. a *packing* with respect to M) if $\{\bigcup M_i, \; M_i \in F\} = M$ (resp. $M_i \cap M_j = \emptyset$ for each M_i and M_j of F). If F is both a covering and a packing, then F is said to be a *partition* of M. If $\omega_1, \ldots, \omega_m$ are weights associated with $M_1, \ldots, M_m$, the *set covering* (resp. the *set packing*, resp. the *set partitioning*) problem is to define a covering (resp. a packing, resp. a partition) $F \subseteq \{M_1, \ldots, M_m\}$ such that:

$$\sum_{M_i \in F} \omega_i$$

is minimum (resp. maximum), (resp. maximum (or minimum)). It is clear that these models are combinatorial optimization problems whose basic set is $\{M_1, \ldots, M_m\}$ and the solutions set is the set of covers, packings or partitions of M, respectively. These models are often encountered in resource allocation, facility location, and shape detection problems.

Finally, one last model that was at the origin of several developments in combinatorial optimization is the *knapsack problem*. Given a number b and n objects $o_1, \ldots, o_n$ each having a certain weight $\omega(o_i)$ and value $v(o_i)$, the problem consists of determining a subset of objects $F \subseteq \{o_1, \ldots, o_n\}$ such that $\sum_{o_i \in F} v(o_i) \leq b$ and $\sum_{o_i \in F} \omega(o_i)$ is maximum. Applications of this model can be found in [MAR 90, NEM 88].

Given that the number of solutions of a combinatorial optimization problem is finite, in order to solve the problem we might think of using, for example, an enumerative method consisting of enumerating all the solutions of the problem, calculating the value of each solution, and choosing the best of them. However, even though the number of solutions is finite, it can be exponential, and hence this method can rapidly reach its limits, even for small problems. If we consider this method, for instance for the traveling salesman problem for 30 towns, and using the most powerful computer currently available, more than 10^{10} centuries would be needed to find an optimal solution. This shows that such naïve methods cannot be applied to combinatorial optimization problems.

As a consequence, other, more powerful tools have proved to be necessary for tackling this kind of problem. Linear programing and integer programming are at the root of such tools. A combinatorial optimization problem can always be formulated as an integer program (generally in 0–1). Therefore any relaxation of the problem, obtained by relaxing the integrality constraints and by considering a subset of constraints, is no more than a linear program. If, by solving this program, we obtain a feasible integer solution of the problem, then it is optimal.

At the end of the 1940s, Dantzig [DAN 51, DAN 63] introduced the first algorithm, the simplex method, for solving linear programming problems. This method not only proved to be practically effective, but it constitued a very important basic tool for combinatorial optimization. Indeed, an optimal solution of a given linear program, found using the simplex algorithm, always corresponds to an extreme point of the polyhedron given by the inequalities of the program. In this way, this algorithm allows us to solve any combinatorial optimization problem whose set of solutions corresponds to the vertices of a polyhedron whose description is given by a system of linear inequalities. It follows from this that, given a combinatorial optimization problem where every solution can be represented by a vector of integers, if we can describe the convex hull of these points using a linear system, we thus reduce the problem to solving a simple linear program.

This transformation from optimization on a finite discrete set, which can have a very large number of solutions, to optimization on a convex domain, has been at the origin of an important evolution in combinatorial optimization. Indeed, this reduction has allowed us to introduce a new approach, known as the *polyhedral* approach, for combinatorial optimization problems. This consists of reducing such a problem to the resolution of a linear program by describing the convex hull of its solutions by a linear inequality system. This method, initiated by Edmonds [EDM 65] for the matching problem, was later revealed to be very powerful for optimally solving these problems. In particular, this method allows us to solve effectively a combinatorial optimization problem even if we only have a partial description of the convex hull of its solutions and even if the latter contains an exponential number of constraints. It also allows us to obtain min–max relations and to devise polynomial-time algorithms.

Indeed, as has been shown by Grötschel *et al.* [GRÖ 81], optimizing on a given polyhedron does not depend on the number of constraints of the system describing the polyhedron, but rather on the *separation* problem associated with this system. Given a solution x, this problem consists of determining if x satisfies the system, and, if not, of finding a constraint of the system that is violated by x. Grötschel *et al.* [GRÖ 81] have shown that a polynomial algorithm exists for optimizing a linear function on a polyhedron if and only if a polynomial algorithm exists for solving the separation problem associated with the system that defines the polyhedron. This equivalence between optimization and separation gave a new boost to combinatorial optimization and to polyhedral approaches. Indeed, a combinatorial optimization problem can be solved as a sequence of linear programs, each being obtained by adding a violated constraint. This latter can be obtained by solving a separation problem. If the solution of one of these programs is a solution of the problem, then it is optimal. Therefore, the initial problem can be solved in polynomial time, if the separation of the constraints can be carried out in polynomial time. Such an algorithm is known as a *cutting-plane* algorithm.

Combinatorial optimization problems are generally **NP**-hard. As a consequence, there is little hope of finding a complete description of the polyhedron associated with a system of inequalities. Using a cutting-plane algorithm, it is possible that an optimal solution of the problem will not be obtained. A *branch-and-bound* technique can then be considered to establish an optimal solution. In each iteration of the method, new constraints can be generated to further tighten the linear relaxation, and improve the bound on the optimal value at each node of the resolution tree. This method, introduced by Padberg and Rinaldi [PAD 91] and known as the *branch-and-cut method*, has been successfully applied to several combinatorial optimization problems, such as the traveling salesman problem [APP 98, GRÖ 91, LAW 85, PAD 91], the maximum cut problem [BAR 88, SIM 95, JÜN 98], and the survivable network design problem [GRÖ 95, KER 05]. Furthermore, the development of new and powerful computational tools has also enabled an important evolution in this method to take place in recent years. This method is now the most effective, and the most widely used, tool for solving combinatorial optimization problems. In this chapter, we present these techniques and discuss certain applications.

In section 10.2, we introduce the basic elements of polyhedral theory. In section 10.3, we study the relationship between combinatorial optimization and linear programming. In section 10.4, we introduce some proof techniques for polyhedra, in particular we introduce methods for proving that a given inequality defines a facet of a combinatorial polyhedron, and that a linear system describes an integer polyhedron. In section 10.5, we discuss integer polyhedra and min–max relationships in combinatorial optimization. We study, in particular, totally unimodular matrices, totally dual integral systems, and blocking and antiblocking polyhedra. In section 10.6, we introduce cut, and branch-and-cut, methods. We also discuss the relationship between separation and optimization. In sections 10.7 and 10.8, we introduce some applications of these techniques to the maximum cut and survivable network design problems.

The rest of this section is devoted to some definitions and notations. We consider non-directed graphs. A graph will be expressed by $G = (V, E)$, where V is the set of *vertices* and E is the set of *edges*. If e is an edge between two vertices u and v, then we denote it by $e = uv$. A path between two vertices u and v in G is a sequence of vertices and edges $(v_0, e_1, v_1, e_2, v_2, \ldots, v_{l-1}, e_l, v_l)$, where $u = v_0$, $v = v_l$ and $e_i = v_{i-1}v_i$ for $i = 1, \ldots, l$ and $v_0, \ldots, v_l$ are distinct vertices of V. Vertices u and v are said to be the *extremities* of the path. A path will be given by its set of edges $(e_1, \ldots, e_l)$. Two paths between two vertices u and v are said to be *edge-disjoint* (resp. *vertex-disjoint*) if they do not have any edges (resp. vertices, except for u and v) in common. A *cycle* is a path whose extremities coincide. If $F \subseteq E$, then $V(F)$ will refer to the set of vertices of the edges of F. If $S \subseteq V$, we will denote by $E(S)$ the set of edges having their extremities in S.

Let $E = \{e_1, \ldots, e_n\}$ be a finite set. If $F \subseteq E$ and $x = (x(e),\ e \in E) \in \mathbb{R}^E$, then we express by $x(F)$ the sum $\sum_{e \in F} x(e)$. If a and x are vectors of $\mathbb{R}^E$, we denote the sum $\sum_{e \in E} a(e)x(e)$ by ax. Therefore the inequality $\sum_{e \in E} a(e)x(e) \leq \alpha$ is written as $ax \leq \alpha$.

10.2. Polyhedra, faces and facets

In this section, we present some definitions and basic properties of polyhedral theory. In particular, we discuss the description of a polyhedron by its facets.

10.2.1. *Polyhedra, polytopes and dimension*

DEFINITION 10.1.– *A polyhedron $P \subseteq \mathbb{R}^n$ is the set of solutions of a finite system of linear inequalities, that is:*

$$P = \{x \in \mathbb{R}^n \ : \ Ax \leq b\}$$

where A is a matrix $m \times n$, $b \in \mathbb{R}^m$, and m and n are two positive integers.

A bounded polyhedron is known as a *polytope*. In other words, a polyhedron $P \subseteq \mathbb{R}^n$ is a polytope if $x^1, x^2 \in \mathbb{R}^n$ exists such that $x^1 \leq x \leq x^2$ for all $x \in P$.

Note that any polyhedron P is *convex*, that is if x^1 and x^2 are any two points of P, then $\lambda x^1 + (1 - \lambda)x^2$ is also a point of P for any $0 \leq \lambda \leq 1$.

Throughout this chapter, we consider rational polyhedra, that is polyhedra for which the coefficients of the system $Ax \leq b$ are all rational. If A is a matrix $m \times n$, we use A_i (resp. A^j) to refer to the i-th row (resp. j-th column) of A for $i = 1, \ldots, m$ (resp. $j = 1, \ldots, n$).

DEFINITION 10.2.– *Points $x^1, \ldots, x^k$ of $\mathbb{R}^n$ are said to be linearly (resp. affinely) independent if the system:*

$$\sum_{i=1}^{k} \lambda_i x^i = 0 \quad (\text{resp.} \sum_{i=1}^{k} \lambda_i x^i = 0 \text{ and} \sum_{i=1}^{k} \lambda_i = 0)$$

allows a unique solution $\lambda_i = 0$ for $i = 1, \ldots, k$.

A set of linearly (resp. affinely) independent points will also be said to be linearly (resp. affinely) independent. Let us note that a linearly independent set of points is affinely independent, however the converse is not true. The following proposition, which is easy to verify, establishes the relationship between linear independence and affine independence.

PROPOSITION 10.1.– *The following two assertions are equivalent:*

1) $x^1, \ldots, x^k$ are affinely independent;

2) $x^2 - x^1, \ldots, x^k - x^1$ are linearly independent.

DEFINITION 10.3.– *A polyhedron P is said to be k-dimensional if the maximum number of affinely independent points of P is $k + 1$. We then write dim(P) $= k$.*

Note that if P is a polyhedron of $\mathbb{R}^n$, then $\dim(P) \leq n$. If $P \subseteq \mathbb{R}^n$ and $\dim(P) = n$, then P is said to be *full-dimensional.*

DEFINITION 10.4.– *A point $x \in \mathbb{R}^n$ is said to be a linear combination of points $x^1, \ldots, x^k$ of $\mathbb{R}^n$ if k scalars $\lambda_1, \ldots, \lambda_k$ exist such that $x = \sum_{i=1}^k \lambda_i x^i$. If, furthermore, $\sum_{i=1}^k \lambda_i = 1$ (resp. $\lambda_i \geq 0$ for $i = 1, \ldots, k$ and $\sum_{i=1}^k \lambda_i = 1$), then x is said to be an affine combination (resp. a convex combination) of these points.*

DEFINITION 10.5.– *Given a set of points $S \subset \mathbb{R}^n$, the affine hull (resp. the convex hull) of S (see Figure 10.1) is the set of points that can be expressed as an affine (resp. convex) combination of points of S.*

COMMENT 10.1.– If 0 does not belong to the affine hull of points $x^1, \ldots, x^n$, then $x^1, \ldots, x^n$ are affinely independent if and only if they are linearly independent.

The convex hull of a set of points S can also be seen as the smallest convex set containing S. We will later express the convex hull of a set of points S by $conv(S)$.

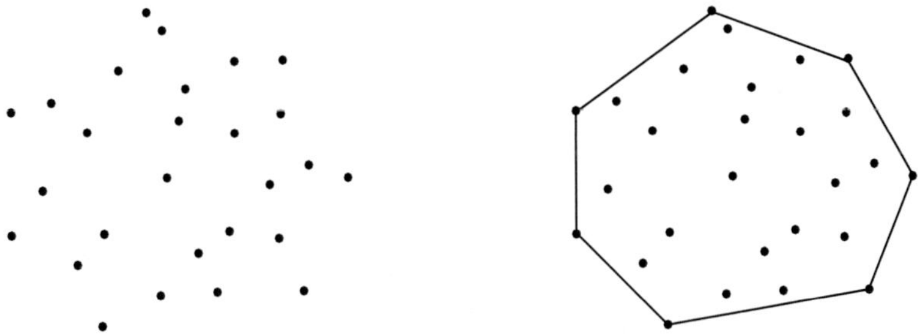

Solutions of S Convex hull of S

Figure 10.1. *Convex hull*

The following proposition establishes the relationship between optimizing on S and optimizing on the convex hull of S.

PROPOSITION 10.2.– *Let $S \subseteq \mathbb{R}^n$ be a set of points and ω a vector of $\mathbb{R}^n$. Thus*

$$\max\{\omega x \ : \ x \in S\} = \max\{\omega x \ : \ x \in conv(S)\}$$

Proof. Let $\bar{x} \in S$ and $x^* \in conv(S)$ such that $\omega\bar{x} = \max\{\omega x \ : \ x \in S\}$ and $\omega x^* = \max\{\omega x \ : \ x \in conv(S)\}$. Since $\bar{x} \in S$, and therefore $\bar{x} \in conv(S)$, then $\omega\bar{x} \leq \omega x^*$. Furthermore, using linear programming, we can assume that x^* is an extreme point of $conv(S)$, and consequently $x^* \in S$. This means that $\omega x^* \leq \omega\bar{x}$, and so $\omega x^* = \omega\bar{x}$. ∎

Proposition 10.2 establishes the link between combinatorial optimization and linear programming. An optimal solution of a combinatorial optimization problem can be obtained by solving the linear program induced by the convex hull of its solutions.

Let us now consider a polyhedron $P \subseteq \mathbb{R}^n$, and let us assume that P is described by a system of m_1 inequalities and m_2 equations, that is P is expressed as:

$$P = \left\{ \begin{array}{lll} x \in \mathbb{R}^n : & A_i x \leq b_i, & i = 1, \ldots, m_1 \\ & B_j x = d_j, & j = 1, \ldots, m_2 \end{array} \right\} \qquad [10.2]$$

This implies that for any inequality $A_i x \leq b_i$, $i \in \{1, \ldots, m_1\}$, there exists a solution $\tilde{x}$ of P such that $A_i \tilde{x} < b_i$. In what follows, we will denote by A and B the matrices whose rows are A_i, $i = 1 \ldots, m_1$ and B_j, $j = 1, \ldots, m_2$, respectively, and by b and d the vectors $(b_1, \ldots, b_{m_1})^T$ and $(d_1, \ldots, d_{m_2})^T$ (for a given matrix A, A^T will denote the transpose of A).

DEFINITION 10.6.– *A point $x^* \in P$ is called an interior point of P if $A_i x^* < b_i$ for $i = 1, \ldots, m_1$.*

PROPOSITION 10.3.– *If P is non-empty, then P contains an interior point.*

Proof. For $i = 1, \ldots, m_1$, let $x^i \in P$ such that $A_i x^i < b_i$. Let $\bar{x} = \frac{1}{m_1}\sum_{i=1}^{m_1} x^i$. We have $A_i\bar{x} < b_i$ for $i = 1, \ldots, m_1$ and $B_j\bar{x} = d_j$ for $j = 1, \ldots, m_2$. Therefore $\bar{x}$ is an interior point of P. ▮

Given a matrix M, the *rank* of M, denoted by rank(M), is the maximum number of linearly independent rows (or columns) of M.

COMMENT 10.2.– If M is a matrix of rank r, and $\{x \in \mathbb{R}^n \ : \ Mx = l\} \neq \emptyset$ for a certain vector l, then the maximum number of affinely independent points in the set $\{x \in \mathbb{R}^n \ : \ Mx = l\}$ is $n - r + 1$.

PROPOSITION 10.4.– *If $P \neq \emptyset$, then dim(P) $= n-$rank(B).*

Proof. Let $k = \text{rank}(B)$. From comment 10.2, there are $n-k+1$ points $x^1, \ldots, x^{n-k+1}$ affinely independent of the system $Bx = 0$. Since $P \neq \emptyset$, from proposition 10.3, P contains an interior point, say $\bar{x}$, so $A_i\bar{x} < b_i$ for $i = 1, \ldots, m_1$. Hence a scalar $\epsilon \neq 0$ exists, sufficiently small, such that $A_i\bar{x} + \epsilon A_i x^j \leq b_i$, for $i = 1, \ldots, m_1$ and $j = 1, \ldots, n-k+1$. Let us consider the points $y^j = \bar{x} + \epsilon x^j$ for $j = 1, \ldots, n-k+1$. We have $y^j \in P$ for $j = 1, \ldots, n-k+1$. Furthermore, since $x^1, \ldots, x^{n-k+1}$ are affinely independent, $y^1, \ldots, y^{n-k+1}$ are also affinely independent, which implies that $\dim(P) \geq n - k$.

Also, as $P \subseteq \{x \in \mathbb{R}^n \ : \ Bx = d\}$, from comment 10.2, $\dim(P) \leq n - k$. Consequently, $\dim(P) = n - k$. ▮

From proposition 10.4, it followsthat a polyhedron is full-dimensional if and only if it contains a point that satisfies all the constraints of the polyhedron with strict inequality.

10.2.2. *Faces and facets*

Let us consider a polyhedron $P \subseteq \mathbb{R}^n$, and let us assume that P is described by the system [10.2]. In this section, we study the inequalities of this system which are essential to the description of P.

DEFINITION 10.7.– *A linear inequality $ax \leq \alpha$ is said to be valid for a polyhedron $P \in \mathbb{R}^n$ if it is satisfied by every point of P, that is $P \subseteq \{x \in \mathbb{R}^n \ : \ ax \leq \alpha\}$.*

DEFINITION 10.8.– *If $ax \leq \alpha$ is a valid inequality, then the polyhedron:*

$$F = \{x \in P : \ ax = \alpha\}$$

is called a face of P, and we say that F is defined by the inequality $ax \leq \alpha$.

By convention, the empty set and the polytope P itself are considered as faces of P. A face of P is said to be *proper* if it is non-empty and different from P (see Figure 10.2).

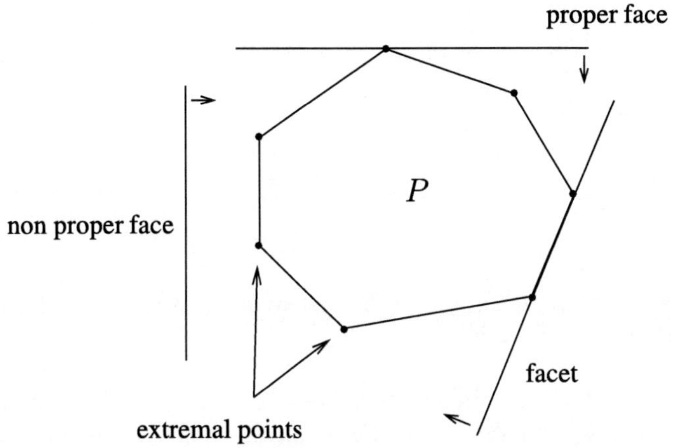

Figure 10.2. *Faces and facets*

PROPOSITION 10.5.– *A non-empty subset F of P is a face of P if and only if a subsystem $A'x \leq b'$ of $Ax \leq b$ exists such that $F = \{x \in P \ : \ A'x = b'\}$.*

Proof. ($\Longrightarrow$) Let us assume that F is a face of P. So an inequality $ax \leq \alpha$, valid for P, exists such that $F = \{x \in P \ : \ ax = \alpha\}$. Let us consider the linear program

$$\max\{ax \ : \ x \in P\} \qquad [10.3]$$

The optimal solutions of the program [10.3] are precisely the elements of F. Let (y^1, y^2) be a dual optimal solution of the program [10.3], where y^1 and y^2 are the dual vectors corresponding to the systems $Ax \leq b$ and $Bx = d$, respectively. Let $A'x \leq b'$ be the subsystem of $Ax \leq b$ whose dual variables have a strictly positive value. From the conditions of complementary slackness in linear programming, we have $F = \{x \in P \ : \ A'x = b'\}$.

($\Longleftarrow$) Let us assume that a subsystem $A'x \leq b'$ of $Ax \leq b$ exists such that $F = \{x \in P : A'x = b'\}$. So for every point $x \in P \setminus F$, there is at least one constraint among the inequalities of $A'x \leq b'$ that is not equally satisfied by x. Let us consider the inequality $ax \leq \alpha$ obtained by adding the inequalities of $A'x \leq b'$. It is clear that $ax = \alpha$ for every $x \in F$ and $ax < \alpha$ for every $x \in P \setminus F$. ∎

DEFINITION 10.9.– *A point x of a polyhedron P is said to be an extreme point or vertex of P if there are not two solutions x^1 and x^2 of P, $x^1 \neq x^2$, such that $x = \frac{1}{2}x^1 + \frac{1}{2}x^2$.*

PROPOSITION 10.6.– *A point $\bar{x} \in P$ is an extreme point of P if and only if $\bar{x}$ is a 0-dimensional face.*

Proof. Let $\bar{P}$ be the polyhedron obtained from P by transforming every inequality equally satisfied by $\bar{x}$ into an equation. Therefore $\bar{P}$ can be written as:

$$\bar{P} = \{x \in \mathbb{R}^n : \bar{A}x \leq \bar{b}, \ \bar{B}x = \bar{d}\}$$

where $\bar{B}x = \bar{d}$ (resp. $\bar{A}x \leq \bar{b}$) is the system of equations (resp. inequalities) of $\bar{P}$. We therefore have $\bar{A}\bar{x} < \bar{b}$.

($\Longrightarrow$) If $\bar{x}$ is not a 0-dimensional face, then $\dim(\bar{P}) > 0$, and, therefore, from proposition 10.4, $\text{rank}(\bar{B}) < n$. Consequently, there is a point $y \neq 0$ such that $\bar{B}y = 0$. Since $\bar{A}\bar{x} < \bar{b}$, there exists $\epsilon \neq 0$, sufficiently small, such that $\bar{A}\bar{x} + \epsilon\bar{A}y \leq \bar{b}$, and so $\bar{x}^1 = \bar{x} + \epsilon y \in \bar{P}$. Moreover, ϵ can be chosen such that $\bar{x}^2 = \bar{x} - \epsilon y \in P$. We then have $\bar{x} = \frac{1}{2}(\bar{x}^1 + \bar{x}^2)$, which implies that $\bar{x}$ is not an extreme point.

($\Longleftarrow$) If $\bar{x}$ is a 0-dimensional face, and therefore $\bar{P}$ is 0-dimensional, from proposition 10.4, we have $\text{rank}(\bar{B}) = n$. Thus $\bar{x}$ is the unique solution of the system $\bar{B}x = \bar{d}$. Let $\bar{x} = \frac{1}{2}(x^1 + x^2)$, with $x^1, x^2 \in P$. Since x^1 and x^2 are solutions of $\bar{B}x = \bar{d}$, we therefore must have $\bar{x} = x^1 = x^2$, and hence $\bar{x}$ is an extreme point of P. ∎

It is clear that if F is a proper face of P, then $\dim(F) \leq \dim(P) - 1$. In the following, we will see that the inequalities that are necessary to the description of P are those that define maximal faces (in the sense of inclusion).

DEFINITION 10.10.– *A proper face of P is said to be a facet if $\dim(F) = \dim(P) - 1$ (see Figure 10.2).*

PROPOSITION 10.7.– *Assume that $P \neq \emptyset$. Then every constraint $A_i x \leq b_i$ that does not define a facet of P is redundant.*

Proof. Let us assume that $A_i x \leq b_i$ is essential in the description of P. We will show that $A_i x \leq b_i$ defines a facet. Indeed, if $A_i x \leq b_i$ is essential in P, then there must exist a point $x^* \in \mathbb{R}^n \setminus P$ such that:

$$\begin{aligned} &A_j x^* \leq b_j, \quad \forall j \in \{1,\ldots,m_1\} \setminus \{i\} \\ &A_i x^* > b_i \\ &Bx^* = d \end{aligned}$$

Since $P \neq \emptyset$, from proposition 10.3, P contains an interior point, let us say $\hat{x}$. Therefore $A_i\hat{x} < b_i$. Let z be a point on the segment between x^* and $\hat{x}$ such that $A_i z = b_i$. So $z = \lambda\hat{x} + (1-\lambda)x^*$ with $0 < \lambda < 1$. Furthermore, we have:

$$\begin{aligned} &A_j z < b_j, \quad \forall j \in \{1,\ldots,m_1\} \setminus \{i\} \\ &A_i z = b_i \\ &Bz = d \end{aligned}$$

This implies that z belongs to the set $F = \{x \in P \,:\, A_i x = b_i\}$, the face of P defined by $A_i x \leq b_i$. Note that the system given by the equations of F is $\{A_i x = b_i,\ Bx = d\}$. Since $\hat{x} \in P \setminus F$, A_i is linearly independent of the rows of B. So, $\text{rank}\begin{pmatrix} A_i \\ B \end{pmatrix} = \text{rank}(B) + 1$. Hence $\dim(F) = \dim(P) - 1$, and, consequently, F is a facet of P. ∎

In what follows, we need theorem 10.1, known as Farkas' lemma, which is one of the fundamental results in mathematical programming. For the proof, see [COO 98].

THEOREM 10.1.– (Farkas' lemma for inequalities). *Given a matrix $m \times n$ A and a vector $b \in \mathbb{R}^m$, the system $Ax \leq b$ has a solution if and only if there does not exist a vector $y \geq 0$ of $\mathbb{R}^m$ such that $yA = 0$ and $yb < 0$.*

COROLLARY 10.1.– (Farkas' lemma). The system $Ax = b$ allows a solution (resp. a positive solution) if and only if there does not exist a vector y such that $yA = 0$ and $yb < 0$ (resp. $yA \geq 0$ and $yb < 0$).

The following proposition shows that a facet of P must be defined by at least one of the inequalities of P.

PROPOSITION 10.8.– *For each facet F of P, one of the inequalities defining F is necessary in the description of P.*

Proof. Let us assume that F is defined by each of the inequalities $A_i x \leq b_i$, for $i \in I$, where I is a subset of $\{1,\ldots,m_1\}$. Let $\tilde{P}$ be the polyhedron obtained from P by removing all the inequalities $A_i x \leq b_i$, $i \in I$. We will show that $\tilde{P} \setminus P \neq \emptyset$, which shows that at least one of the constraints $A_i x \leq b_i,\ i \in I$ must appear in the description of P.

Let us consider a constraint $A_i x \leq b_i$ among those that define F. Since F is a facet of P, we have $F \neq \emptyset$. From proposition 10.3, F must then contain an interior point, say x^0. Since $F \neq P$, A_i is independent of the rows of B. Therefore the system $yB = A_i$ does not have a solution. From corollary 10.1, the system $\{Bx = 0,\ A_i x > 0\}$ has a solution, say x^1. Since $A_k x^0 < b_k$ for $k \in \{1, \ldots, m_1\} \setminus I$, $\epsilon > 0$ exists such that $A_k x^0 + \epsilon A_k x^1 \leq b_k$ for $k \in \{1, \ldots, m_1\} \setminus I$. Let $x^2 = x^0 + \epsilon x^1$, hence $A_k x^2 \leq b_k$ for $k \in \{1, \ldots, m_1\} \setminus I$. Since $A_i x^1 > 0$ and $A_i x^0 = b_i$, we have $A_i x^2 = A_i x^0 + \epsilon A_i x^1 > b_i$, from which $x^2 \notin P$. So, since $Bx^1 = 0$, we also have $x^2 \in \tilde{P}$, and, consequently, $x^2 \in \tilde{P} \setminus P$. ∎

DEFINITION 10.11.– *Two valid inequalities $a_1 x \leq \alpha_1$ and $a_2 x \leq \alpha_2$ are said to be equivalent if there exist $\lambda > 0$ and a vector μ of $\mathbb{R}^{m_2}$ such that:*

$$a_2 = \lambda a_1 + \mu B \qquad [10.4]$$

$$\alpha_2 = \lambda \alpha_1 + \mu d \qquad [10.5]$$

PROPOSITION 10.9.– *Assume that $P \neq \emptyset$. If two valid inequalities of P define the same facet, then they are equivalent.*

Proof. Let $a_1 x \leq \alpha_1$ and $a_2 x \leq \alpha_2$ be two valid inequalities that define the same facet F of P. First note that if system [10.4] has a solution, then this latter also satisfies [10.5]. Indeed, since $F \neq \emptyset$, there is a solution $\tilde{x}$ in F such that $a_1 \tilde{x} = \alpha_1$ and $a_2 \tilde{x} = \alpha_2$. We also have $B\tilde{x} = d$. Thus $\alpha_2 = a_2 \tilde{x} = \lambda a_1 \tilde{x} + \mu B \tilde{x} = \lambda \alpha_1 + \mu d$.

In the following, we will show that if system [10.4] has a solution, then $\lambda > 0$. Indeed, if $\lambda \leq 0$, then $\lambda a_1 x \geq \lambda \alpha_1$ for every $x \in P$, and, consequently, for every $x \in P$, we have:

$$\begin{aligned} a_2 x &= \lambda a_1 x + \mu B x \\ &\geq \lambda \alpha_1 + \mu d \\ &= \alpha_2 \end{aligned}$$

Since $a_2 x \leq \alpha_2$, it follows that $a_2 x = \alpha_2$. But this implies that $P = F$, which contradicts the fact that F is a proper face. Let us now assume that the system [10.4] does not have a solution. Thus by corollary 10.1, a solution $\bar{x} \in \mathbb{R}^n$ exists such that:

$$\begin{aligned} a_1\bar{x} &= 0 \\ B\bar{x} &= 0 \\ a_2\bar{x} &> 0 \end{aligned}$$

Let $I \subset \{1, \dots, m_1\}$ such that $A_i x \leq b_i$ defines F for every $i \in I$. Note that from proposition 10.8, $I \neq \emptyset$. Since F is a facet, and, consequently, $F \neq \emptyset$, from proposition 10.3, F contains an interior point, say x^*. Therefore $A_k x^* < b_k$ for every $k \in \{1, \dots, m_1\} \setminus I$. Let $\epsilon > 0$ such that $A_k x^* + \epsilon A_k \bar{x} \leq b_k$ for every $k \in \{1, \dots, m_1\} \setminus I$. Let $\hat{x} = x^* + \epsilon\bar{x}$. Since $a_2\bar{x} > 0$ and $a_2 x^* = \alpha_2$, we have $a_2\hat{x} = a_2 x^* + \epsilon a_2\bar{x} > \alpha_2$. As a consequence $\hat{x} \notin P$. Since $P \neq \emptyset$, let y be an interior point of P. We therefore have $a_1 y < \alpha_1$, and so $y \notin F$. Let z be the point of F on the segment between y and $\hat{x}$. Then $z = \nu y + (1 - \nu)\hat{x}$ for a certain $0 < \nu < 1$. Furthermore, we have $a_1 z = \nu a_1 y + (1 - \nu) a_1 \hat{x} < \nu\alpha_1 + (1 - \nu)\alpha_1 = \alpha_1$, which is impossible.

As a consequence, system [10.4] has a solution. As has been shown above, this solution also satisfies system [10.5], and we have$\lambda > 0$. ∎

From proposition 10.9, for every facet of P, one and only one inequality that defines F is necessary in the system that describes P. As a consequence of propositions 10.7, 10.8 and 10.9, we have theorem 10.2.

THEOREM 10.2.– *System [10.2] that defines P is minimal if and only if the rows of B are linearly independent and every inequality $A_i x \leq b_i$, $i = 1, \dots, m_1$, defines a distinct facet of P.*

If P is a full-dimensional polyhedron, then, from proposition 10.9, two constraints induce the same facet if and only if one is a positive multiple of the other. We therefore have corollary 10.2.

COROLLARY 10.2.– If P is a full-dimensional polyhedron, then a unique minimal linear system (apart from multiplications by positive scalars) exists that describes P. Furthermore, every constraint of this system defines a distinct facet of P.

Another important class of faces is that of *minimal* faces, that is the faces that strictly do not strictly contain another face. We have the following result.

PROPOSITION 10.10.– (Hoffman and Kruskal [HOF 56]). *A non-empty subset F of P is a minimal face of P if and only if a subsystem $A'x \leq b'$ of $Ax \leq b$ exists such that $F = \{x \in \mathbb{R}^n \ : \ A'x = b', \ Bx = d\}$.*

Proof. ($\Longrightarrow$) Let us assume that F is a minimal face of P. Then from proposition 10.5, a subsystem $A'x \leq b'$ of $Ax \leq b$ exists such that $F = \{x \in P \;:\; A'x = b'\}$. We choose $A'x \leq b'$ in such a way that it is maximal. Let $A''x \leq b''$ be a minimal subsystem of $Ax \leq b$ such that $F = \{x \in \mathbb{R}^n \;:\; A'x = b',\; A''x \leq b'',\; Bx = d\}$. We will show that $A''x \leq b''$ does not contain any inequality.

Let us assume by contradiction that $A''x \leq b''$ contains an inequality $a''x \leq \alpha''$. Since $a''x \leq \alpha''$ is not redundant in the system describing F, from proposition 10.7, the set $F' = \{x \in \mathbb{R}^n \;:\; A'x = b',\; A''x \leq b'',\; a''x = \alpha'',\; Bx = d\}$ is a facet of F. From proposition 10.5, F' is also a face of P, which contradicts the fact that F is a minimal face of P.

($\Longleftarrow$) Suppose now that $\emptyset \neq F = \{x \in \mathbb{R}^n \;:\; A'x = b',\; Bx = d\} \subseteq P$ for a certain subsystem $A'x \leq b'$ of $Ax \leq b$. From proposition 10.5, F is a face of P. Moreover, F contains strictly no faces of P. Hence, F is a minimal face of P. ∎

COROLLARY 10.3.– Every non-empty minimal face of P is of dimension

$$n - \text{rank}\begin{pmatrix} A \\ B \end{pmatrix}$$

The non-empty minimal faces of a polytope are extreme points.

Given a set of points $S = \{x^1, \ldots, x^p\}$, we may need to determine if a given point x^0 belongs to $conv(S)$. This problem reduces to verifying if x^0 is a convex combination of the points $x^1, \ldots, x^p$, that is whether $\lambda_1, \ldots, \lambda_p$ exist such that:

$$\begin{array}{l} \sum_{i=1}^{p} \lambda_i = 1 \\ \sum_{i=1}^{p} \lambda_i x^i = x^0 \\ \lambda_i \geq 0, \quad \text{for } i = 1, \ldots, p \end{array} \qquad [10.6]$$

Proposition 10.11 shows that if system [10.6] does not have a solution, then an inequality (a hyperplane) must exist that separates x^0 and $conv(S)$, that is an inequality $ax \leq \alpha$ such that $ax \leq \alpha$ for every $x \in conv(S)$ and $ax^0 > \alpha$.

PROPOSITION 10.11.– *If $x^0 \in \mathbb{R}^n \setminus conv(S)$, then an inequality exists that separates x^0 and $conv(S)$.*

Proof. If $x^0 \in \mathbb{R}^n \setminus conv(S)$, then system [10.6] does not have a solution. By Farkas' lemma, a scalar $\alpha \in \mathbb{R}$ and a vector $y \in \mathbb{R}^n$ exist such that:

$$\begin{array}{ll} yx^i + \alpha \geq 0 & \text{for } i = 1, \ldots, p \\ yx^0 + \alpha < 0 & \end{array}$$

Let $a = -y$. Since $ax^i \leq \alpha$ for $i = 1, \ldots, p$, from proposition 10.2 it follows that $ax \leq \alpha$ for every $x \in conv(S)$. Since $ax^0 > \alpha$, constraint $ax \leq \alpha$ then separates x^0 and $conv(S)$. ∎

DEFINITION 10.12.– *A cone is the set of solutions of a homogeneous finite system, that is a system in the form $Ax \leq 0$. A cone C is said to be generated by a set of points $\{x^1, \ldots, x^k\}$ if every point x of C can be expressed in the form $\sum_{i=1}^k \lambda_i x_i$ with $\lambda_i \geq 0$ for $i = 1, \ldots, k$.*

THEOREM 10.3.– (Minkowski [MIN 96], Weyl [WEY 50]). *A set $C \subseteq \mathbb{R}^n$ is a cone if and only if C is generated by a finite set of points.*

Proof. ($\Longrightarrow$) Let us assume that C is the set of solutions of a finite system $Ax \leq 0$. We will show that C can be generated by a finite set of points. The proof is by induction on the number of constraints in the system that defines C. If the system has no constraints (that is $C = \mathbb{R}^n$), then the points given by the unit vectors and the vector whose components are all equal to -1 generate C. Let us now assume that we have a finite set S' of points that generate $C' = \{x \in \mathbb{R}^n \; : \; A'x \leq 0\}$ and let us assume that $Ax \leq 0$ is obtained from $A'x \leq 0$ by adding a constraint $ax \leq 0$. Let S'_0, S'_+, S'_- be the subsets of points x of S' such that $ax = 0, ax > 0$ and $ax < 0$, respectively. Note that S'_0, S'_+, S'_- form a partition of S'. For every pair (x, x'), such that $x \in S'_+$ and $x' \in S'_-$, let us consider the vector:

$$y_{x,x'} = (ax)x' - (ax')x \qquad [10.7]$$

Therefore $ay_{x,x'} = 0$ and $A'y_{x,x'} = (ax)A'x' - (ax')A'x \leq 0$. Let $S = S'_- \cup S'_0 \cup \{y_{x,x'} \; : \; x \in S'_+, x' \in S'_-\}$. It then follows that $S \subseteq \{x \in \mathbb{R}^n \; : \; A'x \leq 0, ax \leq 0\}$. We will show in what follows that S generates C. For this, let us consider a solution x^* of C. Since $C \subseteq C'$, then $x^* = \sum_{x \in S'} \lambda_x x$, where $\lambda_x \geq 0$ for every $x \in S'$. The solution x^* can also be written as:

$$x^* = \textstyle\sum_{x \in S'_+} \lambda_x x + \sum_{x \in S'_0} \lambda_x x + \sum_{x \in S'_-} \lambda_x x \qquad [10.8]$$

If $\lambda_x = 0$ for every $x \in S'_+$, since $S'_0 \cup S'_- \subseteq S$, x^* is therefore generated by elements of S. If $x \in S'_+$ with $\lambda_x > 0$ exists, then $\tilde{x} \in S'_-$ with $\lambda_{\tilde{x}} > 0$ must exist. Otherwise, we would have $ax^* > 0$, which would contradict the fact that $x^* \in C$. From expression [10.7], we have:

$$(ax)\tilde{x} + (-a\tilde{x})x - y_{x,\tilde{x}} = 0 \qquad [10.9]$$

By subtracting [10.9], multiplied by an appropriate coefficient, from [10.8] we obtain a positive combination for x^* in which at least one scalar among $\lambda_x, \lambda_{\tilde{x}}$ is reduced to zero. By repeating this operation a certain number of times (at most $|S'_+| + |S'_-|$ times), we obtain a positive combination of points of S for x^*, which implies that S is a generator for C.

($\Longleftarrow$) Let us assume that C is generated by a finite set S. We will show that C is the set of solutions of a finite system of the form $Ax \leq 0$. Let $U = \{a \in \mathbb{R}^n : ax \leq 0, \forall\, x \in S\}$. Therefore U is the set of solutions of a homogeneous system. From the first part of the proof, a subset U' of U must exist such that every solution of U can be expressed as a positive linear combination of points of U'. Let A be the matrix whose rows are the vectors a such that $a \in U'$. We have $C = \{x \in \mathbb{R}^n : Ax \leq 0\}$. Indeed, it is clear that $C \subseteq \{x \in \mathbb{R}^n : Ax \leq 0\}$. Let $\tilde{x} \notin C$. Then the linear system $\sum_{x \in S} \lambda_x x = \tilde{x}$, $\lambda_x \geq 0$ for $x \in S$ does not have a solution. Thus from Farkas' lemma, a vector $a \in \mathbb{R}^n$ exists such that $a\tilde{x} > 0$ and $ax \leq 0$ for every $x \in S$. Consequently, $a \in U$. Since $a\tilde{x} > 0$, $\tilde{x}$ cannot then be a solution of the system $Ax \leq 0$. ∎

10.3. Combinatorial optimization and linear programming

As we saw in the previous sections, there is a strong relationship between combinatorial optimization and linear programming. Indeed, any combinatorial optimization problem can be reduced to solving a linear program. In this section, we discuss this relationship and applications of polyhedral theory in combinatorial optimization.

10.3.1. *Associated polytope*

Consider a combinatorial optimization problem of type [10.1], associated with a family $\mathcal{F}$ of subsets of a finite basic set $E = \{e_1, \ldots, e_n\}$. We can associate with every solution $F \in \mathcal{F}$ of [10.1], the vector $x^F \in \{0,1\}^E$ given by:

$$x^F(e) = \begin{cases} 1 & \text{if } e \in F \\ 0 & \text{if not} \end{cases}$$

The vector x^F is called *the incidence vector* (or the *representative vector*) of F. Let

$$\mathcal{S} = \{x^F : F \in \mathcal{F}\}$$

be the set of incidence vectors of the solutions of [10.1]. So [10.1] is equivalent to the problem:

$$\max\{\omega x : x \in \mathcal{S}\} \qquad [10.10]$$

From proposition 10.2, the optimal value of [10.10] is equal to that of the problem:

$$\max\{\omega x \; : \; x \in conv(S)\} \qquad [10.11]$$

Theorem 10.4 shows that the two problems are equivalent.

THEOREM 10.4.– *A (non-empty) set of points $P \subseteq \mathbb{R}^n$ is a polytope if and only if there exists a set of points S such that $P = conv(S)$.*

Proof. ($\Longrightarrow$) Let us assume that P is a polytope. Let S be the set of its extreme points. It is clear that $conv(S) \subseteq P$. Let us assume that there exists a point $x^0 \in P \setminus conv(S)$. From proposition 10.11, an inequality $ax \leq \alpha$ exists such that $ax \leq \alpha$ for every $x \in conv(S)$ and $ax^0 > \alpha$. Let $\alpha^* = \max\{ax \; : \; x \in P\}$, and let $F = \{x \in P \; : \; ax = \alpha^*\}$. Note that F is a non-empty face of P. Since $x^0 \in P$, we must have $\alpha < \alpha^*$. But this implies that F does not contain any extreme points of P, which would contradict corollary 10.3.

($\Longleftarrow$) Let $S = \{x^1, \ldots, x^p\}$ be a finite set of points of $\mathbb{R}^n$ and let $P = conv(S)$. We will show that P is the set of solutions of a finite linear system. To do this, let us consider the set $T \subseteq \mathbb{R}^{n+1}$ given by the points $(\lambda, y) \in \mathbb{R} \times \mathbb{R}^n$ such that:

$$\begin{array}{l} -1 \leq \lambda \leq 1 \\ -\mathbf{1} \leq y^T \leq \mathbf{1} \\ yx^i \leq \lambda, \quad \text{for } i = 1, \ldots, p \end{array}$$

where $\mathbf{1}$ denotes the vector of $\mathbb{R}^n$ whose components are all equal to 1. It is clear that T is a polytope. Let $(\lambda_1, y^1), \ldots, (\lambda_t, y^t)$ be the extreme points of T. From the first part of this proof, $T = conv(\{(\lambda_1, y^1), \ldots, (\lambda_t, y^t)\})$. We will show that P is the set of solutions of the system:

$$y^i x \leq \lambda_i, \qquad \text{for } i = 1, \ldots, t. \qquad [10.12]$$

To do this, we first show that every point of P is a solution of system [10.12]. Indeed, if $\bar{x} \in P$, then $\bar{x} = \mu_1 x^1 + \ldots + \mu_p x^p$ for certain scalars $\mu_1, \ldots, \mu_p \geq 0$ such that $\sum_{i=1}^p \mu_i = 1$. Therefore $y^i \bar{x} = \mu_1 y^i x^1 + \ldots + \mu_p y^i x^p \leq \mu_1 \lambda_i + \ldots + \mu_t \lambda_i = \lambda_i$ for $i = 1, \ldots, t$. This implies that $\bar{x}$ is a solution of system [10.12].

Let us now consider a solution $\tilde{x}$ of system [10.12]. If $\tilde{x} \notin P$, then from proposition 10.11, an inequality $ax \leq \alpha$ exists such that $ax \leq \alpha$ for every $x \in P$ and $a\tilde{x} > \alpha$. By dividing the inequality by an appropriate coefficient, we can assume that $\mathbf{-1} \leq a^T \leq \mathbf{1}$ and $-1 \leq \alpha \leq 1$. Therefore $(\alpha, a) \in T$, and consequently (α, a)

can be written as $(\alpha, a) = \sum_{i=1}^{t} \gamma_i(\lambda_i, y^i)$ for certain scalars $\gamma_1, \ldots, \gamma_t$ such that $\sum_{i=1}^{p} \gamma_i = 1$. So, $a\tilde{x} = \sum_{i=1}^{t} \gamma_i y^i \tilde{x} \leq \sum_{i=1}^{t} \gamma_i \lambda_i = \alpha$. Since $a\tilde{x} > \alpha$, we obtain a contradiction. Consequently, $\tilde{x} \in P$, which ends the proof of the theorem. ∎

From theorem 10.4, $conv(\mathcal{S})$ is a polytope, and can therefore be described by a finite system of linear inequalities. As a consequence, problem [10.11] is nothing but a linear program. By the simplex algorithm, an optimal solution of [10.11] can be taken among the extreme points of $conv(\mathcal{S})$. Since these latter are precisely the solutions of $\mathcal{S}$, it follows that problems [10.10] and [10.11] are equivalent. Therefore any combinatorial optimization problem can be reduced to a linear program by describing the convex hull of its solutions using a linear system. This convex hull is called the *polytope associated* with the problem (or the *solution polytope* of the problem).

Problem [10.10] can be formulated as an integer program of the form:

$$\max \omega x$$

$$Ax \leq b \qquad [10.13]$$

$$\mathbf{0} \leq x \leq \mathbf{1} \qquad [10.14]$$

$$x \text{ integer} \qquad [10.15]$$

Here, $\mathbf{0}$ is the vector whose components are all equal to 0. Constraints [10.14] are called *trivial inequalities*. The polytope given by inequalities [10.13] and [10.14] generally contains fractional extreme points. Therefore, the linear relaxation, obtained by removing the integrality constraints [10.15], can have a fractional optimal solution. As a consequence, other constraints would be necessary to characterize the polytope associated with the problem and formulate it as a linear program.

As a consequence of the above development, we obtain a general method for solving combinatorial problems. This method, called the *polyhedral approach*, can be summarized as follows:

1) represent the elements of $\mathcal{F}$ by a set $\mathcal{S}$ of 0–1 vectors;

2) define the polytope P, the convex hull of the points of $\mathcal{S}$;

3) determine a complete system that describes P;

4) apply linear programming to solve the problem.

Step 3 is the crucial step of the method. Although the above results imply that there is a finite linear system that describes P, they do not say how to obtain such a system.

The *dominant* of a polyhedron $P \subseteq \mathbb{R}^n$, denoted by dom(P), is the polyhedron that consists of the points x such that $x \geq x'$ for certain $x' \in P$, that is

$$\text{dom}(P) = P + \mathbb{R}^n_+$$

It is clear that $P \subseteq \text{dom}(P)$ and dom(P) is unbounded. We have the following interesting algorithmic property.

COMMENT 10.3.– If $\omega \in \mathbb{R}^n_+$, then $\min\{\omega x \ : \ x \in P\} = \min\{\omega x \ : \ x \in \text{dom}(P)\}$.

Consequently, given a combinatorial optimization problem of the form $\min\{\omega x \ : \ x \in P\}$, where P is its associated polyhedron, instead of studying P, we can consider its dominant. This latter is generally simpler to characterize.

Given a set of points $S \in \mathbb{R}^n$, we say that an inequality $ax \leq \alpha$ is valid for S if it is satisfied for every point of S.

PROPOSITION 10.12.– *An inequality $ax \leq \alpha$ is valid for S if and only if it is valid for $conv(S)$.*

Proof. Since $S \subset conv(S)$, $ax \leq \alpha$ is valid for S if it is valid for $conv(S)$. Let us assume that $ax \leq \alpha$ is valid for S, and let us consider a solution $x \in conv(S)$. Therefore $x = \sum_{i=1}^k \lambda_i x^i$, where $x^i \in S$, $\lambda_i \geq 0$ for $i = 1, \ldots, k$ and $\sum_{i=1}^k \lambda_i = 1$ for a certain integer k. Consequently, $ax = \sum_{i=1}^k \lambda_i (ax^i) \leq \sum_{i=1}^k \lambda_i \alpha = \alpha$. ∎

PROPOSITION 10.13.– *If F is a non-empty $(p-1)$-dimensional face of $conv(S)$, then there are p affinely independent points in $S \cap F$.*

Proof. By definition, there are p affinely independent points $x^1, \ldots, x^p$ in F. If $x^i \in S$ for $i = 1, \ldots, p$, then the assertion is proven. Let us assume that $x^1 \notin S$. Then $x^1 = \sum_{j=1}^k \lambda_j \bar{x}^j$, where $\bar{x}^j \in S$, $\lambda_j \geq 0$ for $j = 1, \ldots, k$, and $\sum_{j=1}^k \lambda_j = 1$ for a certain integer k. Furthermore, the points $\bar{x}^j$, $j = 1, \ldots, k$ are all in F. Indeed, this is clear if $F = conv(S)$. If F is defined by a constraint $ax \leq \alpha$ valid for $conv(S)$, then $ax^1 = \alpha$. Since $a\bar{x}^j \leq \alpha$ for $j = 1, \ldots, k$, it follows that $a\bar{x}^j = \alpha$ for $j = 1, \ldots, k$, and thus $\bar{x}^j \in F$ for $j = 1, \ldots, k$. Since $x^1, \ldots, x^p$ are affinely independent, $t \in \{1, \ldots, k\}$ exists such that $\bar{x}^t, x^2, \ldots, x^p$ are affinely independent. Now the proof can be completed by repeating this process for every point $x^i \notin S$. ∎

From propositions 10.12 and 10.13, it is enough to consider the points of S to establish the validity of a constraint for $conv(S)$ or the dimension of a face of $conv(S)$.

10.3.2. *Extreme points and extreme rays*

Let us consider a polyhedron $P = \{x \in \mathbb{R}^n \ : \ Ax \leq b\}$, where A is an $m \times n$ matrix and b a vector of $\mathbb{R}^m$. Let us note that there may be inequalities in $Ax \leq b$ that are equally satisfied by the solutions of P.

DEFINITION 10.13.– *Let $P^0 = \{r \in \mathbb{R}^n \ : \ Ar \leq 0\}$. The points of $P^0 \setminus \{0\}$ are called rays of P.*

It is easy to see that a point $r \in \mathbb{R}^n$ is a ray of P if and only if for every point $x \in P$, $\{y \in \mathbb{R}^n : y = x + \lambda r,\ \lambda \in \mathbb{R}_+\} \subseteq P$.

DEFINITION 10.14.– *A ray r of P is said to be an extreme ray if two rays, $r^1, r^2 \in P^0$, $r^1 \neq \lambda r^2$, do not exist for every $\lambda \in \mathbb{R}_+$, such that $r = \frac{1}{2}r^1 + \frac{1}{2}r^2$.*

In what follows, we give some properties of extreme points and extreme rays. These will allow us to describe any polyhedron by its extreme points and extreme rays.

PROPOSITION 10.14.– *A point $r \in P^0$ is an extreme ray of P if and only if the set $\{\lambda r : \lambda \in \mathbb{R}_+\}$ is a 1-dimensional face of P^0.*

Proof. Let $\bar{A}$ be the matrix whose rows are the rows A_i of A such that $A_i r = 0$. If $\{\lambda r : \lambda \in \mathbb{R}_+\}$ is a 1-dimensional face of P^0, then from proposition 10.4, $\text{rank}(\bar{A}) = n-1$. Therefore any solution of the system $\bar{A}y = 0$ is of the form $y = \mu r$, $\mu \in \mathbb{R}$. If $r = \frac{1}{2}r^1 + \frac{1}{2}r^2$, where $r^1, r^2 \in P^0$, then $\bar{A}r^1 \leq 0$ and $\bar{A}r^2 \leq 0$. But this implies that $\bar{A}r^1 = \bar{A}r^2 = 0$. Consequently $r^1 = \lambda_1 r$ and $r^2 = \lambda_2 r$ for $\lambda_1, \lambda_2 \in \mathbb{R}$. Thus $r^1 = \lambda r^2$ with $\lambda = \frac{\lambda_1}{\lambda_2} \in \mathbb{R}$, and therefore r is an extreme ray.

If $r \in P^0$ and $\text{rank}(\bar{A}) < n-1$, then $r' \neq \lambda r$, $\lambda \in \mathbb{R}$ exists such that $\bar{A}r' = 0$. Let $\tilde{A}$ be the complementary submatrix of $\bar{A}$ in A. Note that $\tilde{A}r < 0$. Consequently $\epsilon \neq 0$ exists such that $\tilde{A}r + \epsilon\tilde{A}r' \leq 0$ and $\tilde{A}r - \epsilon\tilde{A}r' \leq 0$. So $r^1 = r + \epsilon r'$ and $r^2 = r - \epsilon r'$ are rays of P^0. Since $r = \frac{1}{2}r^1 + \frac{1}{2}r^2$ and $r^1 \neq \lambda r^2$ for every $\lambda \in \mathbb{R}_+$, r cannot be an extreme ray. ∎

THEOREM 10.5.– *For every extreme point x^* of P, $c \in \mathbb{Z}^n$ exists such that x^* is the unique optimal solution of the program max$\{cx : x \in P\}$.*

Proof. Let $I = \{i \in \{1, \ldots, m\} : A_i x^* = b_i\}$. Let $c^* = \sum_{i \in I} A_i$. Since P is rational, an integer $t > 0$ exists such that $c = tc^* \in \mathbb{Z}^n$. Since x^* is an extreme point of P, and from proposition 10.6, x^* is a 0-dimensional face, for every point $x \in P \setminus \{x^*\}$, there exists $i \in I$ such that $A_i x < b_i$. Therefore for all $x \in P \setminus \{x^*\}$, we have $cx = \sum_{i \in I} tA_i x < \sum_{i \in I} tb_i = \sum_{i \in I} tA_i x^* = cx^*$. ∎

THEOREM 10.6.– *If rank$(A) = n - k$ and $P \neq \emptyset$, then P contains a k-dimensional face, and does not contain any face of a lower dimension.*

Proof. Let $F \neq \emptyset$ be a face of P and let us denote by I the set of indices $i \in \{1, \ldots, m\}$, such that $A_i x < b_i$ for a certain $x \in F$. Since $\text{rank}(A) = n - k$, then the matrix of the equations of F, given by A_i such that $i \in \{1, \ldots, m\} \setminus I$, is of rank less than or equal to $n - k$. From proposition 10.4, $\dim(F) \geq k$. Suppose now that F is of minimum dimension. If $\dim(F) = 0$, then, from proposition 10.6, F is reduced to an extreme point of P and the assertion is proven. Let us assume that $\dim(F) > 0$. Let $\hat{x}$

be an interior point of F. Since $\dim(F) > 0$, there must exist a further point $y \in F$. Consider the straight line Δ going through the points $\hat{x}$ and y. It contains the points $z(\lambda) = \lambda y + (1 - \lambda)\hat{x}$ for $\lambda \in \mathbb{R}$. Assume that Δ intersects one of the hyperplanes $\{x \in \mathbb{R}^n : A_i x = b_i\}$, $i \in I$. Let $\lambda^* = \min\{|\lambda^i| : i \in I, A_i z(\lambda^i) = b_i\}$, and let $i^* \in I$ such that $\lambda^* = |\lambda^{i^*}|$. Since $\hat{x}$ is an interior point of F, and, consequently, $A_{i^*}\hat{x} < b_{i^*}$, it follows that $\lambda^* \neq 0$. Consider the set $F^* = \{x \in F : A_{i^*}x = b_{i^*}\}$. It is clear that F^* is a face of P. Also, since $z(\lambda^{i^*}) \in F^* \setminus F$, F^* is a face of lower dimension, a contradiction.

Consequently, Δ does not intersect any of the hyperplanes $\{x \in \mathbb{R}^n : A_i x = b_i\}$, $i \in I$. This implies that $\Delta \subseteq P$, that is $A(\lambda y + (1-\lambda)\hat{x}) \leq b$ for every $\lambda \in \mathbb{R}$. As $A\hat{x} \leq b$, it follows that $A(y - \hat{x}) = 0$ for every $y \in F$. So, $F = \{y \in P : Ay = A\hat{x}\}$. Since $\text{rank}(A) = n - k$, by proposition 10.4, we have $\dim(F) = k$. ∎

THEOREM 10.7.– *Suppose that rank$(A) = n$, and the problem:*

$$\max\{\omega x : x \in P\} \qquad [10.16]$$

has a finite optimal solution. Then there exists an optimal solution of [10.16] that is an extreme point of P.

Proof. The set of optimal solutions of [10.16] is a non-empty face $F = \{x \in P : \omega x = \omega_0\}$. Since $\text{rank}(A) = n$, by theorem 10.6, F contains a 0-dimensional face. By proposition 10.6, it follows that F contains an extreme point. ∎

THEOREM 10.8.– *If rank$(A) = n$ and max$\{\omega x : x \in P\}$ is unbounded, then P has an extreme ray r^* such that $\omega r^* > 0$.*

Proof. Since the linear program $\max\{\omega x : x \in P\}$ does not have a finite optimal solution, by duality in linear programming, the system $\{yA = \omega,\ y \geq 0\}$ does not have a solution. By Farkas' lemma, $r \in \mathbb{R}^n$ exists such that $Ar \leq 0$ and $\omega r > 0$. Consider the linear program:

$$\max\{\omega r : Ar \leq 0,\ \omega r \leq 1\}. \qquad [10.17]$$

The optimal value of [10.17] is then equal to 1. Since this value is bounded and $\text{rank}(A) = n$, by theorem 10.7, [10.17] has an optimal solution that is an extreme point of $\{r : Ar \leq 0,\ \omega r \leq 1\}$. Let r^* be such a point. It is clear that r^* is a point of $P^0 \setminus \{0\}$, and therefore r^* is a ray of P. Furthermore, since r^* is the unique solution of n equations of the system $\{Ar \leq 0,\ \omega r \leq 1\}$, the subsystem of $Ax \leq 0$, equally satisfied by r^*, must be of rank $n - 1$. Otherwise r^* would be zero, which is impossible. Consequently, $\{\lambda r^*,\ \lambda \in \mathbb{R}_+\}$ is a 1-dimensional face of P^0. From proposition 10.14, r^* is then an extreme ray of P. ∎

We can now give the fundamental theorem 10.9, which establishes a description of P in terms of extreme points and extreme rays.

THEOREM 10.9.– (Minkowski [MIN 96]). *If $P \neq \emptyset$ and rank $(A) = n$, then:*

$$P = \left\{ \begin{array}{ll} x \in \mathbb{R}^n : & x = \sum_{i=1}^{p} \lambda_i x^i + \sum_{j=1}^{q} \mu_j r^j,\ \sum_{i=1}^{p} \lambda_i = 1 \\ & \lambda_i \geq 0,\ i = 1, \ldots, p,\ \mu_j \geq 0,\ j = 1, \ldots, q \end{array} \right\}$$

where $x^1, \ldots, x^p$ and $r^1, \ldots, r^q$ are the extreme points and the extreme rays, respectively, of P.

Proof. Let:

$$Q = \left\{ \begin{array}{ll} x \in \mathbb{R}^n : & x = \sum_{i=1}^{p} \lambda_i x^i + \sum_{j=1}^{q} \mu_j r^j,\ \sum_{i=1}^{p} \lambda_i = 1, \\ & \lambda_i \geq 0,\ i = 1, \ldots, p,\ \mu_j \geq 0,\ j = 1, \ldots, q \end{array} \right\}.$$

Since $x^i \in P$ for $i = 1, \ldots, p$, it is clear that $x' = \sum_{i=1}^{p} \lambda_i x^i \in P$, if $\lambda_i \geq 0,\ i = 1, \ldots, p$ and $\sum_{i=1}^{p} \lambda_i = 1$. Also, since $r^j,\ j = 1, \ldots, q$ are rays of P, $x' + \sum_{j=1}^{q} \mu_j r^j \in P$, if $\mu_j \geq 0$ for $j = 1, \ldots, q$. So, $Q \subseteq P$.

Let us now assume that $P \setminus Q \neq \emptyset$, and let $y \in P \setminus Q$. Then there are no $\lambda_i,\ i = 1, \ldots, p$, and $\mu_j,\ j = 1, \ldots, q$, such that:

$$\begin{array}{l} \sum_{i=1}^{p} \lambda_i x^i + \sum_{j=1}^{q} \mu_j r^j = y \\ \sum_{i=1}^{p} \lambda_i = 1 \\ \lambda_i \geq 0, \quad i = 1, \ldots, p \\ \mu_j \geq 0, \quad j = 1, \ldots, q \end{array}$$

From Farkas' lemma, a vector $(a, a_0) \in \mathbb{R}^{n+1}$ exists such that $ax^i - a_0 \leq 0$ for $i = 1, \ldots, p$, $ar^j \leq 0$ for $j = 1, \ldots, q$, and $ay - a_0 > 0$. Consider now the linear program $\max\{ax : x \in P\}$. If this program has a finite optimal solution, then by theorem 10.7, such a solution can be taken among the extreme points of P. Since $y \in P$ and $ax^i < ay$ for every extreme point x^i, this is impossible. If the optimal solution of the above program is unbounded, then by theorem 10.8, an extreme ray r^j exists such that $ar^j > 0$. Again, we have a contradiction. ∎

10.4. Proof techniques

We have seen in section 10.2 that a minimal description of a polyhedron P, using a system of inequalities, can be obtained by characterizing the facets of the polyhedron. By the linear programming duality, this description may allow us to obtain a strong min–max relationship between the optimal solutions of the primal problem $\max\{\omega x : x \in P\}$ and its dual. Furthermore, the use facet defining inequalities in the framework of a cutting-planes method for the primal problem can enable us to speed up the resolution of the problem. Consequently, if P is the polytope associated with a combinatorial optimization problem, given a valid inequality for P, a fundamental question that arises is to determine if this inequality defines a facet of P. Moreover, given a system of inequalities valid for P, it would be interesting to see if the system completely describes P. In this section, we discuss certain proof techniques for these questions.

10.4.1. *Facet proof techniques*

10.4.1.1. *Proof of necessity*

A first technique for proving that a valid constraint $ax \leq \alpha$ defines a facet of P is to show that $ax \leq \alpha$ is *essential* in describing P. In other words, if $P = \{x \in \mathbb{R}^n : Ax \leq b\}$, and if $ax \leq \alpha$ is one of the constraints of $Ax \leq b$, then we must show that there exists a point $\bar{x}$ such that $a\bar{x} > \alpha$, and $\bar{x}$ satisfies all the other inequalities of $Ax \leq b$. This implies that the constraint $ax \leq \alpha$ is necessary to describe P. If P is full-dimensional, from corollary 10.2, it follows that $ax \leq \alpha$ defines a facet. If P is not full-dimensional, then we must also show that $ax \leq \alpha$ is not an equation in the system $Ax \leq b$. To do this, it is enough to establish a solution $\hat{x} \in P$ such that $a\hat{x} < \alpha$.

This technique is generally used for simple constraints. We illustrate this here for the trivial inequalities of the stable set polytope.

A *stable set* in a graph $G = (V, E)$ is a subset of non-adjacent pairwise vertices. If each vertex v of V has a weight $\omega(v)$, *the stable set problem* is to determine a stable set S such that $\sum_{v \in S} \omega(v)$ is maximum. The stable set problem is **NP**-hard even when the weights are all equal to 1.

If $S \subseteq V$ is a subset of vertices, let $x^S \in \mathbb{R}^V$ be the incidence vector of S given by:

$$x^S(v) = \begin{cases} 1 & \text{if } v \in S \\ 0 & \text{if not} \end{cases}$$

Let

$$P(G) = conv\{x^S \ : \ S \subseteq V \text{ is a stable set}\}$$

be the convex hull of the incidence vectors of all the stable sets of G. $P(G)$ is called *the stable sets polytope* of G. The stable set problem in G is therefore equivalent to the linear program $\max\{\omega x : x \in P(G)\}$. The stable set polytope has been widely studied and reported in the literature. Since the problem is **NP**-hard, the polytope $P(G)$ is known explicitly only for certain particular classes of graphs.

If S is a stable set of G, it is clear that x^S satisfies the inequalities:

$$x(u) + x(v) \leq 1 \quad \forall\, uv \in E \qquad [10.18]$$

$$x(v) \geq 0 \quad \forall\, v \in V \qquad [10.19]$$

Also, every integer solution of constraints [10.18] and [10.19] represents a stable set. In what follows, we will show that constraints [10.19] define facets for $P(G)$. For this, we first show that $P(G)$ is full-dimensional.

PROPOSITION 10.15.– *$P(G)$ is full-dimensional.*

Proof. From proposition 10.13, it is enough to show that there exist $n+1$ stable sets whose incidence vectors are affinely independent. Consider the sets $S_0 = \emptyset$ and $S_i = \{v_i\}$ for $i = 1, \ldots, n$. It is clear that these sets induce stable sets. Furthermore, their incidence vectors $x^{S_0}, x^{S_1}, \ldots, x^{S_n}$ are affinely independent. ∎

Consequently, from corollary 10.2, two valid inequalities for $P(G)$ define the same facet if and only if one is a positive multiple of the other.

PROPOSITION 10.16.– *Let $\sum_{v \in V} a(v)x(v) \leq \alpha$ be a valid inequality for $P(G)$, different from inequalities [10.19]. Let us assume that $\sum_{v \in V} a(v)x(v) \leq \alpha$ defines a facet of $P(G)$. Then $a(v) \geq 0$ for every $v \in V$.*

Proof. Assume that a vertex $u \in V$ exists such that $a(u) < 0$. Since $P(G)$ is full-dimensional, and $\sum_{v \in V} a(v)x(v) \leq \alpha$ is different from constraints [10.19], it follows that the facet defined by $\sum_{v \in V} a(v)x(v) \leq \alpha$ is different from the face $\{x \in P : x(u) = 0\}$. From proposition 10.13, there must exist a stable set S of G containing u such that $\sum_{v \in V} a(v)x^S(v) = \alpha$. Let $S' = S \setminus \{u\}$. It is obvious that S' is a stable set. However, we have $\sum_{v \in V} a(v)x^{S'}(v) = \sum_{v \in V} a(v)x^S(v) - a(u) > \alpha$, which is impossible. ∎

PROPOSITION 10.17.– *Constraints [10.19] define facets of $P(G)$.*

Proof. From theorem 10.4 and proposition 10.16, the polytope $P(G)$ can be described by a system of inequalities of the form:

$$\begin{array}{ll} a^i x \leq \alpha_i, & \forall\, i \in I \\ x(v) \geq 0, & \forall\, v \in V \end{array} \qquad [10.20]$$

where I is a set of indices, and $a^i \geq 0$ for every $i \in I$. Let $v \in V$ and $\bar{x} \in \mathbb{R}^V$ such that $\bar{x}(u) = 0$ for every $u \in V \setminus \{v\}$, and $\bar{x}(v) = -1$. Since $a^i \geq 0$ for every $i \in I$, it follows that $\bar{x}$ satisfies all the constraints of system [10.20] except the constraint $x(v) \geq 0$. This shows that this latter is essential in [10.20], and therefore defines a facet of $P(G)$. ∎

10.4.1.2. *Direct proof*

The most direct technique for showing that a valid constraint induces a facet consists of showing that there exist dim(P) affinely independent solutions satisfying the constraint with equality. If P is full-dimensional, we then obtain $ax \leq \alpha$ defines a facet of P. If P is not full-dimensional, as for the first method, we must also show that $ax \leq \alpha$ is not an equation in $Ax \leq b$ by determining a solution of P that satisfies $ax \leq \alpha$ with strict inequality. We will also illustrate this method on the stable set polytope.

A (simple) graph is said to be *complete* if an edge exists between each pair of vertices. If $G = (V, E)$ is a graph, a subset K of vertices of V is called a *clique* of G if it induces a maximal complete subgraph, that is a complete subgraph which is not strictly contained in any complete subgraph. If $S \subseteq V$ is a stable set, it is clear that S cannot intersect a clique in more than one vertex. This implies that the following constraints

$$x(K) \leq 1, \quad \forall\, K \text{ clique of } G \qquad [10.21]$$

are valid for the polytope $P(G)$.

PROPOSITION 10.18.– *Inequalities [10.21] define facets of $P(G)$.*

Proof. Since $P(G)$ is full-dimensional, it is enough to show that there exist $n = |V|$ stable sets of G whose incidence vectors satisfy [10.21] with equality and are affinely independent. Since K is a clique, for every vertex v in $V \setminus K$, a vertex $v' \in K$ exists such that $vv' \notin E$. Let us consider the sets

- $S_v = \{v\}$ for every $v \in K$,
- $S_v = \{v, v'\}$ for every $v \in V \setminus K$.

It is clear that these sets are stable sets of G. Furthermore, their incidence vectors x^{S_v}, $v \in V$ satisfy [10.21] with equality and are linearly independent. Since 0 does not belong to the affine hull of the incidence vectors x^{S_v}, $v \in V$, from comment 10.1, these points are also affinely independent. ∎

Observe that if uv is an edge such that $\{u, v\}$ is contained in a clique K, then inequality [10.18] corresponding to uv is dominated by [10.21].

10.4.1.3. *Proof by maximality*

A final method for showing that a constraint $ax \leq \alpha$, valid for P, defines a facet, consists of proving that the face $F = \{x \in P : ax = \alpha\}$ induced by $ax \leq \alpha$ is not strictly contained in a facet of P. In other words, if $F \subseteq \{x \in P : bx = \beta\}$, where

$bx \leq \beta$ is a constraint that defines a facet of $P(G)$, then $ax \leq \alpha$ must be equivalent to $bx \leq \beta$. If P is full-dimensional, it is enough to show, in this case, that $\rho > 0$ exists such that $b = \rho a$. If P is not full-dimensional, then, as well as the equivalence between $ax \leq \alpha$ and $bx \leq \beta$, we must show that $ax \leq \alpha$ is not an equation in $Ax \leq b$.

Let us again consider the stable set polytope. Let $G = (V, E)$ be a graph. A *wheel* of G is a subgraph $H = (W, T)$, where $W = \{w_0, w_1, \ldots, w_k\}$ and $T = \{w_0 w_i,\ i = 1, \ldots, k\} \cup \{w_i w_{i+1},\ i = 1, \ldots, k-1\} \cup \{w_1 w_k\}$. Assume that $H = (W, T)$ is a wheel of G with even $|W| = k + 1$. Such a wheel is said to be *odd*. It is not hard to see that the following constraint is valid for $P(G)$:

$$\sum_{i=1}^{k} x(w_i) + \frac{k-1}{2} x(w_0) \leq \frac{k-1}{2} \qquad [10.22]$$

PROPOSITION 10.19.– *Suppose that v_0 is not adjacent to any vertex of $V \setminus W$. Then inequality [10.22] defines a facet for $P(G)$.*

Proof. Let us denote constraint [10.22] by $ax \leq \alpha$, and suppose that there is a constraint $bx \leq \beta$ valid for $P(G)$ that defines a facet of $P(G)$ such that $\{x \in P(G) : ax = \alpha\} \subseteq \{x \in P(G) : bx = \beta\}$. Since $P(G)$ is full-dimensional, it is enough to show that $\rho > 0$ exists such that $b = \rho a$. For this, we first show that $a(w_i) = a(w_j)$ for every $i, j \in \{1, \ldots, k\}$. Consider the sets

– $S = \{w_1, w_3, \ldots, w_{k-2}\}$,
– $S' = (S \setminus \{w_1\}) \cup \{w_k\}$.

It is easy to see that these sets are stable sets of G whose incidence vectors satisfy [10.22] with equality. It follows that $bx^S = bx^{S'}$, and, consequently, $b(w_1) = b(w_k)$.

By symmetry, we obtain that there exists $\rho \in \mathbb{R}$ exists such that:

$$b(w_i) = b(w_j) = \rho \text{ for every } i, j \in \{1, \ldots, k\} \qquad [10.23]$$

Since $\{w_0\}$ is a stable set and $ax^{\{w_0\}} = ax^S = \alpha$, then $bx^{\{w_0\}} = bx^S = \beta$. From [10.23], it follows that:

$$b(w_0) = \rho \frac{k-1}{2} \qquad [10.24]$$

Consider a vertex $w \in V \setminus W$. Since w is not adjacent to w_0, the set $S'' = \{w, w_0\}$ is then a stable set of G. Since $ax^{S''} = ax^{\{w_o\}} = \alpha$, and thus $bx^{S''} = bx^{\{w_o\}} = \beta$, it follows that $0 = bx^{S''} - bx^{\{w_o\}} = b(w)$. Since w is arbitrary in $V \setminus W$, we have:

$$b(w) = 0 \quad \text{for every } w \in V \setminus W \qquad [10.25]$$

From [10.23]–[10.25], it follows that $b = \rho a$. Furthermore, for every vertex v of G there is a stable set S of G which contains v such that $ax^S = \alpha$. The face defined by $ax \leq \alpha$ is different from a trivial face $\{x \in P(G) : x(v) = 0\}$. This implies that the face induced by $bx \leq \beta$ is not contained in a trivial face. From proposition 10.16, it follows that $b(v) \geq 0$ for every $v \in V$. Since $bx \leq \beta$ defines a facet of $P(G)$, there must exist at least one vertex $v \in V$ such that $b(v) > 0$. Consequently $\rho > 0$. ∎

10.4.2. *Integrality techniques*

Let $P \subseteq \mathbb{R}^n$ be the solutions polyhedron of a combinatorial optimization problem, and let $Ax \leq b$ be a system of valid inequalities for P. In what follows, we discuss techniques which enable us to show that $Ax \leq b$ completely describes P. For this, we assume that every integer solution of $Ax \leq b$ is a solution of the problem.

10.4.2.1. *Extreme points integrality proof*

A first technique for showing that $Ax \leq b$ describes P consists of proving that the extreme points of the polyhedron $\hat{P} = \{x \in \mathbb{R}^n : Ax \leq b\}$ are all integers. This would imply that $\hat{P} \subseteq P$. Since $P \subseteq \hat{P}$, we therefore have $P = \hat{P}$. To demonstrate this technique, we consider the 2-edge connected subgraph problem.

A graph G is said to be k-edge connected, for fixed $k > 0$, if between each pair of vertices of G, there exist at least k edge-disjoint paths. Given a graph $G = (V, E)$ and a weight function $\omega : E \rightarrow \mathbb{R}$ that associates the weight $\omega(e)$ with each edge $e \in E$, *the k-edge connected subgraph problem* is to determine a k-edge connected subgraph (V, T) of G containing all the vertices of V and such that $\sum_{e \in T} \omega(e)$ is minimum. This problem has applications in the design of reliable telecommunications networks [GRÖ 95, KER 05].

Let $G = (V, E)$ be a graph. If $W \subseteq V$, the set of edges having one extremity in W and the other in $V \setminus W$ is called a *cut*, and denoted by $\delta(W)$. If $W = \{v\}$, we write $\delta(v)$ instead of $\delta(\{v\})$. Theorem 10.10 (Menger's theorem), establishes a relationship between edge-disjoint paths and cuts in a graph.

THEOREM 10.10.– (Menger [MEN 27]). *In a graph G, there are k edge-disjoint chains between two vertices s and t if and only if every cut of G that separates s and t contains at least k edges.*

Let TECP(G) be the convex hull of the incidence vectors of the sets of edges $T \subseteq E$ such that (V, T) is a 2-edge connected subgraph of G. It is easy to see that the following constraints are valid for TECP(G):

$$x(e) \geq 0 \quad \forall\, e \in E \qquad [10.26]$$

$$x(e) \leq 1 \quad \forall\, e \in E \qquad [10.27]$$

$$x(\delta(W)) \geq 2 \quad \forall\, W \subset V,\ W \neq \emptyset \qquad [10.28]$$

From Menger's theorem, every integer solution that satisfies constraints [10.26]–[10.28] induces a 2-edge connected subgraph spanning the vertices of V. In what follows, we denote by $Q(G)$ the polytope given by inequalities [10.26]–[10.28].

A graph G is said to be *series-parallel* if it can be constructed from a graph consisting of two vertices linked by one edge by application of the following operations:

– duplicate an edge (that is add another parallel edge);

– subdivide an edge (that is replace an edge uv with two edges uw and wv, where w is a new vertex).

Series-parallel graphs have the following property.

COMMENT 10.4.– If $G = (V, E)$ is a connected series parallel graph without multiple edges with $|E| \geq 2$, then G contains at least one vertex of degree 2.

In the following, we will show that if $G = (V, E)$ is a series-parallel graph, then constraints [10.26]–[10.28] characterize the polytope TECP(G).

THEOREM 10.11.– [MAH 94] *If G is series-parallel, then TECP(G) $= Q(G)$.*

Proof. Assume by contradiction that there exists a series-parallel graph $G = (V, E)$ such that TECP(G) $\neq Q(G)$. Therefore G is 2-edge connected, otherwise TECP(G) $= Q(G) = \emptyset$. Thus $|E| \geq 2$. Suppose that $|E|$ is minimum, that is for every series-parallel graph $G' = (V', E')$ such that $|E'| < |E|$, we have TECP(G') $= Q(G')$.

Since constraints [10.26]–[10.28] are valid for TECP(G), it is clear that TECP(G) $\subseteq Q(G)$. Since TECP(G) $\neq Q(G)$, there must exist a fractional extreme point $\bar{x}$ of $Q(G)$.

LEMMA 10.1.– *$\bar{x}(e) > 0$ for every $e \in E$*

Proof. Suppose that an edge $e_0 \in E$ exists such that $\bar{x}(e_0) = 0$. Let $\bar{x}' \in \mathbb{R}^{|E|-1}$ be the restriction of $\bar{x}$ on $E \setminus \{e_0\}$, that is the solution defined by $\bar{x}'(e) = \bar{x}(e)$ for

$e \in E \setminus \{e_0\}$. It is clear that $\bar{x}'$ is an extreme point of $Q(G - e_0)$, where $G - e_0$ is the graph obtained from G by removing e_0. Since $\bar{x}'$ is fractional and $Q(G - e_0) =$ TECP$(G - e_0)$, we have a contradiction. ∎

LEMMA 10.2.– *G is 3-edge connected.*

Proof. Assume that G is not 3-edge connected. Since G is 2-edge connected, then G contains a cut consisting of exactly two edges, say e_1 and e_2. Consequently, $x(e_1) = x(e_2) = 1$. Let $G^* = (V^*, E^*)$ be the graph obtained from G by contracting e_1. Let $x^* \in \mathbb{R}^{|E|-1}$ be the restriction of $\bar{x}$ on E^*. Then x^* is a solution of $Q(G^*)$. Furthermore, x^* is an extreme point of $Q(G^*)$. Indeed, if this is not the case, there must then be two solutions y', y'' of $Q(G^*)$ such that $x^* = \frac{1}{2}(y' + y'')$, hence $y'(e_2) = y''(e_2) = 1$. Consider the solutions $y^{*'}$, $y^{*''} \in \mathbb{R}^m$ given by:

$$y^{*'}(e) = \begin{cases} y'(e) & \forall\, e \in E^* \\ 1 & \text{if } e = e_1 \end{cases}$$

and:

$$y^{*''}(e) = \begin{cases} y''(e) & \forall\, e \in E^* \\ 1 & \text{if } e = e_1 \end{cases}$$

It is clear that $y^{*'}$, $y^{*''}$ are solutions of $Q(G)$. We also have $\bar{x} = \frac{1}{2}(y^{*'} + y^{*''})$, a contradiction. Consequently, x^* is an extreme point of $P(G^*)$. Since G^* is series-parallel, $|E^*| < |E|$, and x^* is fractional, this contradicts the minimality of $|E|$. ∎

By lemma 10.1, $\bar{x}(e) > 0$ for every $e \in E$. Since $\bar{x}$ is an extreme point of $Q(G)$, from corollary 10.3, there must be a set of cuts $\{\delta(W_i),\ i = 1, \ldots, t\}$ and a subset of edges $E_1 \subseteq E$ such that $\bar{x}$ is the unique solution of the system:

$$\begin{cases} x(e) = 1, & \forall\, e \in E_1 \\ x(\delta(W_i)) = 2, & i = 1, \ldots, t \end{cases} \qquad [10.29]$$

where $|E_1| + t = |E|$.

LEMMA 10.3.– *Each variable $x(e)$ has a non-zero coefficient in at least two equations of [10.29].*

Proof. It is clear that $x(e)$ must have a non-zero coefficient in at least one equation of system [10.29]. Otherwise every point $\bar{x}'$ such that $\bar{x}'(f) = \bar{x}(f)$ if $f \in E \setminus \{e\}$

and $\bar{x}'(e) = \bar{x}(e) + \epsilon$, where ϵ is a non-zero scalar, would be a solution of [10.29], different from $\bar{x}$, which is impossible. Suppose now that an edge e_0 such that the variable associated with e_0, $x(e_0)$ has a non-zero coefficient in exactly one equation of [10.29]. Without loss of generality, we can assume that between the extremities of e_0 there is only one edge, namely e_0. Let [10.29] be the system obtained from [10.29] by removing the equation containing $x(e_0)$. Let $\bar{x}^0$ be the restriction of $\bar{x}$ given by $\bar{x}^0(e) = \bar{x}(e)$ for every $e \in E \setminus \{e_0\}$. Note that $\bar{x}^0$ is fractional. Indeed, this is clear if $\bar{x}(e_0)$ is integer. Otherwise, since $\bar{x}$ is a solution of system [10.29] whose right-hand side is integer, $\bar{x}$ must have at least two fractional components, and therefore $\bar{x}^0$ is fractional. Furthermore, $\bar{x}^0$ is the unique solution of system [10.29]. Consequently, $\bar{x}^0$ is an extreme point of $Q(G - e_0)$, but as G^0 is series-parallel, from our minimality hypothesis, $Q(G - e_0)$ is integer, a contradiction. ∎

Since G is series-parallel and $|E| \geq 2$, from comment 10.4 and lemma 10.2, it follows that G contains two multiple edges f and g. Also, at most one of the edges f and g may have a fractional value. If this is not the case, since every cut of G contains either both edges f and g or neither, the point $\bar{x}' \in \mathbb{R}^E$ such that $\bar{x}'(f) = \bar{x}(f) + \epsilon$, $\bar{x}'(g) = \bar{x}(g) - \epsilon$ and $\bar{x}'(e) = \bar{x}(e)$ if $e \in E \setminus \{f, g\}$, where $\epsilon > 0$ is a sufficiently small scalar, would be a solution of system [10.29], different from $\bar{x}$, a contradiction. Consequently, we can assume, for example, that $\bar{x}(f) = 1$. By lemma 10.3, there must be a cut $\delta(W_{i^*})$, $i^* \in \{1, \ldots, t\}$ that contains f, and therefore g. Furthermore, $\bar{x}(g)$ is fractional. Indeed, if $\bar{x}(g) = 1$, then from lemmas 10.1 and 10.2, we obtain $2 = \bar{x}(\delta(W_{i^*}) > \bar{x}(f) + \bar{x}(g) = 2$, which is impossible. From lemma 10.3, there must exist a cut $\delta(W_{j^*})$, $j^* \in \{1, \ldots, t\}$ that contains f and g. Let $\delta(W_{i^*})\Delta\delta(W_{j^*})$ be the symmetric difference of the cuts $\delta(W_{i^*})$ and $\delta(W_{j^*})$. (For any two sets I and J, their *symmetrical difference* $I\Delta J$ is defined as $I\Delta J = (I \setminus J) \cup (J \setminus I)$. Note that the symmetric difference of two cuts is a cut.) We therefore have:

$$\begin{aligned} 2 &\leq \bar{x}(\delta(W_{i^*})\Delta\delta(W_{j^*})) \\ &= \bar{x}(\delta(W_{i^*}) \cup \delta(W_{j^*})) - 2\bar{x}(\delta(W_{i^*}) \cap \delta(W_{j^*})) \\ &\leq 4 - 2(\bar{x}(f) + \bar{x}(g)) \\ &< 2 \end{aligned}$$

a contradiction, which ends the proof. ∎

10.4.2.2. *Direct proof*

If the polyhedron is full-dimensional, another more direct technique for showing that a system $Ax \leq b$ valid for P completely describes P is to show that every facet of P is defined by one of the constraints of the system. Since by corollary 10.2 every inequality that defines a facet of P is a positive multiple of a constraint of the system

$Ax \leq b$, we have $P = \{x : Ax \leq b\}$. The method can be presented as follows. We consider an inequality $ax \leq \alpha$ that defines a facet F of P. By using the structure of the extreme points of P (that is the solutions of the underlying combinatorial problem), we establish certain properties related to a, which imply that $ax \leq \alpha$ is a constraint of the system $Ax \leq \alpha$. We will illustrate this method on the matching polytope.

Given a graph $G = (V, E)$, a subset of pairwise non-adjacent edges is called a *matching*. If each edge of G has a certain weight, *the matching problem* in G is to determine a matching whose total weight is maximum. Edmonds [EDM 65] has shown that this problem can be solved in polynomial time. He also produced a linear system that completely describes the associated polytope.

If $G = (V, E)$ is a graph, the *matching polytope* of G, denoted by $P^c(G)$, is the convex hull of the incidence vectors of the matchings of G. It is not difficult to see that if F is a matching of G, then its incidence vector x^F satisfies the following constraints:

$$x(e) \geq 0 \quad \forall\, e \in E \qquad [10.30]$$

$$x(\delta(v)) \leq 1 \quad \forall\, v \in V \qquad [10.31]$$

$$x(E(S)) \leq \frac{|S|-1}{2} \quad \forall\, S \subseteq V,\ |S| \geq 3 \text{ and odd} \qquad [10.32]$$

Theorem 10.12, established by Edmonds [EDM 65], has been proven by Lovász by applying the technique described above [LOV 79].

THEOREM 10.12.– *For every graph $G = (V, E)$, the matching polytope $P^c(G)$ is given by inequalities [10.30]–[10.32].*

Proof. First of all, we can easily verify that $P^c(G)$ is full-dimensional. Indeed, the sets $\{e\}$, $e \in E$ with the empty set form a family of $|E| + 1$ matchings whose incidence vectors are affinely independent. Consequently, from corollary 10.2, two constraints define the same facet of $P^c(G)$ if and only if one is a positive multiple of the other.

Let $ax \leq \alpha$ be a constraint that defines a facet of $P^c(G)$, and let $\mathcal{C}_a$ be the set of matchings of G whose incidence vectors satisfy $ax \leq \alpha$ exactly. Assume that $ax \leq \alpha$ is different from constraints [10.30] and [10.31]. We will show that $ax \leq \alpha$ is necessarily of the type [10.32].

Since $ax \leq \alpha$ is different from inequalities [10.30], then $a(e) \geq 0$ for every $e \in E$. Indeed, if $a(e) < 0$ for a certain edge e, then every matching in $\mathcal{C}_a$ does not contain e, and, consequently, the face defined by $ax \leq \alpha$ is contained in the face

$\{x \in P^c(G) : x(e) = 0\}$. But this implies that $ax \leq \alpha$ and $x(e) \geq 0$ induce the same facet, and therefore one is a positive multiple of the other, a contradiction.

Let G' be the graph induced by the edges $e \in E$ such that $a(e) > 0$ and let S be the set of its vertices. Note that G' is connected. We will show that the incidence vector of every matching of $\mathcal{C}_a$ satisfies:

$$x(E(S)) = \frac{|S| - 1}{2} \qquad [10.33]$$

which implies that $ax \leq \alpha$ defines the same facet as [10.32].

Suppose that a matching M_1 of $\mathcal{C}_a$ exists whose incidence vector does not satisfy [10.33]. Without loss of generality, we can assume that $a(e) > 0$ for every $e \in M_1$. (If M_1 contains edges such that $a(e) = 0$, by removing these edges, we obtain another matching that satisfies $ax \leq \alpha$ with equality and does not satisfy [10.33]). Consequently, there must be two vertices u and v of S that are not incident to any edge of M_1. We can suppose that M_1 is such that the distance in G' (with respect to the weights $a(e)$, $e \in E$) between u and v is minimum. Since $a(e) > 0$ for every edge e of G', the vertices u and v cannot be adjacent in G'. So there is a vertex z, different from u and v, in the shortest path between u and v. Since $ax \leq \alpha$ is different from constraints [10.31], there must be a matching M_2 of $\mathcal{C}_a$ that does not cover z. Otherwise, every matching of $\mathcal{C}_a$ would cover z, and, consequently, the constraints $ax \leq \alpha$ and $x(\delta(z)) \leq 1$ would induce the same face. But this would imply that $ax \leq \alpha$ is a positive multiple of $x(\delta(z)) \leq 1$, a contradiction.

From the choice of M_1, M_2 must cover the two vertices u and v. If M_2 does not cover, for example, u, then u and z would be two vertices not covered by M_2. Since the path between u and z is shorter than that between u and v, this contradicts the choice of M_1. From similar arguments, M_1 must cover z. Thus, in the graph G^* consisting of the edges of $M_1 \cup M_2$, the vertices u, v and z all have a degree 1. Without loss of generality, we can suppose that the connected component of G^* containing u consists of a path Q not going through z. Consider the two matchings:

– $\bar{M}_1 = (M_1 \setminus (M_1 \cap Q)) \cup (M_2 \cap Q)$;

– $\bar{M}_2 = (M_2 \setminus (M_2 \cap Q)) \cup (M_1 \cap Q)$.

Since $\bar{M}_1 \cup \bar{M}_2 = M_1 \cup M_2$, $\bar{M}_1$ and $\bar{M}_2$ are in $\mathcal{C}_a$. But $\bar{M}_2$ does not cover either u or z, which contradicts the choice of the matching M_1. ∎

To show the integrality of a polyhedron $\{x \in \mathbb{R}^n : Ax \leq b\}$, it is also enough to show, using the linear programming duality, that for every vector weight $\omega \in \mathbb{R}^n$, the linear program max $\{\omega x \,:\, Ax \leq b\}$ has an integer optimal solution. Indeed, by theorem 10.5, this implies that the extreme points of $\{x \in \mathbb{R}^n : Ax \leq b\}$ are all integers. This technique has been introduced by Edmonds [EDM 65] to show the integrality of system [10.30]–[10.32] for the matchings of a graph. Other techniques, based on projection, can also be used to show the integrality of a polyhedron [SCH 03].

10.5. Integer polyhedra and min–max relations

As has been highlighted in section 10.4, the principal motivation for proving the integrality of a polyhedron is to establish a combinatorial min–max relation between the optimal solutions of the underlying dual problems. This dual relations in combinatorial optimization has been the subject of extensive studies, which have led to the introduction of new concepts such as totally dual integral systems, and blocking and antiblocking polyhedra. In this section, we introduce these concepts and discuss some applications.

10.5.1. *Duality and combinatorial optimization*

Let:

$$\max\{\omega x \,:\, Ax \leq b,\ x \geq 0\} \qquad [10.34]$$

be a linear program (said to be primal) and:

$$\min\{b^T y \,:\, A^T y \geq \omega^T,\ y \geq 0\} \qquad [10.35]$$

be its dual. From linear programming duality, if one of problems [10.34] or [10.35] has an optimal solution, then the other also has one, and the two optimal solutions have the same value. In other words, if [10.34] has an optimal solution $\bar{x}$, then its dual [10.35] has an optimal solution $\bar{y}$ such that $\omega\bar{x} = b^T\bar{y}$. Thus, if two optimal solutions are known for the primal problem and the dual problem, then we obtain a min–max relation between the optimal solutions of the two problems. This can have interesting applications when the two dual problems have a combinatorial interpretation. Indeed, if the system $Ax \leq b$ describes the solutions polyhedron of a combinatorial optimization problem $\mathcal{P}_1$ and $A^T y \geq \omega^T$ describes that of a combinatorial optimization problem $\mathcal{P}_2$, and if [10.34] has an optimal solution, then we obtain a relation of the form:

$$\max\{\omega(F) \,:\, F \in \mathcal{F}_1\} = \min\{b(F) \,:\, F \in \mathcal{F}_2\} \qquad [10.36]$$

where $\mathcal{F}_1$ and $\mathcal{F}_2$ are the sets of solutions of the problems $\mathcal{P}_1$ and $\mathcal{P}_2$, respectively. Consequently, given a linear program that formulates a combinatorial optimization problem, it will always be useful to study its dual and to inspect the possibility of having a [10.36] type of relation. Since the formulation of a combinatorial optimization problem in terms of a linear program is only possible if we have a complete characterization of the solutions polyhedron using a linear system, intensive research has been done to characterize the matrices whose induced system describes an integer polyhedron. One of these classes is that of matrices called totally unimodular.

10.5.2. *Totally unimodular matrices*

An $m \times n$ matrix A is said to be *totally unimodular* (TU) if the determinant of every square submatrix of A is 0, 1 or -1 (in particular every entry of A must be 0, 1 or -1). Note that if a matrix A is TU, then its transpose A^T is also TU. Theorem 10.13 establishes a fundamental link between totally unimodular matrices and polyhedra.

THEOREM 10.13.– (Hoffman, Kruskal [HOF 56]) *An $m \times n$ matrix A is totally unimodular if and only if for every integer vector $b \in \mathbb{R}^m$, the polyhedron $\{x \in \mathbb{R}^n : Ax \leq b,\ x \geq 0\}$ is integral.*

As a consequence of theorem 10.13, if A is TU and $b \in \mathbb{R}^m$ and $\omega \in \mathbb{R}^n$ are two integer vectors, then programs [10.34] and [10.35] have integer optimal solutions. Consequently, a combinatorial min–max relationship can be established between the optimal solutions of these programs. Theorem 10.14, also by Hoffman and Kruskal [HOF 56], gives a sufficient condition for a polyhedron to be integral.

THEOREM 10.14.– (Hoffman, Kruskal [HOF 56]). *If A is an $m \times n$ matrix TU, and $b \in \mathbb{R}^m$ is an integer vector, then the polyhedron $Ax \leq b$ is integral.*

In the following, we give some examples of totally unimodular matrices. If $G = (V, E)$ is a graph, the vertex-edge incidence matrix of G is the matrix A whose rows correspond to the vertices of G, and whose columns correspond to the edges of G, and such that for a vertex i and an edge e, the entry $A_{i,e}$ is 1 if i is an extremity of e and 0 if it is not. If G is bipartite, then its vertex-edge incidence matrix is TU.

Let $G = (V, E)$ be a bipartite graph. Since constraints [10.30] and [10.31] are valid for the matching polytope $P^c(G)$ in G, we have $P^c(G) \subseteq P^*(G)$, where $P^*(G)$ is the polytope given by inequalities [10.30] and [10.31]. These inequalities can also be written in the form:

$$Ax \leq \mathbf{1}$$

$$x \geq 0$$

where A is the vertex-edge incidence matrix of G (let us remember that $\mathbf{1}$ refers to the vector whose components are all equal to 1). Since A is TU, by theorem 10.13, $P^*(G)$ is integral. Consequently, every extreme point of $P^*(G)$ represents a matching of G, and therefore $P^*(G) \subseteq P^c(G)$. It then follows that $P^*(G) = P^c(G)$.

Let us now consider the maximum cardinality matching problem. Since $P^*(G) = P^c(G)$, this problem is equivalent to the linear program:

$$\max \{\mathbf{1}^T x \ : \ Ax \leq \mathbf{1}, \ x \geq 0\} \qquad [10.37]$$

The dual of [10.37] can be written as:

$$\min \{\mathbf{1}^T y \ : \ A^T y \geq \mathbf{1}, \ y \geq 0\} \qquad [10.38]$$

Note that the variables in [10.38] correspond to the vertices of G and that the constraints correspond to the edges. Since the matrix A is TU, and A^T is also TU, an optimal solution $\bar{y}$ of [10.38] can be supposed to be integer. Therefore $\bar{y}$ is a 0–1 vector, and we can see that in this case $\bar{y}$ represents a set of vertices that covers all the edges of the graph. Such a set is called a vertex cover. From [10.36], we then obtain theorem 10.15.

THEOREM 10.15.– (König [KÖN 31]). *In a bipartite graph, the maximum cardinality of a matching is equal to the minimum cardinality of a vertex cover.*

If $G = (V, E)$ is a directed graph, the vertex-arc incidence matrix of G is the matrix D whose rows correspond to the vertices of G, and columns to the arcs of G such that for a vertex i and an arc e, the entry $D_{i,e}$ is 1 if i is the initial vertex of e, -1 if i is the terminal vertex of e and 0 if not. This matrix is TU, and, as a consequence, we can obtain the famous maximum-flow minimum-cut theorem [FOR 56, FOR 62].

Unfortunately not all the integer matrices (in $-1, +1, 0$) are TU. Integral polyhedra exist for which the matrix of the corresponding system is not TU.

If $\{x \in \mathbb{R}^n \ : \ Ax \leq 0, \ x \geq 0\}$ is an integral polyhedron where A is not TU, although the linear program:

$$\max\{\omega x \ : \ Ax \leq b, \ x \geq 0\} \qquad [10.39]$$

still has an integer optimal solution, its dual may not have one (even if ω is integer). Consequently, no combinatorial min–max relationship can be obtained. A natural question that can therefore be asked is: under what conditions can the dual of [10.39] have an integer optimal solution each time that [10.39] has one for an integer ω? Edmonds and Giles [EDM 77] have studied this question, and have introduced the concept of a totally dual integral system, which we discuss in what follows.

10.5.3. *Totally dual integral systems*

Let A be a rational $m \times n$ matrix and $b \in \mathbb{R}^m$. The linear system $Ax \leq b$ is said to be *totally dual integral* (TDI) if for every integral vector $\omega \in \mathbb{R}^n$ such that the linear program $\max\{\omega x : Ax \leq b\}$ has an optimal solution, the corresponding dual program has an integer optimal solution. A sufficient condition for a polyhedron to be integral is the following.

THEOREM 10.16.– (Edmonds, Giles [EDM 77]). *If $Ax \leq b$ is TDI and b is integer, then the polyhedron $\{x \in \mathbb{R}^n : Ax \leq b\}$ is integral.*

Theorem 10.16 has some important consequences. If $Ax \leq b$ is TDI and b is integer then an optimal solution of the linear program $\max\{\omega x : Ax \leq b\}$ can always be considered integer. Furthermore, if ω is integer, as in this case, an optimal solution of the dual can also be considered integer; this leads to a combinatorial min–max relation between the optimal solutions.

The converse of theorem 10.16 is not true; non-TDI systems that induce integral polyhedra may exist. However, as theorem 10.17 shows, any integral polyhedron can be described by a TDI system.

THEOREM 10.17.– (Giles, Pulleyblank [GIL 79]). *If $P \subseteq \mathbb{R}^n$ is a rational polyhedron, then a linear TDI system $Ax \leq b$ exists, where A is an integer matrix, such that $P = \{x \in \mathbb{R}^n : Ax \leq b\}$. Furthermore, if P is integral, then b can be chosen to be integral.*

From theorem 10.17, it is sometimes necessary to add redundant inequalities to a system for it to become TDI.

Several linear systems, describing integral polyhedra associated with combinatorial optimization problems, have been shown to be TDI, for example the matching polytope.

THEOREM 10.18.– (Cunningham, March [CUN 78]). *The system given by [10.30]–[10.32] is TDI.*

Let $G = (V, E)$ be a graph. A family of odd subsets of vertices of V, $\mathcal{C} = \{S_i : i \in I\}$, where I is a set of indices, is called an *odd cover* of G if for every edge $e \in E$ either $e \in \delta(v)$ for a certain v such that $\{v\}$ is in $\mathcal{C}$, or $e \in E(S_i)$ for a certain S_i such that $|S_i| \geq 3$. We define the weight of $\mathcal{C}$ as being $q + \frac{1}{2}\sum_{i \in I, |S_i| \geq 3}(|S_i| + 1)$, where q is the number of subsets of $\mathcal{C}$ reduced to one single element. We have the following min–max relationship as a corollary of theorem 10.18.

THEOREM 10.19.– (Berge [BER 58]). *The maximum cardinality of a matching in G is equal to the minimum weight of an odd cover of G.*

Note that this theorem generalizes König's theorem 10.15. Extensions of the notion of TDI have also been studied. For more details see Schrijver [SCH 03].

10.5.4. *Blocking and antiblocking polyhedra*

Another aspect of the duality that has been studied in combinatorial optimization is that of blocking and antiblocking polyhedra. These concepts, introduced by Fulkerson [FUL 71, FUL 72], have been the source of several important developments in combinatorial optimization.

10.5.4.1. *Blocking polyhedra*

Let A be a positive $m \times n$ matrix and $w \in \mathbb{R}^m_+$. Consider the linear program:

$$\max \{\mathbf{1}^T x \ : \ A^T x \leq w, \ x \geq 0\} \qquad [10.40]$$

Problem [10.40] is called a *packing problem*. Its dual can be written as:

$$\min \{w^T y \ : \ Ay \geq \mathbf{1}, \ y \geq 0\} \qquad [10.41]$$

Let $\mathcal{B}$ be the polyhedron given by $\{y \in \mathbb{R}^n_+ \ : \ Ay \geq \mathbf{1}\}$. Note that $\mathcal{B}$ is bounded and is full-dimensional. Moreover, the system $Ay \geq \mathbf{1}$ may contain redundant constraints. If $Ay \geq \mathbf{1}$ does not contain any redundant inequalities, then the matrix A is said to be *proper*. A matrix is also said to be proper if it does not contain any rows (that is $\mathcal{B} = \mathbb{R}^N_+$ or $\mathcal{B} = \emptyset$). If A is a proper 0–1 matrix, then the rows of A can be considered as incidence vectors of m non-comparable pairwise subsets of a set of n elements. In this case the family of subsets, represented by the rows of A, is said to be a *clutter*.

Let:

$$\hat{\mathcal{B}} = \{z \in \mathbb{R}^n_+ \ : \ z^T x \geq 1, \ \forall \, x \in \mathcal{B}\} \qquad [10.42]$$

The set $\hat{\mathcal{B}}$ is called a *blocking* of $\mathcal{B}$. Note that if $\mathcal{B} = \emptyset$, then $\hat{\mathcal{B}} = \mathbb{R}^n_+$, and if $\mathcal{B} = \mathbb{R}^n_+$, then $\hat{\mathcal{B}} = \emptyset$. Consequently, $\hat{\mathcal{B}}$ can be seen as the dual of $\mathcal{B}$ and *vice versa*. Fulkerson [FUL 71] established the following relationships between $\mathcal{B}$ and its blocking $\hat{\mathcal{B}}$.

THEOREM 10.20.– (Fulkerson [FUL 71]). *Let A be a proper matrix whose rows are $a^1, \dots, a^m$. Let $b^1, \dots, b^r$ be the extreme points of $\mathcal{B}$, and let B be the $q \times n$ matrix whose rows are $b^1, \dots, b^q$. Then*

1) $\hat{\mathcal{B}} = \{x \in \mathbb{R}^n_+ \, : \, Bx \geq \mathbf{1}\}$,

2) B is proper,

3) the extreme points of $\hat{\mathcal{B}}$ are $a^1, \ldots, a^m$.

Note that from assertion 1, $\hat{\mathcal{B}}$ is a polyhedron. Also, from 1 and 3 we have $\hat{\hat{\mathcal{B}}} = \mathcal{B}$. Consequently, given a matrix A, we can construct the blocking matrix B of A such that the blocking polyhedron associated with one of the matrices has as extreme points the rows of the other. The polyhedra $\hat{\mathcal{B}}$ and $\mathcal{B}$ therefore play a symmetric role: they form what we call a *blocking pair of polyhedra*. The pair of matrices A and B is said to be a *blocking pair of matrices*. Note that if A is an integer matrix and the extreme points of $\mathcal{B}$ are all integer, then the blocking of $\mathcal{B}$, $\hat{\mathcal{B}}$, has all its extreme points as integer.

Let A be an $m \times n$ matrix and B a $q \times n$ matrix. Suppose that A and B are proper. Let $a^1, \ldots, a^m$ and $b^1, \ldots, b^q$ be the rows of A and B, respectively. We say that the *min–max* relation is satisfied for A and B (in this order) if for every vector $w \in \mathbb{R}^n_+$, the packing problem [10.40] has an optimal solution $\bar{x}$ such that:

$$\mathbf{1}^T \bar{x} = \min \{b^j w; \; j = 1, \ldots, q\} \qquad [10.43]$$

If A and B form a blocking pair of matrices, then the vectors $b^1, \ldots, b^q$ are the extreme points of the blocking $\mathcal{B}$ of A. Since the polyhedron $\{x \in \mathbb{R}^m : A^T x \leq w, \; x \geq 0\}$ is bounded for every $w \in \mathbb{R}^n_+$, program [10.40] has an optimal solution for every $w \in \mathbb{R}^n_+$. Thus the dual program [10.41] also has an optimal solution for every $w \in \mathbb{R}^n_+$, and [10.43] is therefore satisfied. The converse is also true as theorem 10.21 shows.

THEOREM 10.21.– (Fulkerson [FUL 71]). *The min–max relation is satisfied for two matrices A and B (in this order) if and only if A and B form a blocking pair of matrices.*

Theorem 10.21 gives a characterization of blocking pairs of matrices. A direct consequence of this theorem is that the [10.43] relation is satisfied for two matrices A and B, so it is also satisfied for B and A. This theorem constitutes a fundamental tool for establishing combinatorial dual relations between the solutions of blocking polyhedra.

The most important case of blocking matrices is when the matrix A is integer (in 0–1) and the packing problem may have an integer optimal solution when w is integer. Indeed, in this case, a combinatorial min–max relation can be directly established. The typical example of blocking matrices satisfying this property is that of matrices representing minimal paths between two vertices s and t of a graph and the minimal cuts separating these vertices.

Let $G = (V, E)$ be a graph and s and t be two vertices of V. Let A be the matrix whose rows are the incidence vectors of the minimal paths between s and t. Let B be the matrix whose rows are the incidence vectors of the minimal cuts separating s and t. Note that the two matrices A and B are 0–1 matrices. Let $w \in \mathbb{R}^n_+$, where $n = |E|$. By considering w as a capacity vector associated with the edges of G, the problem [10.40] corresponding to A and w is nothing but the maximum flow problem between s and t in G with respect to the capacity vector w. From Fulkerson's max-flow min-cut theorem, we know that the maximum value of a flow between s and t is equal to the minimal capacity of a cut separating s and t, where the capacity of a cut is the sum of the capacities of the edges of the cut. Consequently, the min–max relation [10.43] is satisfied for A and B, which implies that A, B is a blocking pair.

If $w = \mathbf{1}^T$, the programs [10.40] and [10.41] for A can be written as:

$$\max \{\mathbf{1}^T x \ : \ A^T x \leq \mathbf{1}, \ x \geq 0\} \qquad [10.44]$$

and:

$$\min \{\mathbf{1}^T y \ : \ Ay \geq \mathbf{1}, \ y \geq 0\}. \qquad [10.45]$$

Note that the variables in [10.44] correspond to the minimal paths of G between s and t, and those in [10.45] correspond to the edges of G. Observe that the packing problem [10.44] has an integer optimal solution. Such a solution represents a packing of paths between s and t, that is a set of pairwise disjoint paths between s and t.

Since A and B form a blocking pair, from theorem 10.20, the extreme points of the polyhedron $\{y \in \mathbb{R}^n_+ \ : \ Ay \geq \mathbf{1}\}$ are precisely the minimal cuts separating s and t. From [10.43], we thus obtain the following min–max relation, which is nothing but theorem 10.10, Menger's theorem.

THEOREM 10.22.– *The minimum number of edges in a cut separating* s *and* t *is equal to the maximum number of pairwise disjoint paths between* s *and* t.

In a similar way, by using the fact that the min–max relationship is also satisfied for B and A, we obtain theorem 10.23.

THEOREM 10.23.– *The minimum number of edges in a chain between* s *and* t *is equal to the maximum number of pairwise disjoint cuts separating* s *and* t.

Other examples of blocking pairs and matrices are given in [FUL 71, SCH 03].

10.5.4.2. *Antiblocking polyhedra*

Let A be a positive matrix $m \times n$, and $w \in \mathbb{R}_+^m$. Let us consider the linear program:

$$\min \{\mathbf{1}^T x \ : \ A^T x \geq w, \ x \geq 0\} \qquad [10.46]$$

Problem [10.46] is called a *set covering problem.* Its dual can be written as:

$$\max \{w^T y \ : \ Ay \leq \mathbf{1}, \ y \geq 0\} \qquad [10.47]$$

Consider the polyhedron $\mathcal{A} = \{y \in \mathbb{R}_+^n \ : \ Ay \leq \mathbf{1}\}$. The polyhedron $\mathcal{A}$ is bounded if and only if A does not contain a zero column. In what follows, we assume that $\mathcal{A}$ is bounded. Let:

$$\bar{\mathcal{A}} = \{z \in \mathbb{R}_+^n \ : \ z^T y \leq \mathbf{1} \ \forall\, y \in \mathcal{A}\}$$

$\bar{\mathcal{A}}$ is called *the antiblocking of* $\mathcal{A}$. Let $b^1, \ldots, b^q$ be the extreme points of $\mathcal{A}$, and let B be the matrix whose rows are $b^1, \ldots, b^q$. Theorem 10.24 is similar to theorem 10.20.

THEOREM 10.24.– (Fulkerson [FUL 71])

1) B is positive and does not contain any zero columns;

2) $\bar{\mathcal{A}} = \{x \in \mathbb{R}_+^n \ : \ Bx \leq \mathbf{1}\}$;

3) $\bar{\bar{\mathcal{A}}} = \mathcal{A}$.

From 1 and 2, $\bar{\mathcal{A}}$ is a polytope. The pair of polyhedra $\mathcal{A}$ and $\bar{\mathcal{A}}$ is said to be an *antiblocking pair of polyhedra*, and the pair of matrices A and B is said to be an *antiblocking pair of matrices*. Note that we do not have a biunivocal relation here between the rows of A and the extreme points of its antiblocking polyhedron $\bar{\mathcal{A}}$, as is the case for blocking polyhedra. In other words, $\bar{\mathcal{A}}$ can have extreme points that are not rows of A.

Let A and B be two positive matrices $m \times n$ and $q \times n$, respectively. Let $a^1, \ldots, a^m$ and $b^1, \ldots, b^q$ be the rows of A and of B respectively, and let us assume that neither A nor B contains any zero columns. We say that the *min–max* relation is satisfied for A and B (in this order) if for every vector $w \in \mathbb{R}_+^n$, the set covering problem [10.46] has an optimal solution x such that:

$$\mathbf{1}^T x = \max \{b^j w; \ j = 1, \ldots, q\} \qquad [10.48]$$

THEOREM 10.25.– (Fulkerson [FUL 71]). *The min–max relation [10.48] is satisfied for two matrices A and B (in this order) if and only if A and B form an antiblocking pair of matrices.*

If relation [10.48] is satisfied for A and B, from theorem 10.25, it is also satisfied for B and A.

To illustrate this concept, let us consider the matching polyhedron. We have seen in section 10.4 that if $G = (V, E)$ is a graph, the matching polytope of G, $P^c(G)$, is given by inequalities [10.30]–[10.32]. $P^c(G)$ can also be written in the form $P^c(G) = \{x \in \mathbb{R}^E \ : \ Bx \leq \mathbf{1}, \ x \geq 0\}$. If we denote by A the matrix whose rows are the incidence vectors of the matchings in G, the min–max relation [10.48] is nothing but that established by duality. Consequently, A and B form an antiblocking pair of matrices. If $w = \mathbf{1}$, the min–max relation for A and B implies theorem 10.19 (Berge's theorem).

10.6. Cutting-plane method

Given a combinatorial optimization problem, it is generally difficult to characterize the associated polyhedron by a system of linear inequalities. Moreover, if the problem is **NP**-complete, there is very little hope of obtaining such a description. Furthermore, even if it is characterized, the system describing the polyhedron can contain a very large (even exponential) number of inequalities, and therefore cannot be totally used to solve the problem as a linear program. However, by using a *cutting-plane method*, a partial description of the polyhedron can be sufficient to solve the problem optimally. We discuss this method below.

Let us consider a combinatorial optimization problem in the form:

$$\max \{\omega x \ : A\, x \leq b, \ x \text{ integer}\} \qquad [10.49]$$

where A is an $m \times n$ matrix and $b \in \mathbb{R}^m$. Let P be the convex hull of the solutions of [10.49]. From proposition 10.2 and theorem 10.4, the problem [10.49] is equivalent to the program $\max\{\omega x \ : \ x \in P\}$. If the inequalities of the system $Ax \leq b$ are sufficient to describe the polyhedron P, then every extreme point of the polyhedron $\{x \in \mathbb{R}^n \ : \ Ax \leq b\}$ is integral, and, consequently, the problem [10.49] is equivalent to its linear relaxation:

$$\max \{\omega x \ : \ Ax \leq b\} \qquad [10.50]$$

Unfortunately, this is not always the case, and the polyhedron $\{x \in \mathbb{R}^n \ : \ Ax \leq b\}$ can indeed contain fractional extreme points. Consequently, an optimal solution of [10.50], let us say x^*, can be fractional. In this case, there must be a valid constraint for P which is violated by x^*. This can be added to the system $Ax \leq b$ and hence we obtain a "stronger" linear relaxation. Such a constraint is called a *cut*. The name "cut" comes from the fact that the addition of the violated constraint to the system $Ax \leq b$ enables us to "cut" a useless part of the polyhedron of the linear relaxation containing x^*.

The crucial step in a cutting-plane method is the generation of valid constraints for the polytope P. A first technique for identifying valid constraints for integer programs has been introduced by Gomory [GOM 58, GOM 60]. Using this technique, Chvátal [CHV 73] has developed a general procedure for generating valid constraints for a combinatorial polyhedron. The procedure is based on the fact that if p is integer and $p \leq q$, then $p \leq \lfloor q \rfloor$.

10.6.1. *The Chvátal–Gomory method*

Let us consider, for example, the stable set problem. Let $G = (V, E)$ be a graph and C a cycle of G. Let $\{v_1, \ldots, v_k\}$ be the set of vertices of C. Suppose that k is odd. The following constraints are therefore valid for the stable set polytope $P(G)$:

$$\begin{aligned}
&x(v_1) + x(v_2) \leq 1\\
&x(v_2) + x(v_3) \leq 1\\
&\vdots\\
&x(v_k) + x(v_1) \leq 1
\end{aligned}$$

By adding up these inequalities (and dividing by 2), we obtain the constraint

$$\sum_{i=1}^{k} x(v_i) \leq \frac{k}{2}$$

Since $\sum_{i=1}^{k} x(v_i)$ is integer for every solution of the stable set problem, then the inequality:

$$\sum_{i=1}^{k} x(v_i) \leq \lfloor \frac{k}{2} \rfloor = \frac{k-1}{2} \qquad [10.51]$$

is valid for the stable set polytope.

This procedure can be presented in a more general context in the following way. Let y be a positive vector of $\mathbb{R}^m$. Every solution of [10.49] satisfies the inequality:

$$\sum_{j=1}^{n} yA^j x_j \leq yb, \qquad [10.52]$$

obtained as a linear combination of constraints of the system $Ax \leq b$. Since $yA^j - \lfloor yA^j \rfloor \geq 0$ and $x_j \geq 0$ for $j = 1, \ldots, n$, the constraint:

$$\sum_{j=1}^{n} \lfloor yA^j \rfloor x_j \leq yb \qquad [10.53]$$

is also satisfied by the solutions of [10.49]. Since the left-hand side of [10.53] is integer, it follows that the inequality:

$$\sum_{j=1}^{n} \lfloor yA^j \rfloor x_j \leq \lfloor yb \rfloor \qquad [10.54]$$

is valid for P. Inequalities of type [10.54] are called *Chvátal–Gomory inequalities*, and the above procedure is known as the *Chvátal–Gomory method.*

As is stated in theorem 10.26, every valid constraint for P can be obtained using the Chvátal–Gomory method.

THEOREM 10.26.– (Schrijver [SCH 80]). *Let $ax \leq \alpha$ be a valid inequality for $P \neq \emptyset$ with $(a, \alpha) \in \mathbb{Z}^{n+1}$. Then $ax \leq \alpha$ is a Chvátal–Gomory inequality.*

A valid inequality for P, equivalent to (or dominated by) a positive linear combination of $Ax \leq b$ is said to be *of Chvátal rank* 0 with respect to $Ax \leq b$. An inequality $ax \leq \alpha$ valid for P is said to be of *Chvátal rank k* with respect to $Ax \leq b$ if $ax \leq \alpha$ is not of Chvátal rank less than or equal to $k - 1$, and if it is equivalent to (or dominated by) a positive linear combination of inequalities, each one can be obtained using the Chvátal–Gomory method from inequalities of Chvátal rank less than or equal to $k - 1$. In other words, an inequality valid for P is of Chvátal rank k if k applications of the Chvátal–Gomory method are necessary to obtain this inequality. Thus the constraints of Chvátal rank 1 are those which are not of rank 0 but which are either equivalent to or dominated by a positive linear combination of constraints of $Ax \leq b$ and inequalities obtained using the Chvátal–Gomory procedure from constraints of $Ax \leq b$. Note that the constraints [10.53] are of Chvátal rank 1 with respect to $Ax \leq b$. The valid inequalities that contribute to the resolution of a combinatorial problem, in the context of a cutting-plane method, are generally of Chvátal rank ≤ 1.

Theorem 10.26 implies that every constraint that is valid for the polyhedron P is of finite Chvátal rank with respect to $Ax \leq b$. The maximum Chvátal rank with respect to $Ax \leq b$ of a valid constraint (facet) for P is called *the Chvátal rank* of the polyhedron $\{x \in \mathbb{R}^n : Ax \leq b\}$. Note that the rank of $\{x \in \mathbb{R}^n : Ax \leq b\}$ is 0 if and only if $P = \{x \in \mathbb{R}^n : Ax \leq b\}$, that is if $\{x \in \mathbb{R}^n : Ax \leq b\}$ is integral.

Suppose that P is the matchings polytope in a graph $G = (V, E)$, and let $Ax \leq b$ be the system given by inequalities [10.30] and [10.31]. It is not hard to see that inequalities [10.32] are of Chvátal rank 1 with respect to $Ax \leq b$. Since constraints [10.30]–[10.32] completely describe P, it follows that the polyhedron $Ax \leq b$ is of rank 1.

10.6.2. *Cutting-plane algorithm*

The Chvátal–Gomory procedure, presented below, is often used to identify valid inequalities for polytopes associated with combinatorial optimization problems. These constraints can then be used in a *cutting-plane algorithm* to solve the problem. This algorithm can be presented as follows.

Suppose that we have certain classes of inequalities valid for the polytope P associated with problem [10.49], and let $\bar{A}x \leq \bar{b}$ be the system formed by these inequalities. Suppose that this system contains the basic constraints $Ax \leq b$ of the problem. Each iteration of the algorithm consists of solving a linear program defined by a subset of constraints of $\bar{A}x \leq \bar{b}$. If the solution found is integral and satisfies $Ax \leq b$, then it is optimal for the problem. Otherwise, other constraints must be added to the linear program.

We start by considering a subsystem $\bar{A}'x \leq \bar{b}'$ of $\bar{A}x \leq \bar{b}$ containing a reduced number of constraints, and we solve the linear program $\max\{\omega x \ : \ \bar{A}'x \leq \bar{b}'\}$. If the optimal solution of this program, say x^1, is feasible for [10.49] (that is if x^1 is integer and is a solution of the system $Ax \leq b$) then it is optimal for [10.49]. Otherwise, we then determine a constraint of the system $\bar{A}x \leq \bar{b}$, say $a^1x \leq \alpha^1$, that is violated by x^1. We add this constraint to the program and we solve the new linear program $\max\{\omega x \ : \ \bar{A}'x \leq \bar{b}', \ a^1x \leq \alpha^1\}$. If the optimal solution, say x^2, of this program is feasible for [10.49], then it is optimal. Otherwise, we determine a constraint $a^2x \leq \alpha^2$ of $\bar{A}x \leq \bar{b}$ that is violated by x^2. We consider the linear program $\max\{\omega x \ : \ \bar{A}'x \leq \bar{b}', \ a^1x \leq \alpha^1, \ a^2x \leq \alpha^2\}$, and so on. This procedure continues until either we find an integral solution feasible for [10.49] and therefore optimal, or it is no longer possible to generate violated constraints. In this case, we use integer programming techniques, such as the branch-and-bound method, to establish an optimal solution of the problem.

Note that if the system $Ax \leq b$ contains a reduced (polynomial) number of constraints, it can then be used to initialize the cutting-plane algorithm. Therefore the first linear program would be the linear relaxation of the problem. At each step of the algorithm, it is sufficient to check if the optimal solution of the current linear program is integral. If this is the case, then it is optimal. The stable set and maximum matching problems are of this type; the linear relaxation of each problem contains $n + m$ constraints where n (resp. m) is the number of vertices (resp. edges) of the graph.

Let us also note that the optimal solution of the linear relaxation, obtained in each iteration of the algorithm, gives an upper bound to the optimal value of the problem. To obtain a tighter linear relaxation from one iteration to the next and to improve this bound to the maximum, it is useful to introduce violated constraints that define facets of the polyhedron P. Indeed, these allow us to cut to the maximum non-feasible solutions of the problem.

10.6.3. *Branch-and-cut algorithms*

If the cutting-plane algorithm does not allow us to provide an optimal solution of the problem, it is necessary to use a branch-and-bound technique to establish one. This technique allows us to construct a resolution tree where each vertex of the tree corresponds to a subproblem, the initial problem being the one associated with the root. This technique is based on two essential procedures:

– *Branching*: this procedure is simple; it just allows us to divide the problem associated with a given vertex of the tree into two disjoint subproblems by fixing one of the variables x_i to 1 for one of the problems and to 0 for the other. An optimal solution of the problem (corresponding to this vertex of the tree) will thus be optimal for one of the subproblems.

– *Bounding*: the aim of this procedure is to establish an upper bound (lower bound in the case of minimization) for the optimal value of the problem associated with a vertex of the tree.

To solve problem [10.49], we can start by solving a linear relaxation of the problem using a cuts algorithm. If an optimal solution is not found in this phase (called the *cutting phase*), we choose a fractional variable x_i, and we apply the branching procedure. In this way we create two subproblems (two vertices of the solution tree, which we link to the root). We establish an upper bound for each subproblem by solving the relaxed programs. If for one of the subproblems the optimal solution is integer, we stop its exploration. Otherwise, we choose one of the vertices, say S, and we divide the associated problem into two subproblems. We then create two new vertices, say S_1 and S_2, which we link to the *father* vertex S. By repeating this procedure, we construct a tree where the vertices correspond to the subproblems created. If the optimal solution for one of the problems is feasible for the initial problem or not as good as a feasible solution already found, then we stop developing the corresponding vertex. This one is declared *sterile*. At each step, we choose a pending vertex of the tree that is not sterile. We divide the corresponding problem into two subproblems and calculate a bound for each of the subproblems created. The algorithm stops when all the pending vertices of the tree are sterile. In this case, the best feasible solution found is optimal.

To calculate a bound for each vertex of the tree, we can simply solve the linear program obtained from the program of the father vertex by adding either the equation $x_i = 0$, or the equation $x_i = 1$. However, this bound can be weak, and can lead to a very slow resolution process, especially when the problem is large. However, if we add violated constraints to this relaxation, we can obtain better bounds and further accelerate the resolution of the program. *A branch-and-cut algorithm* is a branch-and-bound technique in which we apply the cutting-plane algorithm to calculate the bound of each subproblem. This method, introduced by Padberg and Rinaldi [PAD 91] for the traveling salesman problem, has shown itself to be very effective, and is now widely used to solve combinatorial optimization problems in an exact way.

Note that an important step in each iteration of a branch-and-cut algorithm concerns generating violated constraints. The effectiveness of the algorithm depends very closely on the way in which these constraints are generated. Therefore, since the polytope associated with the underlying combinatorial problem may have an exponential number of facets, we can ask if it will be possible, even in this case, to solve the problem in polynomial time using a branch-and-cut algorithm. As we will see in section 10.6.4, this is possible if the separation problem associated with the associated polytope can be solved in polynomial time.

10.6.4. *Separation and optimization*

One of the major developments in polyhedral approaches during the last 25 years has been the introduction of separation techniques for combinatorial optimization problems. Indeed, these techniques allow us to solve hard problems effectively within the framework of branch-and-cut algorithms. They are a direct consequence of the powerful result of Grötschel *et al.* [GRÖ 81], which establishes an equivalence between the problem of optimization on a polyhedron and the separation problem associated with this polyhedron. In the following, we discuss this strong relationship between separation and optimization and the algorithmic consequences for combinatorial optimization problems.

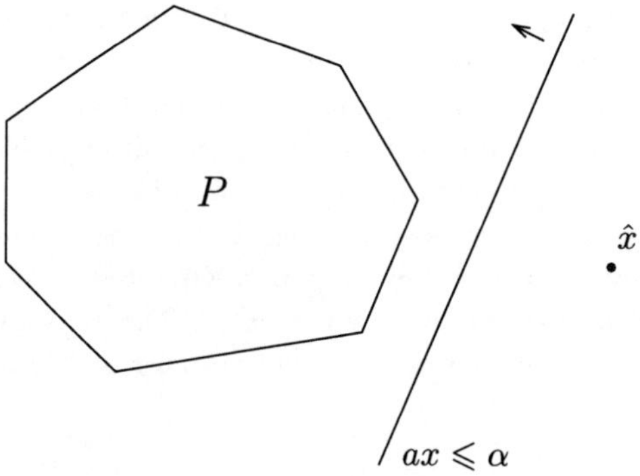

Figure 10.3. *Hyperplane separating P and $\hat{x}$*

The *separation problem* associated with a polyhedron $P \subseteq \mathbb{R}^n$ can be formulated as follows. Given a solution $\hat{x} \in \mathbb{R}^n$, determine whether $\hat{x} \in P$, and, if not, find an inequality $ax \leq \alpha$ valid for P that is violated by $\hat{x}$. In other words, in the case where $\hat{x} \notin P$, the problem consists of finding a hyperplane that separates $\hat{x}$ and P (see Figure 10.3). If P is given by a system of inequalities $Ax \leq b$, then we talk of a separation problem associated with the system $Ax \leq b$.

In a branch-and-cut algorithm, we solve a sequence of separation problems on each vertex of the branching tree, each problem allowing generation of one (or several) violated constraints. The complexity of the algorithm will therefore naturally depend on that of the separation problem of the different classes of inequalities used in the algorithm. Using the ellipsoid method, introduced by Khachian [KHA 79] for linear programming, Grötschel *et al.* [GRÖ 81] have shown that if we can solve the separation problem in polynomial time for a polyhedron P, then we can solve the optimization problem for P, $\max\{\omega x \ : \ x \in P\}$ in polynomial time. The converse of this result is true. If we can optimize in polynomial time, we can also separate in polynomial time. Consequently, the complexity of a cutting-plane algorithm on a polyhedron $\{x \in \mathbb{R}^n \ : \ Ax \leq b\}$ does not depend on the number of constraints in the system $Ax \leq b$ (even if it is exponential), but rather on the complexity of the separation problem that is associated with it. So, to solve an optimization problem of the form $\max\{\omega x \ : \ Ax \leq b\}$, using a cutting-plane algorithm, we do not need to know the system $Ax \leq b$ explicitly. It is sufficient just to be able to verify if a solution $\hat{x}$ satisfies $A\hat{x} \leq b$, and, if not, to determine *one* constraint among $Ax \leq b$ that is violated by $\hat{x}$.

To illustrate this concept, let us again consider the stable set problem in a graph $G = (V, E)$. The formulation of the problem, given by inequalities [10.18] and [10.19], contains a polynomial number of constraints and, consequently, the separation problem corresponding to these constraints can be solved in polynomial time. Let us further consider constraints the [10.51] given below, which are also valid for the stable set polytope.

$$x(V(C)) \leq \frac{|V(C)| - 1}{2}, \quad \forall\, C \text{ odd cycle of } G \qquad [10.55]$$

Constraints [10.55] are called *odd cycle inequalities*. As will be shown below, the branching problem for these constraints can also be solved in polynomial time.

THEOREM 10.27.– (Grötschel *et al.* [GRÖ 88]). *Inequalities [10.55] can be separated in polynomial time.*

Proof. Let $\hat{x} \in \mathbb{R}^V$. Since inequalities [10.18] and [10.19] can be separated in polynomial time, to solve the separation problem for constraints [10.55], we can assume that $\hat{x} \geq 0$ and that $\hat{x}$ satisfies inequalities [10.18]. If $e = ij \in E$, let

$z(e) = 1 - \hat{x}(i) - \hat{x}(j)$. Therefore $z(e) \geq 0$ for every edge $e \in E$. Thus, inequalities [10.55] can be written as:

$$\sum_{e \in C} z(e) \geq 1, \qquad \forall\, C \text{ odd cycle of } G \qquad [10.56]$$

The separation problem for constraints [10.55] with respect to the vector $\hat{x}$ is therefore equivalent to the branching problem for constraints [10.56] with respect to the vector z. Considering $z(e)$ as a weight on edge e, the separation problem of constraints [10.55] is therefore reduced to seeking an odd cycle of minimum weight in a graph having positive weights. This last problem can be solved in polynomial time. To do this, let us consider the bipartite graph $\tilde{G} = (V' \cup V'', \tilde{E})$ obtained from $G = (V, E)$ in the following way: for each vertex v of V, consider two vertices $v' \in V'$ and $v'' \in V''$, and for each edge $uv \in E$, consider the edges $u'v''$ and $u''v'$ in $\tilde{E}$ with the same weight $\tilde{z}(u'v'') = \tilde{z}(u''v') = z(uv)$. Since $\tilde{G}$ is bipartite, every path in $\tilde{G}$ between v' and v'' for every $v \in V$ corresponds to an odd cycle in G of the same weight, and *vice versa*. Consequently, determining an odd cycle in G of minimum weight with respect to z is equivalent to determining a path of minimum weight in $\tilde{G}$ with respect to $\tilde{z}$. Since $\tilde{z} \geq 0$, this last problem can be solved in polynomial time using, for example, Dijkstra's algorithm [DIJ 59]. If C is an odd cycle of G of minimum weight obtained in this way, and if $z(C) \geq 1$, then all the constraints [10.56] are satisfied by z, and therefore [10.55] are satisfied by $\hat{x}$. If $z(C) < 1$, then the constraint of type [10.55] induced by C is violated. ∎

A direct consequence of theorem 10.27 is that the maximum stable set problem can be solved in polynomial time, using a cutting-plane algorithm, in the graph G whose polytope $P(G)$ is described by constraints [10.18], [10.19], and [10.55]. These graphs are called *t-perfect*. A more detailed discussion on these graphs can be found in [SCH 03].

10.7. The maximum cut problem

In this section, we discuss some applications of the techniques presented in the previous sections to two combinatorial optimization problems. The first is *the maximum cut problem*. Given a graph $G = (V, E)$ and weights $(\omega(e),\ e \in E)$ on the edges, this consists of finding a cut $\delta(W)$, $W \subset V$, such that $\sum_{e \in \delta(W)} \omega(e)$ is maximum. This problem has, for the last 20 years, been the object of intensive research [DEZ 97]. It has many applications, for instance in spin glass models in statistical physics, and VLSI circuits.

10.7.1. *Spin glass models and the maximum cut problem*

A *spin glass* is a system obtained by a small dilution (1%) of a magnetic material (iron) in a non-magnetic material (gold). Physicists' interest in this material comes from observing a peak in the curve of what is called the magnetic *susceptibility* as a function of temperature. Such a peak is generally an indication of a transition phase, a change of state of the system, hence the search for models that are likely to explain this phenomenon.

In a spin glass, the magnetic atoms are laid out randomly in space. Between two atoms i, j, there exists an interaction energy:

$$H_{ij} = -J(R)S_i S_j$$

where S_i (S_j) is the spin of the atom i (j), and $J(R)$ is a function that depends on the distance R between the two atoms. To model these systems, physicists have constructed a simplified model: they assume that the spins are situated at the nodes of a regular mesh (instead of being randomly distributed) and are defined by unidimensional vectors (instead of being tridimensional) S_i, taking the values $+1$ and -1. These meshes are generally square or cubic. They further assume that the interactions between the spins only take place between the closest neighbors, and that their energies (J_{ij}) are random variables taking positive or negative values. The interactions then correspond to the links of the mesh.

The energy of the system corresponds to a configuration S of spins (that is an assignment of $+1$ and -1 to the spins), given by:

$$H(S) = -\sum_{ij \in L} J_{ij} S_i S_j \qquad [10.57]$$

where L is the set of the links and J_{ij} the interaction between the spins i and j. The problem that physicists study is to determine a configuration S that minimizes the energy [10.57] of the system. Such a configuration is called *the fundamental state of the system* and the problem is called the *fundamental state problem.* Physicists traditionally use Monte Carlo type heuristics to determine approximate solutions for this problem, even in the case where the mesh is square (planar). As is shown in the following, this problem can be reduced to the maximum cut problem.

We can associate with a spin glass system, a graph $G = (V, E)$ where the vertices correspond to the spins, and two vertices are linked by an edge if there is a link between the spins corresponding to the vertices. We associate the weight $\omega_{ij} = -J_{ij}$

with each link ij. Consequently, the fundamental state problem is equivalent to the program:

$$\min \left\{ H(S) = \sum_{ij \in E} \omega_{ij} S_i S_j \ : \ S_i \in \{-1, 1\}, \ \forall \, i \in V \right\} \qquad [10.58]$$

So, the problem is to determine an assignment of $+1$ and -1 to the vertices of the graph in such a way that $\sum_{ij \in E} \omega_{ij} S_i S_j$ is minimum.

Let S be an assignment of $+1$ and -1 to the vertices of V. Let $V_+ = \{i \in V \ : \ S_i = +1\}$ and $V_- = \{i \in V \ : \ S_i = -1\}$. So:

$$\begin{aligned} H(S) &= \textstyle\sum_{ij \in E} \omega_{ij} S_i S_j = \sum_{i,j \in V_+} \omega_{ij} + \sum_{i,j \in V_-} \omega_{ij} - \sum_{i \in V_+, j \in V_-} \omega_{ij} \\ &= \textstyle\sum_{ij \in E} \omega_{ij} - 2 \sum_{ij \in \delta(V_+)} \omega_{ij}. \end{aligned}$$

Since $\sum_{ij \in E} \omega_{ij}$ is a constant, minimizing $H(S)$ is, as a consequence, equivalent to maximizing $\sum_{ij \in \delta(V_+)} w_{ij}$, the weight of the cut induced by V_+. Therefore, the fundamental state problem reduces to the maximum cut problem in G.

The maximum cut problem was one of the first problems shown to be **NP**-complete. It has been shown that it can be solved in polynomial time in planar graphs [HAD 75] and in graphs that are not contractible to K_5 [BAR 83], that is graphs that cannot be reduced to K_5 (the complete graph over 5 vertices) by contracting and removing edges. In the following, we discuss a polyhedral approach to this problem. Most of the results presented in this section are taken from Barahona and Mahjoub [BAR 86].

10.7.2. *The cut polytope*

Let $G = (V, E)$ be a graph and $\omega \in \mathbb{R}^E$ be a weight vector associated with the edges of G. Let us denote by $P_c(G)$ *the cut polytope of* G, that is the convex hull of the incidence vectors of the cuts of G. The maximum cut problem in G is therefore equivalent to the linear program $\max\{\omega x \ : \ x \in P_c(G)\}$. The polytope $P_c(G)$ is full-dimensional [BAR 85].

Let C be a cycle of G and $F \subseteq C$ such that $|F|$ is odd. Let $\delta(W)$ be a cut of G. Since $\delta(W)$ intersects C in an even number of edges, if $F \subset \delta(W)$, then $\delta(W) \cap (C \backslash F) \neq \emptyset$. Consequently, the following constraints are valid for the polytope $P_c(G)$:

$$x(F) - x(C \setminus F) \leq |F| - 1 \ \ \forall \, C \text{ cycle of } G, \ F \subseteq C, \ |F| \text{ odd} \qquad [10.59]$$

$$0 \leq x(e) \leq 1 \quad \forall\, e \in E \qquad [10.60]$$

It is not hard to see that every integer solution of the system above represents a cut of G. Consequently , these constraints induce an integer formulation of the maximum cut problem. Constraints [10.59] are called *cycle inequalities*. Given a cycle C, a *chord* of C is an edge whose two extremities are in C and are not consecutive when running through C. Theorem 10.28 gives the necessary and sufficient conditions for constraints [10.59] and [10.60] to define facets of $P_c(G)$.

THEOREM 10.28.–

1) An inequality [10.59] defines a facet of $P_c(G)$ if and only if C does not have a chord.

2) An inequality $x(e) \geq 0$ ($x(e) \leq 1$) defines a facet of $P_c(G)$ if and only if e does not belong to a triangle.

Since constraints [10.59] and [10.60] formulate the maximum cut problem as an integer program, and, from theorem 10.28, can define facets, it would be useful to have a polynomial separation algorithm for these constraints. This would allow us to use them efficiently in the context of a cutting-plane method for the problem. It is clear that constraints [10.60] can be separated in polynomial time. In what follows, we show that constraints [10.59] can also be separated in polynomial time.

By changing the variables from $x(e)$ to $1-x(e)$, constraints [10.59] can be written as:

$$\sum_{x \in C \setminus F} x(e) + \sum_{e \in F} (1 - x(e)) \geq 1, \ \forall\, C \text{ cycle of } G,\ F \subseteq C,\ |F| \text{ odd} \qquad [10.61]$$

If $\hat{x} \in \mathbb{R}^n_+$, the separation problem of constraints [10.59] with respect to $\hat{x}$ reduces to checking whether for every C, by associating a weight $1 - \hat{x}(e)$ with an odd number of edges of C and a weight $\hat{x}(e)$ with the other edges of C, the total weight of C is greater than or equal to 1. To solve this problem, we will consider an auxiliary graph.

Let $G' = (V', E')$ be the graph obtained from G in the following way. For every vertex i of G, we consider two vertices i' and i'' in G'. For every edge ij of G, we consider the edges $i'j'$ and $i''j''$ with a weight $x(ij)$ and the edges $i'j''$ and $i''j'$ with a weight $1 - x(ij)$. As we will see, the separation problem of constraints [10.61] reduces to determining a shortest path in G' between two vertices i' and i''. Let us denote by E_{ij} the set of edges $\{i'j', i'j'', i''j', i''j''\}$ for $ij \in E$. Observe that every path in G' between two vertices i' and i'', which uses at most one edge from each set E_{ij}, corresponds to a cycle in G going through the vertex i. Denote by V_1' (resp. V_2') the set of vertices i' (resp. i'') for i in V. Note that an edge e in G' has a weight

$1 - \hat{x}(e)$ if and only if e is between V_1' and V_2'. Let Λ be a path of G' between a vertex i' and a vertex i''. Λ then uses an odd number of edges e with a weight $1 - \hat{x}(c)$. Furthermore, if the weight of Λ is strictly less than 1, then Λ cannot use two edges of types $i'j''$ and $i''j'$. If Λ goes through two edges of type $i'i''$ and $j'j''$, then there must exist a path $\Lambda' \subset \Lambda$ linking either i' and i'' or j' and j'', and using only one single edge among $i'j''$ and $i''j'$. This implies that if Λ is a shortest path between two vertices i' and i'' having a weight strictly less than 1, then Λ can be chosen in such a way that it intersects each E_{ij} in at most one edge, and, consequently, it corresponds to a cycle C in G. By considering F as being the set of the edges e of Λ having a weight $1 - \hat{x}(e)$, then C and F in this case induce a constraint of type [10.61] violated by $\hat{x}$. Furthermore, it is easy to see that every pair C, F, where C is a cycle of G and F is an odd subset of C, that induces a violated constraint of the type [10.61], corresponds to a path between two vertices i' and i'' of G' using at most one edge of each E_{ij} and of weight strictly less than 1.

Consequently, to separate constraints [10.59], we calculate in G' a shortest path between each pair of vertices i', i'', and consider the shortest among all these paths. If the weight of this latter is greater than or equal to 1, then all the constraints [10.59] are satisfied by $\hat{x}$. Otherwise, a violated constraint has been found. Since the weights in G' are all positive, calculating a shortest path between two vertices can be done in $O(n^2)$ (where $n = |V|$). The separation of constraints [10.59] can therefore be done in $O(n^3)$.

Since constraints [10.59] and [10.60] can be separated in polynomial time, the maximum cut problem can therefore be solved in polynomial time for graphs whose cuts polytope is given by these constraints. Theorem 10.29 characterizes these graphs.

THEOREM 10.29.– *Constraints [10.59] and [10.60] completely describe the polytope $P_c(G)$ if and only if G is not contractible to K_5.*

From theorem 10.29, the maximum cut problem can be solved in polynomial time by a cutting-plane algorithm in graphs that are not contractible to K_5. Since planar graphs are part of this class, this theorem also implies that the maximum cut problem can be solved in polynomial time by a cutting-plane algorithm in these graphs.

A graph is called a *bicycle p-wheel* if it consists of a cycle of length p and of two vertices adjacent between themselves and adjacent to every vertex of the cycle.

THEOREM 10.30.– *Let (W, T) be a bicycle $(2k+1)$-wheel, $k \geq 1$, contained in G. Then the inequality:*

$$x(T) \leq 2(2k+1) \qquad [10.62]$$

defines a facet of $P_c(G)$.

Gerards [GER 85] has shown that the separation problem of constraints [10.62] can be reduced to a polynomial sequence of shortest path problems and can therefore be solved in polynomial time.

THEOREM 10.31.– *Let $K_p = (W, T)$ be a complete subgraph of G of order p. So the inequality:*

$$x(T) \leq \lfloor \frac{p}{2} \rfloor \lceil \frac{p}{2} \rceil \qquad [10.63]$$

is valid for $P_c(G)$. Furthermore, it defines a facet of $P_c(G)$ if and only if p is odd.

Inequalities [10.63] can be separated in polynomial time if p is fixed.

Branch-and-cut algorithms based on these classes of facets (and on other families of valid constraints) have been developed to solve planar and non-planar instances of the spin glass fundamental state problem [BAR 88, JÜN 98, LIE 03, SIM 95]. This approach has proved to be the most effective for this problem.

10.8. The survivable network design problem

With the introduction of optical technology, the telecommunications field has seen substantial evolution in recent years. Indeed, fiber-optics offer large transmission capacity, and thus allow the transfer of great quantities of information. Therefore, current networks tend to have a sparse topology (almost a tree). However, the failure of one or more links (or nodes) of a telecommunications network can have catastrophic consequences if the network is not in a position to be able to provide rerouting paths. So, designing a sufficiently survivable network, that is one which can continue to work in the event of a failure, has today become one of the objectives of telecommunications operators.

Survivability is generally expressed in terms of connectivity in the network. We ask that between each pair of nodes, depending on their importance in the network, a minimum number of disjoint paths exist, in such a way that in the case of failure, there is always at least one path that allows traffic to flow between the nodes. If the design of each link in the network carries a certain cost, the problem that is then posed is of designing a network whose topology satisfies the survivability conditions and is of minimum cost.

This problem has attracted much attention in recent years. Several methods of solving the problem have been developed, in particular polyhedral techniques. These have been effective for optimally solving instances of great size [GRÖ 92a, GRÖ 92b, KER 04]. In what follows, we discuss these techniques for certain variants of the problem. First of all we give a formulation of the problem in terms of graphs.

10.8.1. *Formulation and associated polyhedron*

To model survivability in networks, Grötschel and Monma [GRÖ 90] have introduced the idea of type of connectivity of vertices. If $G = (V, E)$ is the graph that represents the nodes and all the possible links between the nodes, we assume that a positive integer $r(s)$ is associated with each vertex $v \in V$. This integer is called *the connectivity type* of s and it represents the minimum number of links that must link the vertex s to the end network. We say that a subgraph satisfies the *conditions of edge survivability* (resp. *node survivability*) if for every pair of nodes s, t of V, there exist at least:

$$r(s,t) \;=\; \min\{r(s), r(t)\}$$

edge (or node) disjoint paths between s and t. A subgraph that satisfies the conditions of edge survivability (or node survivability) is said to be an *edge-survivable subgraph* (or *node-survivable subgraph*). If each edge e of the graph has a cost $c(e)$, the *edge-survivable network design problem* (or *node-survivable network design problem*) is to determine an edge-survivable (or node-survivable) subgraph whose edges' total cost is minimum. We will denote this problem by ESNDP (or NSNDP).

Several combinatorial problems, well known in the literature, are special cases of this general model:

– If the connectivity type is equal to 1 for two given vertices s and t of the graph and 0 for the others, the problem is nothing but the shortest path problem between s and t.

– If the connectivity type is 1 for a subset of vertices (said to be *terminals*) and 0 for the others, the problem is nothing but the Steiner tree problem, and if the connectivity types are all 1, we have the spanning tree problem.

– If for each vertex the connectivity type is equal to k, for fixed k, then it is the k-edge (node)-connected subgraph.

As some of these problems are **NP**-hard, the ESNDP (NSNDP) problem is also **NP**-hard.

In the following, we assume that there are at least two vertices in the graph with a maximum connectivity type. The problem can always be reduced to this case. We also restrict ourselves to the edge-survivability case. Most of the results concerning this problem can easily be extended to the node-survivability case.

For $W \subseteq V$, let:

$$\begin{aligned} r(W) &= \max\{r(s) : s \in W\} \\ con(W) &= \min\{r(W), r(V \setminus W)\} \end{aligned}$$

According to Menger's theorem 10.10, the ESNDP is equivalent to the following integer program:

$$\min \sum_{e \in E} c(e)x(e)$$

$$0 \leq x(e) \leq 1, \quad \forall\, e \in E \qquad [10.64]$$

$$x(\delta(W)) \geq con(W) \quad \forall\, W \subseteq V, \emptyset \neq W \neq V \qquad [10.65]$$

$$x(e) \in \{0, 1\} \quad \forall\, e \in E \qquad [10.66]$$

Constraints [10.65] are called *cut inequalities.*

Let $G = (V, E)$ be a graph and $r = (r(v),\ v \in V)$ be a vector of connectivity types. Let:

$$\text{ESNDP}(G) = conv\{x \in \mathbb{R}^E\ :\ x \text{ satisfies } [10.64] - [10.66]\}$$

be the convex hull of the solutions of the ESNDP problem. ESNDP(G) is called *edge-survivable subgraph polytope.*

Given a graph $G = (V, E)$ and a vector $r = (r(v),\ v \in V)$, we say that an edge $e \in E$ is *essential* if the graph $G - e$ is not edge-survivable. The essential edges must therefore belong to every solution of the problem. If E^* denotes the set of essential edges in G, then we have the following result.

PROPOSITION 10.20.– (Grötschel, Monma [GRÖ 90]). *The dimension of ESNDP(G) is equal to $|E| - |E^*|$.*

10.8.2. *Valid inequalities and separation*

To lighten the presentation, we consider the *weak survivability* case, that is $r(v) \in \{0, 1, 2\}$ for each vertex $v \in V$. Most of the results established in this case can easily be extended to the general case. Furthermore, given that the survivability conditions in this case induce a topology that has shown itself to be sufficiently effective in practice, this variant has been intensively investigated.

A first class of valid inequalities for the polytope ESNDP(G) is that given by the cut inequalities [10.65]. The separation problem for these constraints is equivalent to the minimum cut problem with positive weights on the edges, and can therefore be solved in polynomial time. Given a solution $\hat{x} \in \mathbb{R}^E$, we associate with each edge e a capacity $\hat{x}(e)$, and we calculate the maximum flow between each pair of vertices. From the max-flow min-cut theorem, if for a pair s, t, the flow between s and t is strictly less than $\min\{r(s), r(t)\}$, then the constraint [10.65] induced by the minimum cut between s and t is violated by $\hat{x}$. This separation can be implemented in $O(n^4)$ using the Gomory–Hu tree [GOM 61].

10.8.2.1. *Multicut inequalities*

Let $(V_1, \ldots, V_p)$ be a partition of V, that is $V_i \neq \emptyset$ for $i = 1, \ldots, p$, $V_i \cap V_j = \emptyset$ for every $i, j \in \{1, \ldots, p\}$ $i \neq j$, and $V_1 \cup \ldots \cup V_p = V$. If $con(V_i) = 1$ for $i = 1, \ldots, p$, the graph obtained from a solution of the ESNDP by contracting the subgraphs induced by V_i, $i = 1, \ldots, p$ must be connected. Consequently, the following constraints are valid for the polytope ESNDP(G):

$$x(\delta(V_1, \ldots, V_p)) \geq p - 1, \quad \text{for every partition } (V_1, \ldots, V_p) \text{ such that } con(V_i) = 1,\ i = 1, \ldots, p \qquad [10.67]$$

where $\delta(V_1, \ldots, V_p)$ is the set of edges between the elements of the partition. Inequalities [10.67] are called *multicut inequalities.* In [GRÖ 90], Grötschel and Monma have shown that inequalities [10.67] and inequalities [10.64] are sufficient to describe the polytope ESNDP(G) when $r(v) = 1$ for every vertex $v \in V$. Nash-Williams [NAS 61] has shown that these inequalities, with the positivity constraints $x(e) \geq 0$ for every $e \in E$, characterize the dominant of ESNDP(G) in this case.

Cunningham [CUN 85] has shown that if $r(v) = 1$ for every $v \in V$, the separation problem of constraints [10.67] can be reduced to a sequence of $|E|$ minimum cut problems, and can therefore be solved in polynomial time. In [BAR 92], Barahona has shown that the problem can be reduced to $|V|$ minimum cut problems, and therefore can be solved in $O(n^4)$.

10.8.2.2. *Partition inequalities*

Grötschel *et al.* [GRÖ 92b] have introduced a class of inequalities called partition inequalities, which generalize the cut constraints [10.65]. These inequalities can be presented as follows. Let $(V_1, \ldots, V_p)$, $p \geq 3$ be a partition of V such that $1 \leq con(V_i) \leq 2$ for $i = 1, \ldots, p$. Let $I_2 = \{i : con(V_i) = 2,\ i = 1, \ldots, p\}$. *The partition inequality* induced by $(V_1, \ldots, V_p)$ is given by:

$$x(\delta(V_1, \ldots, V_p)) \geq \begin{cases} p - 1 & \text{if } I_2 = \emptyset \\ p & \text{if not} \end{cases} \qquad [10.68]$$

Note that if $p = 2$, constraint [10.68] is nothing but a cut constraint [10.65]. So, if the connectivity types are all equal to 2, constraint [10.68] is redundant with respect to the cut constraints $x(\delta(V_i)) \geq 2$, $i = 1, \ldots, p$.

The separation problem of constraints [10.68] is generally **NP**-hard. Kerivin and Mahjoub [KER 02] have shown that, if $r \in \{1, 2\}^V$, the separation problem of the constraints:

$$x(\delta(V_1, \ldots, V_p)) \geq p \quad \text{if } I_2 \neq \emptyset \qquad [10.69]$$

where $(V_1, \ldots, V_p)$ is a partition of V, can be reduced to the minimization of a submodular function, and can therefore be solved in polynomial time (a function $f : 2^S \longrightarrow \mathbb{R}$, where S is a finite set, is said to be *submodular* if $f(A \cup B) + f(A \cap B) \leq f(A) + f(B)$ for all subsets A and B of S). Recently, Barahona and Kerivin [BAR 04] have given a polynomial combinatorial algorithm to separate constraints [10.69] in this case. As a consequence of the complexities of the separation problems of [10.67] and [10.69], the separastion problem of constraints [10.68] is polynomial if $r \in \{1, 2\}^V$.

10.8.2.3. *F-partition inequalities*

Let us now assume that the connectivity types are all equal to 2, that is $r(v) = 2$ for every $v \in V$. A class of valid inequalities for the ESNDP(G) in this case has been introduced by Mahjoub [MAH 94] as follows. Consider a partition $(V_0, V_1, \ldots, V_p)$ of V and let $F \subseteq \delta(V_0)$ be of odd cardinality. By adding up the following valid inequalities:

$$x(\delta(V_i)) \geq 2, \quad \forall\, i = 1, \ldots, p$$
$$-x(e) \geq -1, \quad \forall\, e \in F$$
$$x(e) \geq 0, \quad \forall\, e \in \delta(V_1) \setminus F$$

we obtain:

$$2x(\Delta) \geq 2p - |F|$$

where $\Delta = \delta(V_0, \ldots, V_p) \setminus F$. By dividing by 2 and rounding up the second member to the next integer, we obtain:

$$x(\Delta) \geq p - \lfloor \frac{|F|}{2} \rfloor. \qquad [10.70]$$

Inequalities [10.70] are called *F-partition inequalities*. Note that these inequalities are of Chvátal rank 1 with respect to the system given by the trivial constraints and the cut constraints. Let us also note that if $|F|$ is even, inequality [10.70] can be obtained from the trivial inequalities and the cut inequalities.

Inequalities [10.70] are a special case of a more general class of valid inequalities given by Grötschel *et al.* [GRÖ 92b] for the ESNDP(G). Kerivin *et al.* [KER 04] have considered a subclass of F-partition constraints called *odd wheel inequalities* and have given sufficient conditions for these to define facets. They have also extended these inequalities for the case where the connectivity types are 1 or 2 for each vertex of the graph.

The separation problem for the F-partition constraints is still an open problem. However, if F is fixed, as has been shown by Baïou *et al.* [BAÏ 00], the problem can be solved in polynomial time.

10.8.3. *A branch-and-cut algorithm*

By using the results given above, Kerivin *et al.* [KER 04] have developed a branch-and-cut algorithm for the ESNDP when the connectivity types are 1 or 2 for each vertex of the graph. This problem, which we will denote by $(1,2)$-ESNDP, has important practical applications as has been mentioned in [MON 89].

The algorithm starts by solving the linear program reduced to trivial constraints and cut constraints associated with the vertices of the graph, that is:

$$\begin{aligned} &\text{Min } cx \\ &x(\delta(v)) \geq r(v), \quad \forall\, v \in V \\ &0 \leq x(e) \leq 1, \quad \forall\, e \in E \end{aligned} \qquad [10.71]$$

If the optimal solution, say $\bar{x}$, of this program is a solution of the $(1,2)$-ESNDP, that is it is integral and satisfies the cut constraints, then it is optimal. In general, the solution $\bar{x}$ is not feasible for the $(1,2)$-ESNDP, and, consequently, with each iteration of the branch-and-cut algorithm, it is necessary to generate valid constraints for the $(1,2)$-ESNDP that are violated by $\bar{x}$. These constraints are generated among the cut constraints [10.65], the multicut constraints [10.67], the partition constraints [10.68], the F-partition constraints [10.70] and the following constraints, which are a direct extension of constraints [10.70] when $r(v) \in \{1,2\}$ for every $v \in V$:

$$x(\delta(V_0,\ldots,V_p)) \geq p - \lfloor \frac{p_1 + |F|}{2} \rfloor \qquad [10.72]$$

where p_1 is equal to the number of elements, with the exception of V_0, of the partition $(V_0,\ldots,V_p)$ having a connectivity type equal to 1.

The separation procedures used in the branch-and-cut algorithm are either exact or heuristic depending on the class of inequalities. In certain cases, separation can be done in polynomial time, but the use of a heuristic can be a lot more effective in accelerating the separation process, and thus the resolution of the problem.

The separation of the cut inequalities is done by calculating the Gomory–Hu tree using Goldberg and Tarjan's max-flow algorithm [GOL 88]. For separating the partition constraints [10.68], Kerivin *et al.* have developed a heuristic where they consider two cases, depending on the right-hand side of the inequality. Their heuristic is based on Barahona's algorithm [BAR 92] for multicut constraints.

For separating the F-partition inequalities [10.70] and [10.72], Kerivin *et al.* propose two heuristics [KER 04]. The first is based on the concept of critical extreme points introduced by Fonlupt and Mahjoub [FON 99]. The procedure consists of applying some graph reduction operations, and of looking for odd cycles formed by edges whose value is fractional. If such a cycle is found, then an F-partition constraint violated by $\bar{x}$ is detected. The second heuristic allows us to transform cuts, with as many edges $e \in E$ with $\bar{x}(e) = 1$ as possible, into F-partitions. This is done by calculating a Gomory–Hu tree with respect to the weights $(1 - \bar{x}(e),\ e \in E)$. If $\delta(W)$ is a cut of the Gomory–Hu tree, the F-partition constraint is therefore generated by considering the partition given by W and the vertices in $V \setminus W$, and by choosing a set of edges $F \subseteq \delta(W)$ (the same process can be applied for the partition induced by $V \setminus W$ and the vertices in W).

The numerical results presented in [KER 04], concerning uniform instances where $r(v) = 2$ for every $v \in V$ having up to 417 vertices and instances with $r \in \{1,2\}^V$ having up to 101 vertices, show that these different classes of inequalities are useful for solving the problem optimally. In particular, in the case where $r(v) = 2$ for every $v \in V$, the F-partition inequalities seem to play a decisive role in establishing the optimal solution. They allow us to improve the bound at the root of the branching tree, and to thus decrease its size. When $r \in \{1,2\}^V$, the partition constraints [10.68] seem to be the most important deciding factor in solving the problem. Like the F-partitions constraints for the uniform case, these constraints allow us to significantly reduce the bound at the root, and to considerably reduce the size of the branching tree.

10.9. Conclusion

In this chapter we have discussed polyhedral approaches in combinatorial optimization. We have closely examined polyhedron descriptions in terms of facets and in terms of extreme points and their implementations in the context of a branch-and-cut technique. As has been highlighted, these approaches have been shown to be powerful in exactly solving difficult combinatorial optimization problems. As examples, we have presented applications to spin glass in statistical physics and to survivable network design problems, for which these approaches are particularly effective.

For certain problems, it is possible that a branch-and-cut algorithm does not give an optimal solution even after an enormous computation time. It is then useful in such cases to have an approximate solution with a relatively very small error. For this, it is useful to be able to compute, in each iteration of the algorithm, a feasible solution of the problem. This gives (in the maximization case) a lower bound on the optimal value. This bound with that given by the linear relaxation allows us to calculate the relative error of the current solution. One of the most commonly used techniques for calculating feasible solutions is that known as *random rounding*. The method consists of transforming a fractional solution into a feasible integer solution by randomly changing the fractional components of the solution into integer values. Other heuristic and metaheuristic methods can also be used to calculate feasible solutions.

In some cases, the constraints of the problem being studied can be given in an explicit way, while the number of variables can be very large (exponential). For example, this is the case in multiflow problems where the constraints are the flow conservation equalities and the capacity constraints, and the variables represent all the paths in the network. In this type of situation, polyhedral approaches are generally combined with column generation methods [WOL 98]. These consist of solving a sequence of problems each having a reduced number of variables.

Finally, polyhedral approaches can also be used in the context of *lift-and-project* methods. These techniques, introduced by Balas *et al.* [BAL 96], consist of considering a formulation of the problem (and a linear description of the associated polyhedron) in a higher dimensional space, and of using cutting-plane and projection techniques to solve the problem.

10.10. Bibliography

[APP 98] APPLEGATE D., BIXBY R., CHVÁTAL V., COOK W., "On the solution of traveling salesman problems", *Proceedings of the International Congress of Mathematicians, Berlin 1998-Volume III: Invited Lectures, Documenta Mathematica Extra Volume ICM 1998 III*, Fischer G., Rehmann U., p. 645–656, 1998.

[BAÏ 00] BAÏOU M., BARAHONA F., MAHJOUB A.R., "Separation of partition inequalities", *Mathematics of Operations Research*, vol. 25, p. 243–254, 2000.

[BAL 76] BALAS E., PADBERG M.W., "Set partitioning: a survey", *SIAM Review*, vol. 18, p. 710–760, 1976.

[BAL 96] BALAS E., CERIA S., CORNUÉJOLS G., "Mixed $0-1$ programming by lift-and-project in a branch-and-cut framework", *Management Science*, vol. 42, p. 1229–1246, 1996.

[BAR 83] BARAHONA F., "The max-cut problem on graphs not contractible to K_5", *Operations Research Letters*, vol. 2, p. 107–111, 1983.

[BAR 85] BARAHONA F., GRÖTSCHEL M., MAHJOUB A.R., "Facets of the bipartite subgraph polytope", *Mathematics of Operations Research*, vol. 10, p. 340–358, 1985.

[BAR 86] BARAHONA F., MAHJOUB A.R., "On the cut polytope", *Mathematical Programming*, vol. 36, p. 157–173, 1986.

[BAR 88] BARAHONA F., GRÖTSCHEL M., JÜNGER M., REINELT G., "An application of combinatorial optimization to statistical physics and circuit layout design", *Operations Research*, vol. 36, p. 493–513, 1988.

[BAR 92] BARAHONA F., "Separating from the dominant of the spanning tree polytope", *Operations Research Letters*, vol. 12, p. 201–203, 1992.

[BAR 04] BARAHONA F., KERIVIN H., "Separation of partition inequalities with terminals", *Discrete Optimization*, vol. 1, no. 2, p. 129–140, 2004. (Also available from IBM as research report NRC22984), 2003.

[BER 58] BERGE C., "Sur le couplage maximum d'un graphe", *Comptes Rendus de l'Académie des Sciences Paris, séries 1, Mathématiques*, vol. 247, p. 258–259, 1958.

[CHV 73] CHVÁTAL V., "Edmonds polytopes and a hierarchy of combinatorial problems", *Discrete Mathematics*, vol. 4, p. 305–337, 1973.

[COO 98] COOK W., CUNNINGHAM W., PULLEYBLANCK W.R., SCHRIJVER A., *Combinatorial Optimization*, Wiley, 1998.

[COR 01] CORNUÉJOLS G., *Combinatorial Optimization, Packing and Covering*, Society for Industrial and Applied Mathematics, Philadelphia, PA

[CUN 78] CUNNINGHAM W.H., MARSH A.B., "A primal algorithm for optimum matching", *Mathematical Programming Study*, vol. 8, p. 50–72, 1978.

[CUN 85] CUNNINGHAM W.H., "Optimal attack and reinforcement of a network", *Journal of the ACM*, vol. 32, p. 185–192, 1985.

[DAN 51] DANTZIG G.B., "Maximization of a linear function of variables subject to linear inequalities", *Activity Analysis of Production and Allocation. Proceedings of the Conference on Linear Programming*, Chicago, IL, p. 359–373, 1951.

[DAN 63] DANTZIG G.B., *Linear Programming and Extensions*, Princeton University Press, Princeton, NJ, 1963.

[DEZ 97] DEZA M., LAURENT M., *Geometry of Cuts and Metrics*, Springer, Berlin, 1997.

[DIJ 59] DIJKSTRA E., "A note on two problems in connexion with graphs", *Numerisches Mathematik*, vol. 1, p. 269–271, 1959.

[EDM 65] EDMONDS J., "Maximum matching and a polyhedron with $0-1$ vertices", *Journal of Research of the National Bureau of Standards*, vol. 69, num. B, p. 125–130, 1965.

[EDM 77] EDMONDS J., GILES R., "A min-max relation for submodular functions on graphs", *Annals of Discrete Mathematics*, vol. 1, p. 185–204, 1977.

[FON 99] FONLUPT J., MAHJOUB A.R., "Critical extreme points of the 2-edge connected spanning subgraph polytope", *Proceedings IPCO'99*, LNCS 1610, p. 166–183, 1999.

[FOR 56] FORD L.R., FULKERSON D.R., "Maximal flow through a network", *Canadian Journal of Mathematics*, vol. 8, p. 399–404, 1956.

[FOR 62] FORD L. R., FULKERSON D.R., *Flows in Networks*, Princeton University Press, Princeton, NJ, 1962.

[FUL 71] FULKERSON D.R., "Blocking and anti-blocking of polyhedra", *Mathematical Programming*, vol. 1, p. 168–194, 1971.

[FUL 72] FULKERSON D.R., "Anti-blocking polyhedra", *Journal of Combinatorial Theory Series B*, vol. 12, p. 50–71, 1972.

[GER 85] GERARDS A.M.H., "Testing the odd-bicycle wheel inequalities for the bipartite subgraph polytope", *Mathematics of Operations Research*, vol. 10, p. 359–360, 1985.

[GIL 79] GILES R., PULLEYBLANK W.R., "Total dual integrality and integer polyhedra", *Linear Algebra and Its Applications*, vol. 25, p. 191–196, 1979.

[GOL 88] GOLDBERG A.V., TARJAN R.E., "A new approach to the maximum-flow problem", *Journal of the ACM*, vol. 35, p. 921–940, 1988.

[GOM 58] GOMORY R.E., "Outline of an Algorithm for Integer Solutions to Linear Programs", *Bulletin of the American Mathematical Society*, vol. 64, p. 275–278, 1958.

[GOM 60] GOMORY R.E., "Solving linear programming problems in integers", *Combinatorial Analysis*, p. 211–216, R.E. BELLMAN and M. HALL, JR., eds., American Mathematical Society, 1960.

[GOM 61] GOMORY R.E., HU T.C., "Multi-terminal network flows", *Journal of the Society for Industrial and Applied Mathematics*, vol. 9, p. 551–570, 1961.

[GRA 95] GRAHAM R.L., GRÖTSCHEL M., LOVÁSZ L., *Handbook of Combinatorics, Volume II, Part IV: Applications*, Elsevier, North Holland, The MIT Press, Cambridge, MA, 1995.

[GRÖ 81] GRÖTSCHEL M., LOVÁSZ L., SCHRIJVER A., "The ellipsoid method and its consequences in combinatorial optimization", *Combinatorica*, vol. 1, p. 70–89, 1981.

[GRÖ 88] GRÖTSCHEL M., LOVÁSZ L., SCHRIJVER A., *Geometric Algorithms and Combinatorial Optimization*, Springer, Berlin, 1988.

[GRÖ 90] GRÖTSCHEL M., MONMA C.L., "Integer polyhedra associated with certain network design problem with connectivity constraints", *SIAM Journal on Discrete Mathematics*, vol. 3, p. 502–523, 1990.

[GRÖ 91] GRÖTSCHEL M., HOLLAND O., "Solution of large-scale symmetric travelling salesman problems", *Mathematical Programming*, vol. 51, p. 141–202, 1991.

[GRÖ 92a] GRÖTSCHEL M., MONMA C.L., STOER M., "Computational results with a cutting plane algorithm for designing communication networks with low-connectivity constraints", *Operations Research*, vol. 40, p. 309–330, 1992.

[GRÖ 92b] GRÖTSCHEL M., MONMA C.L., STOER M., "Facets for polyhedra arising in the design of communication networks with low-connectivity constraints", *SIAM Journal on Optimization*, vol. 2, num. 3, p. 474–504, 1992.

[GRÖ 95] GRÖTSCHEL M., MONMA C.L., STOER M., "Design of survivable networks", *Handbook in Operations Research and Management Science*, vol. 7, p. 617–671, M.O. BALL *et al.* eds., North-Holland, 1995.

[HAD 75] HADLOK M., "Finding a maximum cut of a planar graph in polynomial time", *SIAM Journal on Computing*, vol. 4, p. 221–225, 1975.

[HOF 56] HOFFMAN A.J., KRUSKAL J., "Integer boundary points of convex polyhedra", *Linear Inequalities and Related Systems*, p. 223–246, H.W. KUHN and A.W. TUCKER, eds., Princeton University Press, Princeton, NJ, 1956.

[JÜN 98] JÜNGER M., RINALDI G., "Relaxation of the max-cut problem and computation of spin glass ground states", *Operations Research Proceedings*, p. 74–83, P. KISCHKA, H. W. LORENZ, U. DERIGS, W. DOMSCHKE, P. KLEINSCHMIDH and R. MÖRING, eds., Springer, Berlin, 1998.

[KER 02] KERIVIN H., MAHJOUB A.R., "Separation of the partition inequalities for the (1,2)-survivable network design problem", *Operations Research Letters*, vol. 30, p. 265–268, 2002.

[KER 04] KERIVIN H., MAHJOUB A.R., NOCQ C., "(1,2)-Survivable networks: facets and branch-and-cut", *The Sharpest Cut: The Impact of Manfred Padberg and his work*, p. 121–152, MPS-SIAM, 2004.

[KER 05] KERIVIN H., MAHJOUB A.R., "Design of survivable networks: a survey", *Network*, vol. 46, p. 1–21, 2005.

[KHA 79] KHACHIAN L.G., "A polynomial algorithm in linear programming", *Soviet Mathematics Doklady*, p. 191–194, 1979.

[KÖN 31] KÖNIG D., "Graphok és matrixok", *Matematikai és Fizikai Lapok*, vol. 38, p. 116–119, 1931.

[LAW 85] LAWLER E.L., LENSTRA J.K., RINNOOY KAN A.H.G., SHMOYS D.B., *The traveling Salesman Problem – A Guided Tour of Combinatorial Optimization*, Wiley, 1985.

[LIE 03] LIERS F., JÜNGER M., REINELT G., RINALDI G., "Computing exact ground states of hard ising spin glass by branch-and-cut", *Preprint*, 2003.

[LOV 79] LOVÁSZ L., "Graph theory and integer programming", *Discrete Optimization I, Annals of Discrete Mathematics 4*, p. 146–158, P.L. HAMMER, E.L. JOHNSON and B.H. KORTE, eds., 1979.

[MAH 94] MAHJOUB A.R., "Two-edge connected spanning subgraphs and polyhedra", *Mathematical Programming*, vol. 64, p. 199–208, 1994.

[MAR 90] MARTELLO S., TOTH P., *Knapsack Problems: Algorithms and Computer Implementation*, Wiley, 1990.

[MEN 27] MENGER K., "Zur Allgemeinen Kurventheorie", *Fundamanta Mathematicae*, vol. 10, p. 96–115, 1927.

[MIN 96] MINKOWSKI H., *Geometrie der Zahlen (Erste Lieferung)*, Teubner, Leipzig, 1896.

[MON 89] MONMA C.L., SHALLCROS D.F., "Methods for designing communication networks with certain two-connected survivability constraints", *Operations Research*, vol. 37, p. 531–541, 1989.

[NAS 61] NASH-WILLIAMS C.S.J., "Edge-disjoint spanning trees of finite graphs", *Journal of the London Mathematical Society*, vol. 36, p. 445–450, 1961.

[NEM 88] NEMHAUSER G.L., WOLSEY L.A., *Integer and Combinatorial Optimization*, Wiley, 1988.

[PAD 91] PADBERG M., RINALDI G., "A branch-and-cut algorithm for the resolution of large-scale symmetric traveling salesman problems", *SIAM Review*, vol. 33, p. 60–100, 1991.

[SCH 80] SCHRIJVER A., "On cutting planes", *Annals of Discrete Mathematics*, vol. 9, p. 291–296, 1980.

[SCH 03] SCHRIJVER A., *Combinatorial Optimization, Polyhedra and Efficiency*, Springer, Berlin, 2003.

[SIM 95] SIMONE C.D., DIEHL M., JÜNGER M., MUTZEL P., REINELT G., RINALDI G., "Exact ground states in spin glasses: New experimental results with a branch-and-cut algorithm", *Journal of Statistical Physics*, vol. 80, p. 487–496, 1995.

[WEY 50] WEYL H., "Elementare Theorie der konvexen Polyeder", *Commentarii Mathematici Helvetici, 7 (1935) 290–306*, p. 3–18, Translated into English: "The elementary theory of convex polyhedra", in *Contributions to the Theory of Games I*, H.W. KUHN and A.W. TUCKER, eds., Princeton University Press, Princeton, NJ, 1950.

[WOL 98] WOLSEY L.A., *Integer Programming*, Wiley, 1998.

Chapter 11

Constraint Programming

11.1. Introduction

Constraint programming [MIL 04, TSA 93, VAN 89] is a problem-solving and optimization method that guarantees a clear separation between:

– a precise definition of the problem to be solved in terms of constraints to be satisfied and an objective function to be optimized;

– a set of operators that allow us to satisfy constraints by making decisions;

– a set of propagation procedures that allow us to establish the consequences of a decision on the decisions to come, and also on the objective function;

– a set of heuristics that allow us to choose and to order the various possible decisions at each stage of the solution.

A simple example of a disjunctive scheduling problem [CAR 88] allows us to demonstrate this separation. The definition of the problem includes:

– a set of operations of known duration to be ordered;

– a set of precedence constraints between operations, which are transformed into inequalities $\mathit{start}(op_i) + \mathit{duration}(op_i) \leq \mathit{start}(op_j)$, where $\mathit{start}(op_i)$ is a variable that refers to the start date of the operation op_i;

– a set of resource-sharing constraints between operations, expressed as "($\mathit{start}(op_i) + \mathit{duration}(op_i) \leq \mathit{start}(op_j)$) or ($\mathit{start}(op_j) + \mathit{duration}(op_j) \leq \mathit{start}(op_i)$)", each resource under consideration being unique and able to perform one operation at a time.

Chapter written by Claude LE PAPE.

The definition of the problem also includes an objective cost function to be minimized. This cost function can be extremely simple, for example the total duration of the schedule, that is the difference between $max_i(start(op_i) + duration(op_i))$ and $min_j(start(op_j))$, or a weighted sum of manufacturing delays in relation to fixed deadlines. It can also consist of a complex combination of several conflicting criteria (commitment of critical resources, meeting deadlines, reducing stocks, etc.).

Solving the problem involves a unique *decision operator*, which replaces a disjunctive resource-sharing constraint with one of the two inequalities making up the disjunction. With each application of this operator, a decision branch is created in such a way as to enable exploration of the two alternatives, "op_i before op_j" on the one hand, "op_j before op_i", on the other.

The *propagation* allows us to establish, before or after each decision branch, a set of inequalities to add in order to restrict the search space. Notably, if an inequality that appears in a disjunctive constraint cannot be part of a solution (or of a better solution than the best solution already known), the other inequality of the disjunction can be added. A disjunctive constraint stays disjunctive as long as its two inequalities remain possible. It is replaced by a single inequality if only one of the two inequalities is still possible. A failure and therefore a return to an anterior decision branch occur if its two inequalities become impossible [LEP 85a, LEP 85b, LEP 88].

Finally, a *heuristic* specifies how to choose, at each decision branch, the first disjunctive constraint to satisfy, and which of this constraint's two inequalities to try first.

A heuristic search procedure capable of providing an optimal ordering is derived from the four "ingredients" above. What is important is that the distinction made between these ingredients allows us to define them and to modify them separately, almost independently of each other. Many advantages come directly from this:

– Firstly, this break down often facilitates the addition of new types of constraints to an existing application. In the best case, a few lines of code are enough to add these additional constraints, which gives great *robustness* concerning possible evolutions of the problem under consideration.

– Being able to consider propagation independently of the rest of the system allows us to re-use previously defined propagation procedures, which have been tested in terms of efficiency, and carefully optimized. The use of shared libraries of constraints [PUG 92] – and of propagation procedures associated with these constraints – is a very important *efficiency* factor.

– Being able to consider decision-making operators and/or heuristics independently of the rest of the system allows us to test multiple combinations of them. As a consequence it is easy to combine various sets of operators and of heuristics, and to diversify the search for solutions and thus ensure a greater *robustness* concerning the numerical characteristics of the instances of the problem to be solved. It is also easy to distinguish different solution contexts necessitating the use of different solution strategies, for example for a global optimization on the one hand and for an incremental modification with bounded calculation times on the other.

– Finally, in many cases, constraint programming allows us to assimilate results, ideas or techniques from other domains of operations research: bounds and cuts deduction, local optimizations, metaheuristics.

Each of the following sections looks in detail at the definition of the problem (section 11.2), the choice of decision operators appropriate for a given problem (section 11.3), the propagation of the constraints (section 11.4) and the choice of one or several search heuristics (section 11.5), and how the assimilation of various techniques around a kernel of constraint programming allows us to attain the levels of efficiency and robustness necessary to implement industrial applications.

11.2. Problem definition

One of the most important advantages of constraint programming is its capacity to take into account extremely diverse types of variables and types of constraints.

Notably, constraint programming is not limited to using numerical variables. Let us consider, for example, an airline rostering problem. For each flight, this consists of selecting a group of persons (pilots, flight attendants) having well-defined characteristics. In constraint programming, this group can be represented by a single set variable, that is a variable whose value in a given solution is the set of persons assigned to the flight under consideration [PUG 91]. Different types of constraints between such variables can then be defined in an extremely compact way: the fact that two or several sets must be disjoint; that one set must be included in another; that the number of elements in a set that speak Portuguese must be greater than or equal to 2; etc. Extremely efficient propagation methods can then be associated with these constraints. For example, given the cardinalities and the possible elements of several disjoint sets, an assignment problem may be solved to verify that a solution exists that satisfies these constraints [REG 96].

Another interesting example is that of routing several communications in a telecommunications network [BER 02, CHA 04]. For each communication, a unique variable represents the path followed in the network. It is then easy to express constraints on the number of arcs or on the length in kilometers of this path. Actually, such a variable of type "path" can be implemented in the form of a set variable representing the set of arcs of this path. Let us consider, for example, a complete graph comprising n nodes N_1, N_2, ..., N_n, and $n(n-1)$ directed arcs between these nodes. A capacity $C(N_i, N_j)$ and a quantity of traffic to clear $Q(N_i, N_j)$ on a unique simple path (to be established) of length less than or equal to 3 arcs are given for each pair of nodes (N_i, N_j). Moreover, to facilitate rerouting part of the communications in the case of a failure, it is required that for every pair of nodes (N_i, N_j), the path taken from N_i to N_j and the path taken from N_j to N_i are not both direct. A classical modeling of this problem would introduce $O(n^4)$ Boolean variables X_{ijkl}, of value equal to 1 if and only if the path taken to clear the traffic between N_i and N_j goes through the arc (N_k, N_l). In constraint programming, it is possible to use a set variable $E(N_i, N_j)$ representing the set of arcs of the chosen path between N_i and N_j and an integer variable $L(N_i, N_j)$ representing the length by the number of arcs of this path. A specific constraint linking $E(N_i, N_j)$ and $L(N_i, N_j)$ specifies that the set $E(N_i, N_j)$ is a simple path of length $L(N_i, N_j)$ between N_i and N_j. The length limit is then expressed by reducing the domain of $L(N_i, N_j)$ to the interval $[1, 3]$. The constraint designed to facilitate rerouting is expressed as "$(N_i, N_j) \notin E(N_i, N_j)$ or $(N_j, N_i) \notin E(N_j, N_i)$". Finally, for every arc (N_k, N_l), a constraint $\Sigma_{(N_i,N_j)|(N_k,N_l)\in E(N_i,N_j)} Q(N_i, N_j) \leq C(N_k, N_l)$ allows us to force adherence to the capacity limit.

Because of the many existing modeling possibilities, it is extremely important in constraint programming for the model to be well defined. The best model in fact depends on many factors: compactness of the representation, elimination of symmetries, efficiency of propagation algorithms, heuristic guidance possibilities through the results of propagation. Appreciating all of these elements may require a certain amount of experience.

11.3. Decision operators

The most general decision operator obviously consists of assigning a value to a variable of the stated problem or, alternatively, eliminating this value from the domain of possible values of this variable. For example, if a variable V has four possible values 1, 2, 3 and 4, the decisions $V = 1$, $V = 2$, $V = 3$, $V = 4$, $V \neq 1$, $V \neq 2$, $V \neq 3$, $V \neq 4$, may be considered. This can be done in various ways:

– Create four branches corresponding to each of the equalities $V = 1$, $V = 2$, $V = 3$ and $V = 4$. The order of branch exploration is to be chosen heuristically. For example, if V has a negative coefficient assigned to it in an objective function to be minimized, the value 4 may be preferred. But if several variables V, W, X, Y and Z must take distinct values and 2 is not a possible value for W, X, Y and Z, then the value 2 may also be preferred for V since the decision $V = 2$ does not eliminate any possibilities for W, X, Y and Z.

– Choose one of these four values v, either because it is of greater interest from the point of view of the objective function, or because $V = v$ has, on the face of it, more chance of being part of a global solution of the stated problem. Two branches can then be created, the first one corresponding to the addition of $V = v$ and the second to the addition of $V \neq v$. The advantage of this type of branching is that the propagation of the inequality $V \neq v$ can provide useful information about other variables and even, as a result of the propagation, about the variable V itself. Let us assume, for example, a problem contains the constraints ($V = 4$ or $W = 2$), $X \leq 2W - 1$, $V < X$. The addition of $V \neq 4$ allows us to deduce $W = 2$, then $X \leq 3$, then $V < 3$, and therefore to eliminate the value 3 from the domain of V. Thus the branch $V = 3$ will never be explored.

– In a similar way, it is possible to choose a value v to avoid. The first branch then corresponds to the addition of $V \neq v$ and the second to the addition of $V = v$.

In many cases, however, it is neither necessary nor sensible to proceed by assigning and eliminating values. Let us consider, for example, the case of an integer variable that must take its value from an interval $[v_{min}, v_{max}]$. Let us assume, moreover, that this variable is used, with a positive coefficient, in a linear cost function to be minimized. In such circumstances, it seems on the face of it to be sensible to first try the minimal value v_{min}, that is to explore as a priority a subtree in which the variable is set to its minimal value v_{min}. In case of failure, or backtracking designed to optimize the global value of the cost function, the value v_{min} is eliminated from the domain of possible values of the variable, which leads us to try the new minimal value $v_{min} + 1$, and so on. From this ensues an exploration of very similar subtrees: a large part of the work accomplished to explore the subtree corresponding to the value v_{min} is repeated to explore the subtree corresponding to the value $v_{min} + 1$. It may be much more efficient to work using a dichotomy: create a subtree in which the domain of the variable is limited to $[v_{min}, (v_{min} + v_{max})/2[$, then a subtree in which it is limited to $[(v_{min} + v_{max})/2, v_{max}]$.

In the case of a set variable or of a variable of type "path" (see section 11.2), it is easy to set the value of the variable to a set or to a given path. The inverse constraint, that is forbid a given value, is, however, difficult to use: to rebranch, we would need to, for example, find the most promising path among those that have not yet been eliminated. It is often easier to make a more global decision: deciding that the set will or will not contain a certain element or that the path will or will not go through a certain vertex or a certain arc. Let us note that with regard to inclusion, such a choice in fact consists of either increasing the lower bound of the domain of the variable (by imposing an element), or reducing its upper bound (by forbidding an element). Actually, it is very common in constraint programming to branch by modifying the lower or upper bound of the domain of a variable.

A variable representing a cost function to be minimized is a particular case, since each time that an improving solution is found, it is possible to restrict the following searches to the solutions of strictly lower cost than the cost of this solution. A dichotomic search is often used: in a first subtree, the domain of the variable is restricted to $[v_{min}, (v_{min} + v_{max})/2[$. If a solution of cost v is obtained, v_{max} is replaced with v and the process iterated. If not, it is v_{min} that is replaced with $(v_{min} + v_{max})/2$

and the process is also iterated. To be sure of obtaining the optimal solution, it is obviously necessary to continue exploring each subtree until proof of the existence or the non-existence of a solution; this is not always possible, for obvious reasons of complexity.

In the introduction, we have deliberately chosen an example of a problem for which it is not sensible to try assigning a value to each variable. The decision operators must be defined in such a way as to progress efficiently towards a quality solution (or to proving that there is none). In the case of a disjunctive scheduling problem, if the cost function increases with the start dates of the operations, it is enough to order the operations that are in conflict for the assignment of a resource. When all the conflicts have been resolved, there is no longer a disjunctive constraint and it is enough to propagate the precedence constraints to establish the earliest start and end dates for all the operations – and therefore the cost of the solution. In the case of certain non-increasing cost functions, for example when they concern minimizing the sum of the differences with given ideal end dates, it is also possible to use a linear program to establish the best possible assignment of start and end dates for a given sequencing. Coupling constraint programming with other techniques, linear programming in this case, allows us to explore the search space in an appropriate way.

11.4. Propagation

Propagation is a key element of constraint programming procedures. It allows us to reduce the search space to be explored and deduces information allowing us to guide the search towards quality solutions.

To guarantee the robustness of an application when adding new constraints, it is important that each type of constraint is propagated independently of the others. In general, a lack of global "vision" ensues that must, for the sake of efficiency, be compensated by a significant propagation of the main constraints of the problem.

In most cases, the tools available on the market maintain "arc-consistency" [BES 95, MAC 77, MOH 86, MON 74, VAN 92] or "arc-B-consistency" [LHO 93] of the constraints. A constraint $C(V_1, \ldots, V_n)$, concerning the variables $V_1, \ldots, V_n$, will be said to be arc-consistent if and only if for every i, $1 \leq i \leq n$, and for every value v_i in the domain of possible values of V_i, values $v_1, \ldots, v_{i-1}, v_{i+1}, \ldots, v_n$ in the domains of $V_1, \ldots, V_{i-1}, V_{i+1}, \ldots, V_n$ exist such that $C(v_1, \ldots, v_n)$ is satisfied. Maintaining arc-consistency consists of removing values of the domains of the variables until the condition expressed above is satisfied. Arc-B-consistency or "arc bounds consistency" is a weaker condition which only imposes the existence of $v_1, \ldots, v_{i-1}, v_{i+1}, \ldots, v_n$ for the smallest and the largest value in the domain of V_i.

Let us consider, for example, the case of three variables V, W and X, of possible values 1, 2, 3, 4 and 5, and the three constraints ($V = 4$ or $W = 2$), $X \leq 2W - 1$ and

$V < X$. It is easy to verify that the first two constraints are arc-consistent: whatever the value chosen for V, there exists a value for W that satisfies ($V = 4$ or $W = 2$); whatever the value chosen for W, there exists a value for V that satisfies ($V = 4$ or $W = 2$); whatever the value chosen for X, there exists a value for W that satisfies $X \leq 2W-1$; whatever the value chosen for W, there exists a value for X that satisfies $X \leq 2W - 1$. The third constraint $V < X$ is neither arc-consistent nor arc bounds consistent: the value 5 must be removed from the domain of V and the value 1 must be removed from the domain of X. Because of the latter removal, the second constraint $X \leq 2W - 1$ is in turn no longer either arc-consistent or arc bounds consistent: the value 1 must be removed from the domain of W. Arc-consistency of the three constraints is finally re-established with the domains $\{1, 2, 3, 4\}$ for V, $\{2, 3, 4, 5\}$ for W and $\{2, 3, 4, 5\}$ for X. Let us note that the value 3, which is globally impossible for V, is not removed from the domain of this variable.

The concept of arc-consistency is directly linked to the expressions used to represent the problem. Let us assume, for example, that in an assignment problem n variables $V_1, \ldots, V_n$ must take different pairwise values. This can be expressed using $n(n-1)/2$ constraints $V_i \neq V_j$. When the domain of V_i is reduced to one single value, arc-consistency then consists of removing this value from the domain of V_j and *vice versa*. However, if we use a unique constraint *all-different*$(V_1, \ldots, V_n)$, arc-consistency consists of removing from the domain of V_i every value that does not allow us to assign all-different values to the n variables V_j. For example, if the domain of V_k is the set $\{1, \ldots, n\}$ and the domain of the other variables V_i, $i \neq k$, reduces to $\{1, \ldots, n-1\}$, maintaining arc-consistency over *all-different*$(V_1, \ldots, V_n)$ allows us to deduce $V_k = n$, while maintaining arc-consistency over the $n(n-1)/2$ binary constraints $V_i \neq V_j$ does not provide any information for $n > 2$ [REG 94].

Maintaining arc-consistency of a global constraint (*all-different*$(V_1, \ldots, V_n)$, for example) often necessitates the application of reasonably sophisticated operations research or graph theory algorithms. In many cases, it is actually about solving relaxed subproblems of the global problem, which can cause complexity problems. For example, if for a scheduling problem we consider a global resource constraint, which specifies that n non-interruptible operations of given duration cannot overlap one another pairwise in time, the problem of calculating the earliest and latest start and end dates of the operations (arc-B-consistency) is **NP**-complete. It therefore needs to be restricted to an approximative calculation, making a compromise in this way between the interests of the deduced information and the calculation time [BAP 01].

In the case of a set variable, the "bounds" of the domain of possible values of the variable are, on the one hand, the set of elements that must necessarily belong to the set (lower bound for the inclusion relation) and, on the other, the set of elements that may possibly belong to the set (upper bound for the inclusion). The ideas of arc-consistency or arc-B-consistency cannot then, for reasons of efficiency, be used as they are. Let us go back to our routing example. The first type of constraint specifies that the set $E(N_i, N_j)$ is a simple path of length $L(N_i, N_j)$ between N_i and N_j. Enumerating all the paths that satisfy this constraint is obviously out of the question. Nevertheless, a certain number of propagation rules can be put into place:

– Arcs entering N_i are forbidden.

– Arcs leaving from N_j are forbidden.

– If an arc leaving a given node N_k is imposed, then all the other arcs leaving N_k are forbidden.

– If an arc entering a given node N_k is imposed, then all the other arcs entering N_k are forbidden.

– If only one arc leaving N_i is possible, this arc is imposed.

– If only one arc entering N_j is possible, this arc is imposed.

– If an arc entering a given node N_k, $k \neq j$ is imposed and only one arc leaving N_k is possible, this arc is imposed.

– If an arc leaving a given node N_k, $k \neq i$ is imposed and only one arc entering N_k is possible, this arc is imposed.

– If no arc entering a given node N_k, $k \neq i$ is possible, then all the arcs leaving N_k are forbidden.

– If no arc leaving a given node N_k, $k \neq j$ is possible, then all the arcs entering N_k are forbidden.

– The number of imposed arcs of $E(N_i, N_j)$ is a lower bound of $L(N_i, N_j)$.

– The number of possible arcs of $E(N_i, N_j)$ is an upper bound of $L(N_i, N_j)$.

– Let L_{ij} be the length of the shortest path from N_i to N_j in the graph of possible arcs of $E(N_i, N_j)$; L_{ij} is a lower bound of $L(N_i, N_j)$.

– If the value of $L(N_i, N_j)$ is known and equal to the number of imposed arcs of $E(N_i, N_j)$, all non-imposed arcs are forbidden.

– If the value of $L(N_i, N_j)$ is known and equal to the number of possible arcs of $E(N_i, N_j)$, all the possible arcs are imposed.

– If the upper bound of $L(N_i, N_j)$ is 1, the set $E(N_i, N_j)$ is reduced to the singleton (N_i, N_j).

– If the lower bound of $L(N_i, N_j)$ is greater than or equal to 2, the arc (N_i, N_j) is forbidden.

– If in the graph of possible arcs of $E(N_i, N_j)$, a path from N_i to N_k does not exist, all the arcs entering N_k and all the arcs leaving N_k are forbidden.

– If in the graph of possible arcs of $E(N_i, N_j)$ a path from N_k to N_j does not exist, all the arcs entering N_k and all the arcs leaving N_k are forbidden.

– Let (N_k, N_l) be an arc not forbidden by the two previous rules. Let L_{ik} be the length of the shortest path from N_i to N_k in the graph of possible arcs of $E(N_i, N_j)$ and L_{lj} be the length of the shortest path from N_l to N_j in the graph of possible arcs of $E(N_i, N_j)$. If the upper bound of $L(N_i, N_j)$ is strictly less than $L_{ik} + 1 + L_{lj}$, the arc (N_k, N_l) is forbidden.

This list of rules is not restrictive. Nevertheless constraint propagation needs to be restricted to a set of rules that can be applied in a reasonable calculation time. For example, in general it is not desirable to calculate the length of the longest simple path between N_i and N_j in the graph of possible arcs of $E(N_i, N_j)$ (a calculation that would provide an upper bound of $L(N_i, N_j)$). Indeed, establishing whether a simple path of length greater than or equal to p exists between two nodes of a graph constitutes an **NP**-complete problem [GAR 79] which it is not desirable to solve systematically with each modification of the graph.

A special case concerns the propagation of the cost function to be minimized. A constraint programming algorithm will be all the more efficient if an upper bound on this cost function leads to important deductions that allow us to focus the search on a subspace containing good solutions. In the scheduling domain, for example, it is relatively easy to optimize the total duration of the schedule or the maximum delay for a certain number of tasks, since an upper bound on these criteria leads immediately to a bound on the end date of the tasks concerned. This is not the case if the criterion to optimize is a weighted sum of delays, or the number of tasks delayed. It can then be useful to develop a global constraint linking many of the problem variables to the variable representing the cost of a solution [BAP 98, BAP 01].

Another global constraint special case is obtained by grouping all the linear constraints of a problem into a single global constraint. It is then possible to use linear programming tools to establish if this linear system has a solution and what the lower and upper bounds are of certain variables within this linear system. This combination of linear programming and constraint programming sometimes allows us to solve problems that are impossible to deal with efficiently with each of the two techniques taken separately. In general, however, it is expensive in terms of calculation time and therefore not always "viable".

11.5. Heuristics

11.5.1. *Branching*

The most frequently used method for exploring the search space consists of exploring a tree "in depth first". At each node of the tree, a variable and a possible value of this variable are selected and two subtrees are created. In the left-hand subtree, explored first, the chosen value is assigned to the variable. This assignment is propagated and the process is iterated. When this first subtree has been explored completely, a second (right-hand) subtree is explored in turn. In this subtree, the value chosen is removed from the domain of the variable, this domain reduction is propagated and the process iterated.

As previously indicated, the choice of a variable and of a value can be, more generally, replaced by the choice of a constraint to be added in the left-hand subtree, the opposite constraint then being added to the right-hand subtree. In all cases, the choice of the constraint to be added is crucial to the efficiency of the search: indeed, it is desirable to make decisions that allow us to converge quickly on new solutions of the problem. Most often, the chosen strategy consists of establishing a constraint C such that (i) C and $not(C)$ both have a considerable impact on the search space; and (ii) C is much more likely than $not(C)$ to lead to a good solution. These two criteria can sometimes be contradictory.

For example, in the case of a communications routing problem, it is preferable to route the most important traffic first. But the choice of an arc or of a path to route such traffic is not obvious. Indeed, choosing an arc or a set of arcs of large capacity allows us to preserve the usage possibilities of these arcs for routing other communications and therefore, globally, to keep a maximum number of possibilities for routing these other communications. However, choosing an arc or a set of arcs of barely sufficient capacity allows us, by propagating the constraints, to reduce the graph of possible arcs of the other communications, which can allow us to converge more quickly. Lastly, in the case where a cost is associated with the use of the capacity of different arcs, the choice is even more difficult: is it better to choose an expensive path for the communication under consideration, which leaves many possibilities for the other communications, or a more economical path for the communication under consideration, which is very restrictive for the other communications?

Let us note that this type of dilemma is encountered in the context of any tree search procedure. Notably, in the case of integer linear programming, a "good" branching may be, depending on the applications:

– a branching that allows us to heighten the value of the continuous relaxation;

– a branching that allows us to reduce the number of integer variables taking a fractional value in the solution of the continuous relaxation;

– a branching that allows us to apply many local cuts, notably according to reduced costs arguments.

The only distinctive feature of constraint programming in this respect is that the main effect of a choice is obtained by propagation. In the case of constraint programming, propagation plays the role that is played by the solution of continuous relaxation in the case of integer linear programming.

11.5.2. *Exploration strategies*

The advantage of a depth-first search lies firstly in the ability to incrementally update the characteristics of the problem, notably the domains of the variables, both when a constraint is added and when backtracking. A depth-first strategy is generally also considered to be less memory consuming than a breadth-first search. It does, however, have a major disadvantage: it can happen that one of the first decisions made is "bad" and that it is necessary to explore a huge subtree before making the opposite decision. Many strategies have therefore been developed to diversify the search. We will only mention the most important ones here.

If the propagation of an upper bound on the cost function is effective, it may be worthwhile, each time a new solution is found or once every n solutions, to restart a search from the root node, to which a constraint bounding the maximum cost of the solutions sought has been added. Indeed, propagating the addition of this constraint modifies the domains of the set of variables, which can lead the heuristic to explore the search space very differently.

It may also be advantageous to alternate between several heuristics or to start several tree searches in parallel. If the probability of unsuccessfully exploring a large subtree is small, starting several alternating or parallel searches allows us to almost guarantee for certain that new improving solutions will be quickly found.

Another possibility consists of making a type of local search based on constraint programming around the obtained solutions. For example, Kilby, Prosser and Shaw [KIL 00] solve vehicle routing problems in this way with capacity constraints, time windows, and additional constraints such as "client A and client B must be served by the same vehicle" or "client A must be served after client B". These additional constraints are very common in collection and delivery problems. In [KIL 00], "pure" local search algorithms are compared with hybrid algorithms:

– a heuristic constructed by inserting visits in a solution, in which the order of insertion of visits is determined depending on the number of possible positions, after constraint propagation, for the insertion of each visit;

– an existing solution-improving heuristic, which consists of removing part of the visits of the solution, then reinserting them differently using a constraint programming algorithm.

The experimental results show that, when the number of additional constraints increases, hybrid algorithms making use of constraint programming are more efficient.

More generally, it is often effective to keep a large percentage of the decisions made to end up with a solution and to explore the subspace opened by questioning the other decisions [CHA 04, NUI 94]. The decisions kept can be chosen partly at random or deterministically, depending on the problem under consideration. Certain constraint programming tools even offer the possibility of explicitly defining neighborhoods, which are then explored as a tree. Most metaheuristics in the literature (simulated annealing, tabu search, etc.) in this way have their equivalent in constraint programming, which allows us to benefit from both the advantages of local searches for exploring the search space and the advantages of constraint programming for representing (and propagating) extremely varied constraints [SHA 02].

A variant consists of exploring a search tree "in bands" [BEC 00, CAS 01, HAR 95, LEP 99, WAL 97]. This involves exploring a neighborhood of the solution obtained on the left-hand branch as a tree. Many variants exist, which are found to be more or less efficient depending on the probability of the heuristic "making a mistake" in different levels of the search tree. For example, if the mistakes are much more common near the root, it is possible to define a band by limiting the distance to the root of the nodes for which the right-hand subtree is explored.

These diverse strategies can of course be combined, the best combination often depending on the problem under consideration. Thus, the best algorithm known for the pre-emptive *job-shop* problem [LEP 99] combines (i) a tree search procedure "in bands" around the best solution found previously, and (ii) a local optimization operator ("Jackson" derivation) applied to each improving solution until a local optimum is obtained, both implemented on the basis of constraint programming. Experimental results show that constraint propagation is the element that contributes most strongly to the efficiency of the global algorithm, but that exploring the search tree "in bands", focusing this exploration around the best previously found solution, and using the local optimization operator, also contributes to this efficiency.

11.6. Conclusion

In this chapter, we have introduced the basic ideas of constraint programming, highlighting those that, in our experience, have shown themselves to be the most important in practice and which explain the success of constraint programming in domains as varied as products and services configuration, vehicle routing, network design, time-tabling and scheduling, to mention just the principal application domains. It is obviously impossible to be exhaustive, given the abundant theoretical and experimental results made available each year. We advise the reader who wishes to go beyond this introduction to refer to the works [BAP 01, MIL 04, TSA 93, VAN 89].

11.7. Bibliography

[BAP 98] BAPTISTE P., LE PAPE C., PÉRIDY L., "Global constraints for partial CSPs: a case study of resource and due-date constraints", *International Conference on Principles and Practice of Constraint Programming*, Pisa, Italy, 1998.

[BAP 01] BAPTISTE P., LE PAPE C., NUIJTEN W., *Constraint-Based Scheduling*, Kluwer Academic Publishers, 2001.

[BEC 00] BECK J.C., PERRON L., "Discrepancy-bounded depth-first search", *International Workshop on the Integration of AI and OR Techniques in Constraint Programming for Combinatorial Optimization Problems*, Paderborn, Germany, 2000.

[BER 02] BERNHARD R., CHAMBON J., LE PAPE C., PERRON L., RÉGIN J.-C., "Résolution d'un problème de conception de réseau avec Parallel Solver", *Journées Francophones de Programmation Logique et Programmation par Contraintes*, Nice, France, 2002.

[BES 95] BESSIERE C., FREUDER E., RÉGIN J.-C., "Using inference to reduce arc consistency computation", *International Joint Conference on Artificial Intelligence*, Montreal, Quebec, 1995.

[CAR 88] CARLIER J., CHRÉTIENNE P., *Problèmes d'ordonnancement: Modélisation/Complexité/Algorithmes*, Masson, Paris, 1988.

[CAS 01] CASEAU Y., LABURTHE F., LE PAPE C., ROTTEMBOURG B., "Combining local and global search in a constraint programming environment", *Knowledge Engineering Review*, vol. 16, p. 41–68, 2001.

[CHA 04] CHABRIER A., DANNA E., LE PAPE C., PERRON L., "Solving a network design problem", *Annals of Operations Research*, vol. 130, p. 217–239, 2004.

[GAR 79] GAREY M.R., JOHNSON D.S., *Computers and Intractability. A Guide to the Theory of NP-Completeness*, W. H. Freeman, San Francisco, 1979.

[HAR 95] HARVEY W.D., GINSBERG M.L., "Limited discrepancy search", *Fourteenth International Joint Conference on Artificial Intelligence*, Montreal, Quebec, 1995.

[KIL 00] KILBY P., PROSSER P., SHAW P., "A comparison of traditional and constraint-based heuristic methods on vehicle routing problems with side constraints", *Constraints*, vol. 5, p. 389–414, 2000.

[LEP 85a] LE PAPE C., Représentation et satisfaction de contraintes à l'aide de graphes de potentiel, Rapport de recherche, Laboratoires de Marcoussis, 1985.

[LEP 85b] LE PAPE C., SAUVE B., "SOJA: un système d'ordonnancement journalier d'atelier", *Journées internationales sur les systèmes experts et leurs applications*, Avignon, France, 1985.

[LEP 88] LE PAPE C., Des systèmes d'ordonnancement flexibles et opportunistes, PhD thesis, University of Paris XI, 1988.

[LEP 99] LE PAPE C., BAPTISTE P., "Heuristic control of a constraint-based algorithm for the preemptive job-shop scheduling problem", *Journal of Heuristics*, vol. 5, p. 305–325, 1999.

[LHO 93] LHOMME O., "Consistency techniques for numeric CSPs", *International Joint Conference on Artificial Intelligence*, Chambéry, France, 1993.

[MAC 77] MACKWORTH A.K., "Consistency in networks of relations", *Artificial Intelligence*, vol. 8, p. 99–118, 1977.

[MIL 04] MILANO M., *Constraint and Integer Programming: Toward a Unified Methodology*, Kluwer Academic Publishers, 2004.

[MOH 86] MOHR R., HENDERSON T.C., "Arc and path consistency revisited", *Artificial Intelligence*, vol. 28, p. 225–233, 1986.

[MON 74] MONTANARI U., "Network of constraints: fundamental properties and applications to picture processing", *Information Sciences*, vol. 7, p. 95–132, 1974.

[NUI 94] NUIJTEN W., Time and Resource Constrained Scheduling: A Constraint Satisfaction Approach, PhD thesis, Eindhoven University of Technology, 1994.

[PUG 91] PUGET J.-F., ALBERT P., "PECOS: programmation par contraintes orientée objets", *Génie logiciel et systèmes experts*, vol. 23, p. 100–105, 1991.

[PUG 92] PUGET J.-F., "Programmation par contraintes orientée objet", *Douzièmes journées internationales sur les systèmes experts et leurs applications*, Avignon, France, 1992.

[REG 94] RÉGIN J.-C., "A filtering algorithm for constraints of difference in CSPs", *Twelfth National Conference on Artificial Intelligence*, Seattle, Washington, 1994.

[REG 96] RÉGIN J.-C., "Generalized arc-consistency for global cardinality constraint", *Thirteenth National Conference on Artificial Intelligence*, Portland, Oregon, 1996.

[SHA 02] SHAW P., FURNON V., DE BACKER B., "A constraint programming toolkit for local search", VOSS S., WOODRUFF D. L., Eds., *Optimization Software Class Libraries*, Kluwer Acamemic Publishers, 2002.

[TSA 93] TSANG E.P.K., *Foundations of Constraint Satisfaction*, Academic Press, 1993.

[VAN 89] VAN HENTENRYCK P., *Constraint Satisfaction in Logic Programming*, MIT Press, 1989.

[VAN 92] VAN HENTENRYCK P., DEVILLE Y., TENG C.M., "A general arc-consistency algorithm and its specializations", *Artificial Intelligence*, vol. 57, p. 291–321, 1992.

[WAL 97] WALSH T., "Depth-bounded discrepancy search", *International Joint Conference on Artificial Intelligence*, Nagoya, Japan, 1997.

List of Authors

Jérémy BARBAY
Departamento de Ciencias de la Computación
University of Chile
Santiago
Chile

Alain BILLIONNET
CEDRIC-CNAM
Paris
France

Alberto CESELLI
Dipartimento di Tecnologie dell'Informazione
Università degli Studi di Milano
Italy

Irène CHARON
Groupe Mathématiques de l'Informatique et des Réseaux
Département Informatique et Réseaux
ENST
Paris
France

Frédérico DELLA CROCE
Dipartimento di Automatica e Informatica
Politecnico di Torino
Italy

Bruno ESCOFFIER
LAMSADE
University Paris-Dauphine
France

Andrea GROSSO
Dipartimento di Informatica
Università degli Studi di Torino
Italy

Olivier HUDRY
Groupe Mathématiques de l'Informatique et des Réseaux
Département Informatique et Réseaux
ENST
Paris
France

Claude LE PAPE
Schneider Electric
Lyon
France

Irene LOISEAU
Departamento de Computación
Facultad de Ciencias Exactas y Naturales
Universidad de Buenos Aires
Argentina

Nelson MACULAN
Programa de Engenharia de Sistemas e Computação
COPPE
Universidade Federal do Rio de Janeiro
Brazil

Ali Ridha MAHJOUB
LAMSADE
University Paris-Dauphine
France

Vangelis Th. PASCHOS
LAMSADE
University Paris-Dauphine
France

Matteo SALANI
Dipartimento di Tecnologie dell'Informazione
Università degli Studi di Milano
Italy

Olivier SPANJAARD
LIP6
University Pierre et Marie Curie
Paris
France

Pierre TOLLA
LAMSADE
University Paris-Dauphine
France

Index

D, E, F, G

I, L

M, N, O

P, Q, R

S, T, V, W

Summary of Volume 2
Paradigms of Combinatorial Optimization

Chapter 9. Integer Quadratic Knapsack Problems

Chapter 10. Graph Coloring Problems

PART II. NEW APPROACHES

Chapter 11. Polynomial Approximation

Summary of Volume 3
Applications of Combinatorial Optimization